TITANIUM MATRIX COMPOSITES

HOW TO ORDER THIS BOOK

BY PHONE: 800-233-9936 or 717-291-5609, 8AM–5PM Eastern Time

BY FAX: 717-295-4538

BY MAIL: Order Department
Technomic Publishing Company, Inc.
851 New Holland Avenue, Box 3535
Lancaster, PA 17604, U.S.A.

BY CREDIT CARD: American Express, VISA, MasterCard

BY WWW SITE: http://www.techpub.com

TITANIUM MATRIX COMPOSITES
Mechanical Behavior

EDITED BY

Shankar Mall
Professor and Head
Air Force Institute of Technology
Department of Aeronautics and Astronautics
Wright-Patterson Air Force Base, OH 45433

Theodore Nicholas
Senior Scientist
Air Force Wright Laboratory Materials Directorate
Wright-Patterson Air Force Base, OH 45433

TECHNOMIC
PUBLISHING CO., INC.
LANCASTER · BASEL

Titanium Matrix Composites

a **TECHNOMIC**®publication

Published in the Western Hemisphere by
Technomic Publishing Company, Inc.
851 New Holland Avenue, Box 3535
Lancaster, Pennsylvania 17604 U.S.A.

Distributed in the Rest of the World by
Technomic Publishing AG
Missionsstrasse 44
CH-4055 Basel, Switzerland

Printed in the United States of America
10 9 8 7 6 5 4 3 2 1

Main entry under title:
  Titanium Matrix Composites: Mechanical Behavior

A Technomic Publishing Company book
Bibliography: p.
Includes index p. 465

Library of Congress Catalog Card No. 97-61635
ISBN No. 1-56676-567-6

Table of Contents

v

Preface

THE PROPULSION AND airframe components in advanced aerospace systems and several other engineering systems will rely heavily upon high-temperature materials with improved specific properties. One class of materials that has received significant attention in this regard is the silicon carbide fiber reinforced titanium alloy matrix composites, commonly referred to as titanium matrix composites (TMCs). In the 1990s, an intense effort was directed towards the development, understanding, characterization, and life prediction of TMCs. The primary interest arose from the National Aerospace Plane (NASP) Program in the United States where TMC was considered a prime candidate as a structural material for the aircraft. At about the same time, the U.S. Air Force Integrated High Performance Turbine Engine Technology (IHPTET) Program sought to achieve dramatic performance gains in advanced propulsion systems. One approach considered TMCs as materials for several turbine engine components because of their potential to save considerable weight. These and several other programs devoted a considerable amount of time and money towards TMCs, which resulted in major advancements in the state of the art of their performance, characterization, and modeling. Also, several applications of TMCs were successfully fabricated to demonstrate clearly that this class of materials is technically feasible, ready to deliver the projected benefits, and well past the stage of "one shot" applications. In fact, TMCs are consistently identified as one of the enabling material technologies for the twenty-first century.

It is, therefore, appropriate to assemble and document in one source a state-of-the-art review and summary of advancements related to mechanical behavior and related mechanics issues of TMCs under the var-

ix

ious thermomechanical loadings that are currently scattered over hundreds of reports, papers, and unpublished places. With this in mind, the editors contacted various experts, and this first book on TMCs is the result of such contributions, for which the editors are grateful to all authors. While it may appear that each author might have focused to some extent on his/her contributions to TMC technology, it can be assured that a sincere attempt has been made by all authors to represent, as well as possible, the general development in the field. Since TMC technology is so recent and evolving continually, we apologize on behalf of all authors to those whose contributions might have been inadvertently not included in this book. Finally, we present this book as a valuable source for those who desire to get familiar with, enhance knowledge, and pursue research in this rapidly developing field.

S. MALL
T. NICHOLAS

1

Introduction

S. MALL

Department of Aeronautics and Astronautics
Air Force Institute of Technology
Wright-Patterson Air Force Base, OH 45433

T. FECKE AND M. A. FORINGER

Aero-Propulsion and Power Directorate
Wright Laboratory
Wright-Patterson Air Force Base, OH 45433

BACKGROUND

METAL MATRIX COMPOSITES (MMCs) consist of metallic materials reinforced with particulate, flakes, whiskers, continuous fibers, or woven continuous fibers. Metal matrix composite is a quasi-homogeneous structure that produces synergistic mechanical and physical properties better than those of the base metal (matrix material). Principal advantages of metal matrix composites are increased mechanical and physical properties that can be tailored to a specific application, usually at less weight than unreinforced metallic materials. Because most of metallic materials and reinforcements have relatively high operating temperatures, metal matrix composites maintain these advantages at much higher temperatures than is possible with polymer matrix composites. Additionally, because these are metal-based composites, electrical and thermal conductivity can be substantially improved along with better resistance to environmental and thermal instability.

There are two distinct types of composite systems including metal matrix composites, depending on the length-to-diameter ratio of the reinforcement. Continuous fiber-reinforced metal matrix composites have reinforcements with a very large length-to-diameter ratio (i.e., nonbroken filaments or fibers). Discontinuously reinforced metal ma-

1

trix composites have reinforcements with a small length-to-diameter ratio (i.e., chopped fibers, particulate, whisker, or flake fillers). In the latter, only greater stiffness is generally realized, whereas in the former, both the modulus and the strength of the continuous fiber are almost fully translated into the composite. Therefore, continuously reinforced metal matrix composites generally have mechanical and physical properties superior to discontinuously reinforced metal matrix composites. The direction of fibers dictates the properties of a continuously reinforced metal matrix composite. This provides the option to tailor specific properties of metal matrix composites in specific directions, which is not possible with monolithic metallic alloys. The highest values for strength and modulus are obtained in the direction where all reinforcing fibers are straight and parallel. These properties are weaker in other directions. Although the difference in these properties between parallel (also referred to as longitudinal or axial) and perpendicular (or transverse) directions to the fibers is generally very large in polymer matrix composites, it is much less in the metal matrix composites.

The availability of boron fibers (produced by chemical vapor deposition) in the late 1960s stimulated interest in metal matrix composites in the aerospace arena, and the development of lower cost multifilament silicon carbide and alumina fibers and whiskers in the mid to late 1970s excited interest in metal matrix composites in lower cost industries. This interest has resulted in metal matrix composites emerging as advanced materials for the last 30 years for a range of applications in the automotive, aerospace, recreational, and other fields. Examples in automotive applications include cylinder liners, pistons, connecting rods, valve lifters, piston pins, and drive trains. Aerospace engine applications include exhaust nozzles, links, blades, cases and frames, shafts, compressor rotors, spacers, and vanes. Aerospace structural applications include aft fuselage structures, landing and arresting gears, stiffeners, drag brakes, and compression and torque tubes. Applications of metal matrix composites in the recreational industry include golf club heads, masts and spars for boats, and bicycle frames. Other applications are wear-resistant, hard materials for cutting tools, drill bits, valves, gates, electrical connectors, substrates, packing materials, battery plates, superconducting wires, medical implants, and tubing in nuclear plants. Over the last 30 years, there have been many successes, failures, and challenges with metal matrix composites.

With the recent improvements in titanium alloys and silicon carbide fibers, a class of metal matrix composites, consisting of a titanium alloy matrix and continuous silicon carbide fiber reinforcement, is being consistently identified as one of the enabling technologies for many applications in both military and commercial arena to meet the design requirements for materials displaying high strength and stiffness-to-density ratios at moderate to high temperatures (400 to 800°C). This class of metal matrix composites is referred to as titanium matrix composite (TMC). These materials are now being considered as viable engineering materials, well past the stage of "one-shot" applications. To use titanium matrix composites effectively, they must be thoroughly understood. Engineers, designers, and others should be aware of technical challenges along with capabilities and shortcomings involved with the use of this class of composites. The purpose of this book is to establish a basis for this knowledge.

HISTORICAL PERSPECTIVE

The development of metal matrix composites, as mentioned earlier, began after boron fibers were successfully developed in the 1960s [1,2]. These were the first high-strength and high-modulus reinforcing fibers with higher temperature capability that had clear advantage over the contemporary glass fibers for use in the metal matrix composites. The first metal used as a matrix material was aluminum because of its low processing temperature. The earliest applications of boron/aluminum metal matrix composites were structural tubular struts used as the frame and rib truss members in the midfuselage section, as the landing gear drag link of the space shuttle orbiter, and as the fan blade in a jet engine [3].

Boron fibers were also tried with the titanium matrix but were less successful due to the formation of a large reaction zone between the fiber and matrix. This resulted in a strength reduction compared with unreinforced titanium. The controlling factor in the formation of reaction zone was the time at temperature. An attempt to overcome this problem included the vapor deposition of a thin coating to serve as a diffusion barrier to prevent interaction between boron and metals at elevated temperatures [4]. The coating that showed the most success was the silicon carbide (SiC). The resulting fiber is known as borsic (silicon carbide-coated boron fiber). Even with the coating, boron fibers still reacted significantly with the titanium matrix. Beryllium fibers were then tried. Beryllium has the highest modulus-to-density ratio of any metal (over 6 times that for steel or titanium). However, the reaction zone between the beryllium and titanium, although better than the boron/titanium, is still significant. Additionally, beryllium is a toxic substance that increases the complexity of fabrication.

These experiences with coating led to the development of the silicon carbide fiber in the late 1970s, which started to show that titanium matrix composites could have significantly higher mechanical properties than homogeneous titanium. This is primarily the result of a greatly reduced reaction zone between the silicon carbide fiber and the titanium matrix due to the incorporation of the carbon coating on this fiber. The reaction zone for this combination is about 60% less than the beryllium/titanium combination, 70% less than the coated boron/titanium combination, and 85% less than the uncoated boron and titanium combination. The ultimate tensile strength for titanium matrix reinforced with silicon carbide fibers is typically 60% higher than that for homogeneous titanium without any reinforcement. With this breakthrough, titanium matrix composites (TMCs) started to show that they could be used in real-world applications.

APPLICATIONS

The primary driver for the development of TMCs has been the gas turbine engine industry. As designers look for ways to increase the performance of aircraft jet engines, more and more stringent requirements are placed on materials to be used. These include high operating temperatures, long expected structural lifetimes, extreme consequences of structural failures, and the need to reduce weight. The Integrated High Performance Turbine Engine Technology (IHPTET) initiative is one of the big drivers toward the introduction of TMCs in turbine engine components [5]. This joint Department of Defense-NASA-industry effort has the basic goal of doubling engine capability by the year 2003. This ambitious goal will require a significant increase in the operating temperatures within the engine as well as lower weight. TMCs play a major role in a variety of applications to accomplish this goal.

The primary advantage with the use of TMCs is the weight savings in elevated temperature applications. For example, a TMC rotor bling (bladed ring) weighs just 4.6 kg, whereas a more traditional nickel superalloy rotor disk weighs 25 kg (Figure 1) [5]. This type of weight savings is possible because the TMC is strong enough to handle the expected hoop stresses at flying speeds with a smaller cross-sectional area than what is required by the nickel superalloy. Because the primary direction of loading in a spinning ring is in the tangential di-

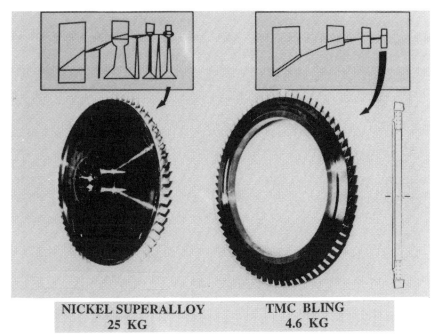

NICKEL SUPERALLOY   **TMC BLING**
25 KG   **4.6 KG**

Figure 1. Comparison of weight between TMC and nickel superalloy blings.

rection, the fibers in the TMC are oriented along the circumference of the ring. This type of application takes advantage of all three of the main strong points of TMCs: their relative low weight, their ability to handle elevated temperatures, and their high strength in the direction of the fiber.

TMCs enable components with an increased hoop radius due to the lower density, which can yield about 30 to 50% reduction in weight of rotating parts in aircraft engines along with the enhanced engine performance, such as impellers, disks, integrally bladed rotors (blotors) or bladed disks (blisks), and bladed rings (blings) now primarily fabricated of nickel based superalloys [6]. These high payoff potentials have resulted in many Department of Defense and aerospace engine industry-sponsored programs to demonstrate feasibility and performance capability of TMC components. A notable demonstration under the sponsorship of the IHPTET program involved two compressor rotors and the spacer between them that were fabricated of a TMC, SCS-6/Ti-6-4 [5]. The weight of these three TMC components was 15.5 kg, which is 78% less than that of the monolithic metallic parts (nickel alloy) that they replaced, i.e., the third and fourth stages, clos-

Figure 2. TMC four-stage compressor rotor for gas turbine engines.

est to the open end, of the four-stage turbine compressor in Figure 2. This reduction of 53.9 kg in the rotating weight improves the engine thrust-to-weight ratio and, ultimately, allows the lightening of engine support structures, further augmenting the weight savings.

TMC shafts for engines have been successfully fabricated since the 1980s (Figure 3) [7–11]. TMC engine shafts are lighter in weight and have the reduced dynamic loads with improved turbomachinery clearance control. A 30% weight reduction from nickel alloy shafts or 40% stiffness increase from titanium shafts can be achieved with TMC shafts. TMC shafts are typically fabricated with angle-ply layups oriented from ±15 to ±45° with respect to the shaft axis. A 127-cm long, 12.1-cm diameter, low-pressure turbine fan shaft for the advanced IHPTET engine was fabricated more recently using 36 plies at ±15° orientation [11]. This was successfully tested up to a runoff limit of 10^5 cycles, which exceeded the predicted capability. Also, TMC tubes up to 150 cm in length have been fabricated which incorporated the monolithic titanium sections into their ends for the ease of welding to the splined or flanged connections [4].

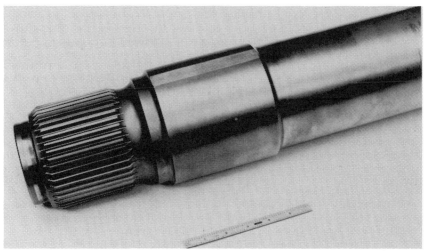

Figure 3. TMC shaft.

Hollow blades for fans and compressors, fabricated of TMCs, are very attractive to engine designers because they reduce the weight by 30% by elimination of shrouds required for vibration control (Figure 4) [12–14]. Use of TMCs also allows more sweep in blades, smaller blade rest chord, and a thinner blade root. One program sponsored by the U.S. Air Force in the early 1980s demonstrated the successful feasibility of a large (46-cm-long), hollow fan blade for the F110 engine

Figure 4. TMC hollow fan blade.

(to power the F-16 fighter plane), which was fabricated by consolidating unidirectional 18 plies with concave and convex airfoil surfaces [12]. This, when tested, met or exceeded all predicted strength, stiffness, and fatigue capabilities.

There are several other applications of TMCs that have been successfully demonstrated in the different components of aerospace engines. TMC ducts or cases offer a 40% greater stiffness and a 50% weight reduction over the current unreinforced titanium structures (Figure 5) [5]. Similarly, TMC links and actuators for moving exhaust flaps offer about 40% weight improvement over IN 718 components [6]. In 1992 TMC compression links (Figure 6) were installed in F110-GE-100 engine exhaust, and flight was tested for 31 hours in the F-16 fighter aircraft. This successful flight testing was preceded by over 2 hours of ground engine tests [15]. Other examples include large flat structures, I-beam sections, box sections, and other structural members, fabricated from TMCs, for the potential F120 engine exhaust structure (Figure 7) [16].

Figure 5. TMC ducts and cases.

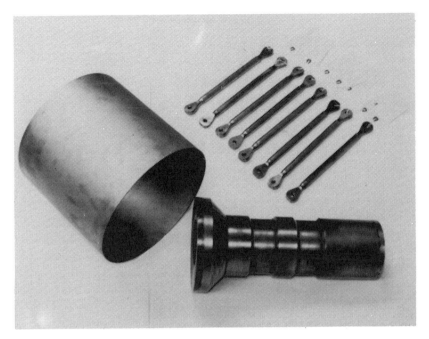

Figure 6. TMC compression links for turbine engine exhaust flaps.

Figure 7. TMC rear frame (exhaust structure).

9

The applications mentioned above have clearly demonstrated that the substantial weight savings are possible with TMCs; however, the high cost of TMCs is the greatest challenge to their widespread use. This has led to the Titanium Matrix Composite Turbine Engine Component Consortium (TMCTECC) initiative, which is now in progress [17]. The primary goal of TMCTECC, a cooperative U.S. Department of Defense-industry effort, is to establish a commercially viable market for TMCs to help drive down the manufacturing cost. The TMCTECC is broken up into three distinct programs. One program concentrates on the TMC-manufacturing process, the second one is to develop a reinforced fan frame for General Electric's GE90 class engine, and the third one is to develop a reinforced hollow fan blade for Pratt & Whitney's PW4084 class engine. Both engines are used in Boeing's 777 commercial airplane. These two engines were selected because there is a possibility of a near term growth of TMCs usage, and these applications require the capabilities that TMCs could address.

The reinforced hollow fan blade portion of the TMCTECC initiative takes advantage of the stiffness properties of TMCs to develop a blade with a TMC insert that is based on the encouraging results of the F110 fan blade [14]. The fibers, in this case, are aligned along the blade in the radial direction. This blade, 102-cm long, uses four to eight plies of TMC on either side of a hollow case region. It is hoped that a stiffer and lighter blade can be produced using TMCs at the comparable cost of a homogeneous titanium hollow blade. One of the major challenges is to design a blade that adequately resists foreign object damage, because it is yet to be shown that the presence of the TMC helps the impact resistance of the blade. Other challenges include the manufacturing of TMC components using the similar infrastructure as needed for the homogeneous titanium blade.

Titanium matrix composites were also under consideration for use in the National Aerospace Plane (NASP) program jointly run by NASA and the U.S. Air Force [18]. This program was intended to develop a reusable space vehicle that could land and take off like an airplane. Because this type of endeavor would require lightweight materials with high mechanical properties and the capability to withstand elevated temperatures, TMCs were considered for use in the aircraft skin. In this case, because the stresses were of multidirectional nature, composite laminates with off-axis plies would be needed to optimize the design of TMC components. Although the NASP program was ultimately canceled, any future development of a hypervelocity vehicle will impose similar requirements on the material to be used. TMCs seem to fit this role perfectly.

With the advent of the U.S. Air Force's F-22 fighter plane, TMCs are finally into the operational world. The engine on the F-22, Pratt & Whitney's F119, includes 35.6-cm-long actuator piston rods for the exhaust nozzle that are constructed of silicon carbide/titanium composite (TMC) [19]. These piston actuators, which offer a greater than 30% weight savings, exceeded all mechanical design requirements and have been qualified for use in the production of F119 engines. Although this application is small in comparison with other activities, nonetheless this is an important achievement for TMCs because it is the first real application in a production aircraft engine. TMC-reinforced actuators are now also being considered for the F-22 airframe applications.

MECHANICAL PROPERTIES

The forthcoming chapters provide, in the great depth, the detailed mechanical behavior of TMCs under the various thermomechanical loading conditions. This section includes an overview of a few basic mechanical properties of TMCs. These properties are strongly dependent on several factors, such as fabrication methods, constituents (type of titanium alloy and fibers), fiber-to-matrix volume ratio, test procedure (load or displacement control mode), and environment. Furthermore, the heat treatment may also influence the mechanical properties of TMCs due to the phase transformation in the titanium matrix. Finally, the statistical considerations should not be ignored. This multitude of factors should be kept in mind during any involvement with mechanical properties as they are presented in this section to show their representative trends or values to highlight the capabilities offered by this class of composites.

There are several titanium matrix composite systems that are either currently available or being developed. Four titanium alloy-based composite systems are discussed here because of the availability of an extensive range of mechanical property data and the fact that they represent broad based trends about the present capabilities of TMCs [20–34]. All these TMCs are reinforced with the continuous silicon carbide fiber, SCS-6, manufactured by Textron Specialty Materials Division. This fiber, which is produced by chemical vapor deposition on a carbon monofilament core, is β-phase silicon carbide having a radially columnar grain structure and a double-pass, carbon-rich outer coating that is added to control reaction with titanium alloy and to prevent fiber damage during normal handling. The fiber has a nominal diameter of 142 μm. The elastic modulus of the fiber is about 400 GPa, and the ultimate tensile strength of the fiber is about 3600 MPa at room temperature. The tensile strength at elevated temperatures is shown in Figure 8.

The matrix materials of the four TMC systems are the conventional titanium alloys, Ti-15-3 (Ti-15 V-3 Cr-3 Al-3 Sn, weight percent), TIMETAL®21S (Ti-15 Mo-2.7 Nb-3Al-0.2 Si, weight percent), Ti-6-4 (Ti-6 Al-4V, atomic percent), and the advanced titanium alloy, titanium aluminide, Ti-24-11 (Ti-24 Al-11 Nb, atomic percent). The fiber volume fraction ranges from 30 to 40%. The microstructure of Ti-15-3 and TIMETAL®21S consists of metastable β-phase, Ti-6-4 consists of α- and β-phases, and Ti-24-11 consists of α_2- and β-phases. TMCs are typically fabricated by stacking alternate layers of fibers and titanium alloy foils (0.1 to 0.15-mm thick) and consolidating them into a multilayer laminate in the desired layup by vacuum hot press (VHP) or hot isostatic press (HIP). More recently, innovative processes, including tape casting, induction plasma deposition, electron beam physical vapor deposition, fiber coating, and fiber or wire co-winding, have been developed to increase flexibility and improve quality of TMCs while reducing their fabrication costs [6].

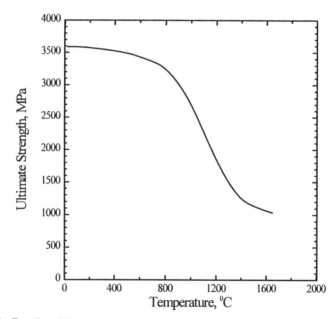

Figure 8. Tensile ultimate strength versus temperature relationship of SCS-6 fiber.

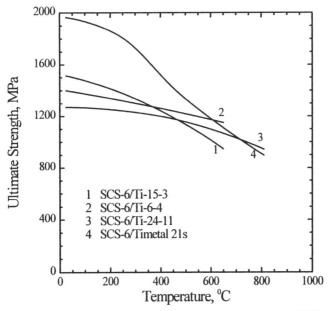

Figure 9. Tensile ultimate strength of four unidirectional TMCs.

The capabilities of the four TMCs in the fiber direction under tensile loading, i.e., the longitudinal tensile strength as a function of temperature, is shown in Figure 9. In turbine engine and aircraft applications, the use of TMCs should yield weight savings. Therefore, specific strength, i.e., strength per density, is an important consideration in these applications. The density of TMCs (4 g/cm^3) is about one-half that of nickel-based superalloys (8 g/cm^3). Figure 10 compares the specific tensile strength of four unidirectional TMCs with a typical superalloy, IN 100, which clearly shows their superiority over the conventional superalloys. Composites are generally used with different layups. The effects of layup and temperature on the tensile strength for two TMCs (with the conventional titanium alloy and the advanced titanium aluminide alloy) are depicted in Figures 11 and 12. As seen in polymer matrix composites, the tensile strength in the longitudinal direction of TMC laminate decreases with the inclusion of off-axis plies. For comparison purposes and to highlight the advantage of reinforcing fibers, tensile strengths of the four titanium alloy matrix materials (without any fiber) as functions of temperature are shown in Figure 13, and these can be compared with those of TMCs in Figure 9.

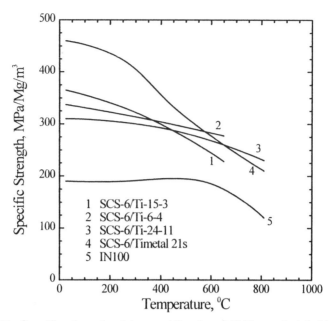

Figure 10. Specific strength of four unidirectional TMCs and nickel-based superalloy (IN 100).

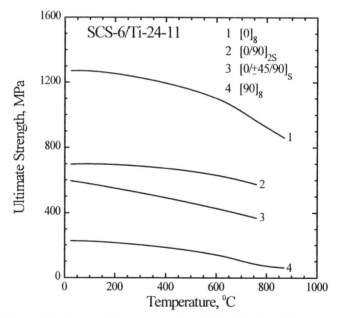

Figure 11. Tensile ultimate strength of SCS-6/Ti-24-11 laminates.

14

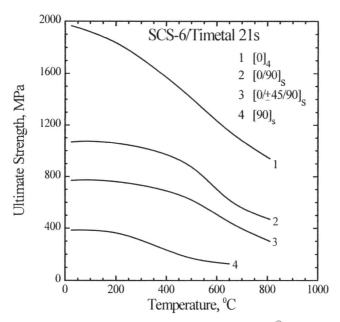

Figure 12. Tensile ultimate strength of SCS-6/TIMETAL®21S laminates.

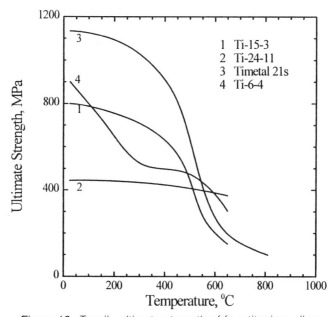

Figure 13. Tensile ultimate strength of four titanium alloys.

The modulus or specific modulus (modulus per density) of materials is also an important consideration in aerospace and engine applications. The specific moduli of the four unidirectional TMCs are compared with a nickel-based superalloy (IN 100) in Figure 14, whereas Figure 15 shows their moduli as a function of temperature. The influence of laminate layups on the longitudinal modulus for the two TMCs, SCS-6/Ti-24-11 and SCS-6/Timetal, are shown in Figures 16 and 17. Other mechanical properties that are critical toward the engineering applications of TMCs include the fatigue life and fatigue crack growth behavior. As shown in Figures 18 and 19, TMCs exhibit these properties in the direction of fiber orientation, which are generally better than superalloys even before considering their density benefits.

The mechanical properties of TMCs are of the anisotropic nature. Transverse properties (strength and modulus) of this class of composites, i.e., for 90° laminates, are lower than superalloys or unreinforced titanium alloys. These characteristics are similar to those of polymer matrix composites. However, it should be reiterated that the ratio of longitudinal to transverse properties in TMCs is much smaller than

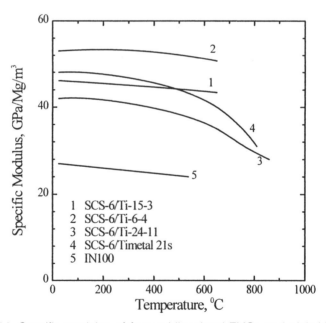

Figure 14. Specific modulus of four unidirectional TMCs and nickel-based superalloy (IN 100).

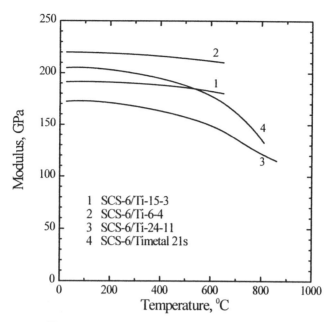

Figure 15. Modulus of four unidirectional TMCs.

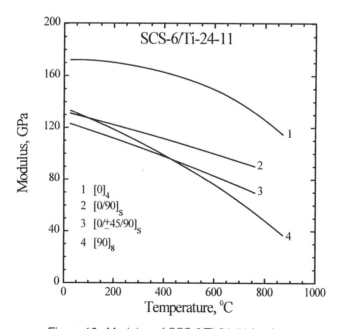

Figure 16. Modulus of SCS-6/Ti-24-11 laminates.

17

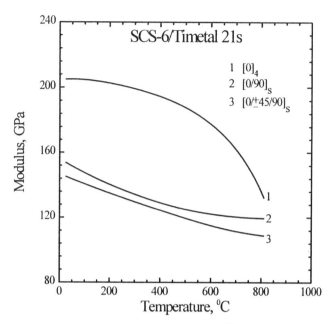

Figure 17. Modulus of SCS-6/TIMETAL®21S laminates.

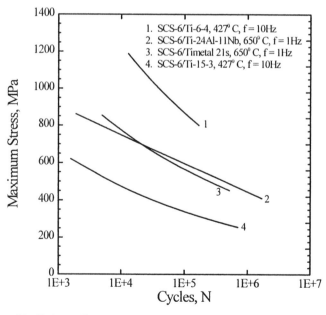

Figure 18. Fatigue life diagrams of four TMCs at elevated temperature.

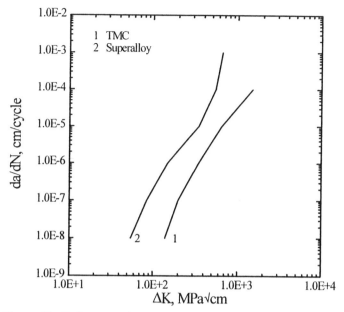

Figure 19. Fatigue crack growth behavior of TMC and superalloy.

its counterpart in polymer matrix composites. Thus, applications of TMCs would require the tailoring of laminates to optimize these features, and design methods that can manage these are available. Many applications of TMCs, as discussed earlier, have used these features.

SUMMARY

Since the inception of titanium matrix composites in the 1980s, considerable development and progress have occurred toward their evaluation and application in aircraft engines and airframe structures. It has been demonstrated that this class of composites is technically feasible material and ready to deliver the projected benefits. TMCs, with continuous silicon carbide fibers as reinforcement, are the most mature fiber reinforced metal matrix composites with very attractive properties that could provide almost a twofold increase in structural efficiency compared with titanium alloys. The fabrication processes of TMCs are ready for production.

There are many challenges that must be overcome before the widespread acceptance of TMCs. Perhaps the most challenging task ahead is the current cost of TMCs being about $11,000 per kilogram which is not competitive with the current superalloys [6]. This cost must be reduced considerably. Part of this problem is the projected consumption of TMCs in the advanced propulsion systems, which is estimated currently at about 5000 kg; therefore, additional applications are needed to achieve the large usage of TMCs to reduce their cost. The TMCTECC program is addressing this issue. Maximizing the composite benefits and development of innovative techniques of fabrication are also needed to meet this and other challenges. This would require a thorough understanding of mechanical behavior and related mechanics issues of titanium matrix composites under the various thermomechanical loadings. Topics in these subject areas are discussed in the forthcoming chapters.

REFERENCES

1. DeBolt, H. 1982. "Boron and Other Reinforcing Agents," *Handbook of Composites,* G. Lubin, ed., New York: Van Nostrand Reinhold Co., Chapter 10.

2. Krukonis, V. J. 1977. "Boron Filaments," J. B. Milweski, and H. S. Katz, ed., *Handbook of Fillers and Reinforcements for Plastics,* New York: Van Nostrand Reinhold Co., Chapter 28.

3. Kinner, W. K. 1980. "Metal-Matrix Composites," *Materials Engineering,* Vol. 91, No. 1, pp. 64–66.

4. Mittnick, M. A. 1990. "Continuous SiC Fiber Reinforced Metals," *Proceedings of 35th International SAMPE Symposium,* Covina, CA, April 2–5, pp. 1372–1382.

5. "IHPTET: Technology Teams in Action," Turbine Engine Division, WL/POT, Wright Laboratory, Wright-Patterson AFB, OH.

6. Singerman, S. A. and Jackson, J. J. 1996. "Titanium Metal Matrix Composites for Aerospace Applications," *Superalloys 1996, Proceedings of Eighth International Symposium on Superalloys,* R. D. Kissinger et al., ed., TMS, Warrendale, PA.

7. Johnson, S. R. and Ravenhall, R. 1987. "Metal Matrix Composite Shaft Research Program," Report AFWAL-TR-87-2007, Wright-Patterson AFB, OH.

8. Profant, D. and Burt, G. 1984. "Manufacturing Methods for Metal Matrix Composite (MMC) Shafts," Interim Reports for Contract F33615-80-C-5176, Wright-Patterson AFB, OH.

9. Gray, D. et al. 1994. "Titanium Metal Matrix Composite Shafts," Report AFWAL-TR-84-4124, Wright-Patterson AFB, OH.

10. Sterling, E. M. and Bell, J. E. 1990. "Advanced High Stiffness Power Turbine Shaft," Report USA AVSCOM TR 90-D-5, U.S. Army, AATD, Fort Eustis, VA.

11. Barnes, R. et al. 1993. "Exoskeletal Structures Program," Report WL-TR-93-2034, Wright-Patterson AFB, OH.

12. Ravenhall, R. 1982. "Advanced Reinforced Titanium Blade Development," Report AFWAL-TR-82-4011, Phase I Final Report, Contract F33615-80-C-3236, Wright-Patterson AFB, OH.

13. Ravenhall, R. 1982. "Advanced Reinforced Titanium Blade Development," Report AFWAL-TM89-357, Phase II and III Final Report, Contract F33615-80-C-3236, Wright-Patterson AFB, OH.

14. Kasperski, D., August 1990–August 1995. "Reinforced Hollow Fan Blade (RHFB) PRDA II," Reports for Contract F33615-90-C-2040, Wright-Patterson AFB, OH.

15. Jackson, J. J., et al. October 1991–September 1993. "Titanium Matrix Composite Engine Components," Reports for Contract F33615-91-C-5728, WL/MLLM, Wright-Patterson AFB, OH.

16. Oliver, R. R., et al. September 1994. "Manufacturing Technology for Metal Matrix Composite Exhaust Nozzle Components," Contract F33615-91-C-5730, WL/MTPM, Wright-Patterson AFB, OH.

17. Anderson, R. et al. November 1994–April 1996. "Titanium Matrix Composite Turbine Engine Component Consortium," Contract F33615-94-2-4439, WL/MTPM, Wright-Patterson AFB, OH.

18. Jackson, J. J. et al. 1991. "Titanium Aluminide Composites," Report 1112, NASP JPO, Wright-Patterson AFB, OH.

19. Tucker, R. D. 1995. "Lightweight Engine Structures and Drum Rotor," Report WL-TR-95-2070, Wright-Patterson AFB, OH.

20. Larsen, J. M., Revelos, W. C., and Gambone, M. L. 1992. "An Overview of Potential Titanium Aluminide Matrix Composites in Aerospace Applications," *Intermetallic Matrix Composites II,* D. B. Miracle, D. L. Anton, and J. A. Graves, ed., MRS Proceedings, Vol. 273, Materials Research Society, Pittsburgh, PA, pp. 3–16.

21. Gambone, M. L. 1989. "Fatigue and Fracture of Titanium Aluminides, Vol. I," U.S. Air Force report WRDC-TR-89-4145.I, Wright-Patterson AFB, OH.

22. Gambone, M. L. 1989. "Fatigue and Fracture of Titanium Aluminides, Vol. II," U.S. Air Force report WRDC-TR-89-4145.II, Wright-Patterson AFB, OH.

23. Russ, S. M., Nicholas, T., Bates, M., and Mall, S. 1991. "Thermomechanical Fatigue of SCS-6/Ti-24A1-11Nb Metal Matrix Composite," *Failure Mechanisms in High Temperature Composite Materials, AD—Vol. 22/AMD—Vol. 122,* G. K. Haritos, G. Newaz, and S. Mall, ed., American Society of Mechanical Engineers, New York, pp. 37–43.

24. Mall, S., Hanson, D. G., Nicholas, T., and Russ, S. M. 1992. "Thermomechanical Fatigue Behavior of a Cross-Ply SCS-6/β21-S Metal Matrix Composite," *Constitutive Behavior of High Temperature Composites, MD—Vol. 41,* B. S. Majumdar, G. M. Newaz, and S. Mall, ed., American Society of Mechanical Engineers, New York, pp. 91–106.

25. Larsen, J. M., Russ, S. M., and Jones, J. W. 1993. "Possibilities and Pitfalls in Aerospace Applications of Titanium Matrix Composites," *Proceedings of NATO AGARD Conference on Characterization of Fibre Reinforced Titanium Metal Matrix Composite,* September 1993, Bordeaux, France, pp 1-1-1-21.

26. Mall, S. and Ermer, P. G. 1991. "Thermal Fatigue Behavior of a Unidirectional SCS-6/Ti-15-3 Metal Matrix Composite," *Journal of Composite Materials,* Vol. 25, No. 12, pp. 1668–1686.

27. Mall, S. and Portner, B. 1992. "Characterization of Fatigue Behavior in Cross-Ply Laminate of SCS-6/Ti-15-3 Metal Matrix Composite at Elevated Temperature," *ASME Journal of Engineering Materials and Technology,* Vol. 114, No. 4, pp. 409–415.

28. Mall, S. and Schubbe, J. J. 1994. "Thermomechanical Fatigue Behavior of a Cross-Ply SCS-6/Ti-15-3 Metal Matrix Composite," *Composites Science and Technology,* Vol. 50, pp. 49–57.

29. Roush, J. T., Mall, S., and Vaught, W. H. 1994. "Thermal-Mechanical Fatigue Behavior of an Angle-Ply SCS-6/Ti-15-3 Metal Matrix Composite," *Composites Science and Technology,* Vol. 52, pp. 47–59.

30. Sanders, B. P. and Mall, S. 1994. "Longitudinal Fatigue Response of a Metal Matrix Composite under Strain Controlled Mode at Elevated Temperature." *Journal of Composites Technology and Research,* Vol. 16, No. 4, pp. 304–313.

31. Boyum, E. A. and Mall, S. 1995. "Fatigue Behavior of a Cross-Ply Titanium Matrix Composite under Tension-Tension and Tension-Compression Cycling," *Materials Science and Engineering,* Vol. A200, pp. 1–11.

32. Hart, K. A. and Mall, S. 1995. "Thermomechanical Fatigue Behavior of a Quasi-Isotropic SCS-6/Ti-15-3 Metal Matrix Composite," *ASME Journal of Engineering Materials and Technology,* Vol. 117, No. 1, pp. 109–117.

33. Sanders, B. P. and Mall, S. 1996. "Transverse Fatigue Response of a Metal Matrix Composite under Strain Controlled Mode at Elevated Temperature, Part I: Experiments," *Journal of Composites Technology and Research,* Vol. 18, No. 1, pp. 15–21.

34. Kraabel, D. L., Sanders, B. P., and Mall, S. 1997. "Tension-Compression Fatigue Behavior of a Unidirectional Titanium Matrix Composite at Elevated Temperature," *Composites Science and Technology* Vol. 57, No. 1, pp. 99–117.

2

Monotonic Response

GOLAM M. NEWAZ

Mechanical Engineering Department
Wayne State University
Detroit, MI 48202

INTRODUCTION

THE DEFORMATION AND failure characteristics of metal matrix composites (MMCs) reinforced with silicon carbide (SiC) fibers, such as SCS-6, are significantly different from the pure matrix material because of the interaction of damage entities with ensuing matrix plasticity. This interaction is complex, and their combined effects on the monotonic behavior of this class of advanced composites remain an important area of continued investigations by many researchers in the field. In pure metals, the deformation characteristics are controlled by matrix plasticity, particularly for deformation in the nonlinear regime. MMCs are typically reinforced with 30–45% by volume of fibers. Because of the processing temperatures to melt the matrix and the differences in thermal expansion coefficients of $9.0 \times 10^{-6}/°C$ and $4.5 \times 10^{-6}/°C$ for the matrix and fiber, respectively, the consolidated composite will develop significant thermal residual stresses. These stresses can have profound influence on the deformation and failure of the composites.

There are several avenues for deformation of the MMCs under consideration. When the brittle fibers and the ductile matrix deform elastically, the average constitutive behavior of the composite is easy to assess. However, when the matrix deformation is influenced by plasticity, the overall composite deformation becomes more complicated, and assessment of the constitutive behavior of the composite becomes difficult. In numerous occasions, continuum plasticity modeling has been used to determine the behavior of the composite based on the fiber and matrix properties [1–6]. These models have shown promise for describing the behavior of unidirectional MMCs in the fiber direction. But for various loading, plasticity-based models have been inadequate because of the complex damage modes involving matrix plasticity and fiber-matrix interfacial debonding [7–9]. These complexities pose significant challenges in determining the true response of MMCs, particularly using analytical methods.

The deformation mechanisms of a metal matrix composite were further investigated by using a combination mechanical measurements and microstructural analysis [10]. For this purpose, quasi-static tension and compression tests were performed on the unidirectional laminate in longitudinal and transverse directions at room temperature. Plasticity and damage were the main deformation modes. Plasticity involved deformation of the matrix between the fibers and Poisson's contraction of the matrix from the fibers. Damage involved fiber matrix debonding, matrix cracking, and fiber cracking. All of these mechanisms were present, and they were related to the appropriate stress and strain characteristics.

In this chapter, we will discuss unidirectional properties of MMCs and their deformation response in tension, compression, and shear with primary focus on room temperature response. Limited discussion on elevated temperature response is forwarded also. Some analytical aspects are discussed along with supporting experimental results to develop a comprehensive understanding of the behavior of continuous fiber MMCs.

COMMON CONSTITUTIVE MODELS AND APPROACHES

There are a number of constitutive models that predict the behavior of the composite based on constituent properties. Two earlier models are that of Dvorak and Aboudi [4,5]. These models assume a regular square array of fibers, that the fibers are continuous and parallel, that all fibers are identical, and that complete bonding exists between constituents.

To compare a few models, the coordinate system for a unidirectional ply shown in Figure 1 will be used. The x_1 axis is the fiber longitudinal direction, and the x_2 and x_3 axes are the transverse and through-the-thickness directions, respectively. In general, each of the models defines the origin of the coordinate system differently. However, x_1 will always be the longitudinal direction, and x_2 and x_3 will always be the transverse and through-the-thickness directions, respectively.

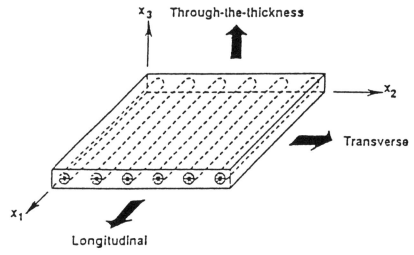

Figure 1. Ply coordinate system.

Vanishing Fiber Diameter Model

The Vanishing Fiber Diameter (VFW) model was first suggested by Dvorak and Bahei-El-Din [1,2] for use as a micromechanics model for metal matrix composites. The VFW model was incorporated into a two-dimensional laminate code called AGLPLY [2]. This code was capable of elastic-plastic analysis. Dvorak and Bahei-El-Din [1–4] used this code to perform numerous studies on the plasticity of laminated composite plates.

MODEL DESCRIPTION

The vanishing fiber diameter model is designed for the predictions, in an average sense, of lamina and laminate properties and stress-strain behavior. The model consists of an elastic-plastic matrix unidirectional reinforced by continuous elastic fibers. Both constituents are assumed to be homogeneous and isotropic. The fibers are assumed to have a very small diameter; although the fibers occupy a finite volume fraction of the composite, they do not interfere with matrix deformation in the transverse and thickness direction. Figure 2(a) shows a schematic of this lamina model. It can also be represented by parallel fiber and matrix bars or plates with axial coupling, as illustrated in Figure 2(b).

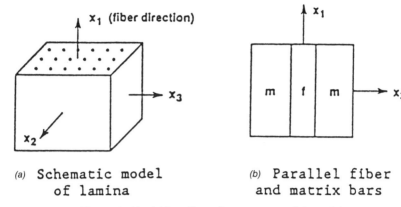

(a) **Schematic model of lamina**

(b) **Parallel fiber and matrix bars**

Figure 2. Vanishing fiber diameter material model.

If the Cartesian coordinates are chosen so that x_1 coincides with the fiber direction, the second-order tensor components of stress and strain, σ and ε, are expressed as:

$$\sigma = [\sigma_{11}\ \sigma_{22}\ \sigma_{33}\ \sigma_{12}\ \sigma_{13}\ \sigma_{23}]^T$$
$$\varepsilon = [\varepsilon_{11}\ \varepsilon_{22}\ \varepsilon_{33}\ \gamma_{12}\ \gamma_{13}\ \gamma_{23}]^T \tag{1}$$

where $\gamma_{ij} = 2\varepsilon_{ij}$ $(i,j = 1,2,3; i \neq j)$ are the engineering shear strain components.

For equilibrium and compatibility, several requirements are imposed on the material model shown in Figure 2. The stress average in each constituent can be related to the overall composite stress in the longitudinal (fiber) direction as follows:

$$\bar{\sigma}_{11} = v_f \sigma_{11}^f + v_m \sigma_{11}^m \tag{2}$$

A bar over a symbol indicates overall composite stress or strain, and the subscripts f and m denote quantities related to the fiber and matrix. The volume fractions v_m and v_f are such that $v_f + v_m = 1$. The other stress components in each constituent were assumed to be uniform and to obey the following equilibrium equations:

$$\bar{\sigma}_{22} = \sigma_{22}^f = \sigma_{22}^m$$
$$\bar{\sigma}_{33} = \sigma_{33}^f = \sigma_{33}^m$$
$$\bar{\sigma}_{12} = \sigma_{12}^f = \sigma_{12}^m \tag{3}$$
$$\bar{\sigma}_{13} = \sigma_{13}^f = \sigma_{13}^m$$
$$\bar{\sigma}_{23} = \sigma_{23}^f = \sigma_{23}^m$$

The only constraint in the model is in the longitudinal (fiber) direction; the matrix and fiber must deform equally. Thus:

$$\bar{\varepsilon}_{11} = \varepsilon_{11}^{f} = \varepsilon_{11}^{m} \tag{4}$$

The other strain components can be related to the overall attainment, as follows:

$$\bar{\varepsilon}_{ij} = v_f \varepsilon_{ij}^{f} + v_m \varepsilon_{ij}^{m} \qquad (ij \neq 11) \tag{5}$$

The fiber material is assumed to be linearly elastic, and its properties may be a function of temperature. In the elastic range, the matrix material is isotropic with thermoelastic properties that are a function of temperature. Because the fibers are elastic up to failure, the inelastic strains of the lamina are caused by matrix deformation. Because the fiber imposes an elastic constraint on the matrix that affects the shape of the lamina yield surface, additional kinematic constraints appear in the hardening rule of the lamina and influence the magnitude of the overall plastic strains. The yield behavior was examined and accounted for in the formulation of the lamina constitutive equations.

Because the VFW model essentially treats the transverse properties as if the fiber and matrix were in series with each other, there is no local constraint in the matrix transverse to the fiber. Thus, the VFW model predicts only longitudinal stresses in the fiber and matrix when a unidirectional ply is subjected to any thermal or mechanical load in the fiber direction.

PROGRAM DESCRIPTION

The AGLPLY program predicts MMC properties using the VFW model. The AGLPLY program is very user-friendly. Any symmetric layup can be modeled. The material properties of both the fiber and matrix are input as a function of user-selected temperatures. For each temperature, a stress-strain relationship is given by specifying an arbitrary number of points on the material stress-plastic strain curve. Intermediate material properties are found by linear interpolation between the given data points. Each ply can consist of a different fiber and matrix combination. Any combination of inplane loads and temperature paths can be specified.

The AGLPLY program assumes that the fiber is always elastic; the matrix, however, may have elastic-plastic properties. The matrix yield is determined by the Von Mises yield criteria. The user has a choice of either the Ziegler or Phillips hardening rule, both of which represent kinematic hardening.

The output of the AGLPLY program consists of overall elastic laminate properties and average stresses and strains for both the fiber and
matrix in each ply. Instantaneous laminate moduli and coefficients of
thermal expansion are also printed for each load or temperature step.

Aboudi Model

Aboudi has developed a continuum model for the prediction of the
average behavior of the fiber-reinforced composites whose constituents are elastoplastic work-hardening materials. The derived
constituent theory is summarized in References [5,6].

MODEL DESCRIPTION
The model is based on the assumption that the continuous fibers extend in the x_1 direction and are arranged in a double-periodic array in
the x_2 and x_3 directions, see Figure 3(a). As a result of this periodic
arrangement, it is sufficient to analyze a representative cell as shown
in Figure 3(b). The model is formulated to predict the average behavior of the composite using a first-order theory in which the displacement in each subcell is expanded linearly in terms of the distances
from the center of the subcell. The effective constitutive relations are
generated by imposing continuity of displacements and tractions
across the boundaries of the individual subcells. Closed-form expressions that relate the overall applied stresses to the stresses in the
fiber and the individual subcell in the matrix [6] predict the effective
response of the composite under arbitrary loading conditions, resulting in an effective transverse isotropic constitutive law.

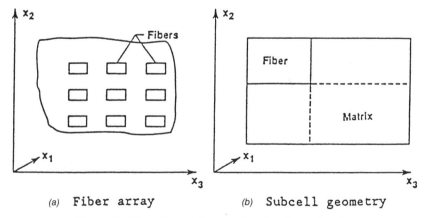

(a) **Fiber array**   (b) **Subcell geometry**

Figure 3. Aboudi model geometry showing subcells.

The present continuum theory assumes both fiber and matrix to be elastic in the linear region end plastic work-hardening in the nonlinear region. The elastic-viscoplastic formulation of Bodner [11] was used to characterize the material in the inelastic region. This unified theory can characterize the plastic behavior of a material with isotropic work-hardening and temperature-dependent properties. Anisotropic hardening with the multidimensional Bauschinger effect can also be incorporated in the unified theory of Bodner [11]. For elastic-plastic materials with isotropic hardening, the plastic strain rates are given by the Prandtl-Reuss flow rule in which plastic incompressibility is assumed.

In the unified theory of Bodner, z_0, z_1, m, and n are parameters that characterize the behavior of the matrix material in the inelastic range. The parameter z_0 is related to the yield stress of a uniaxial stress-strain curve of the material, and z_1 is proportional to the ultimate stress. The material constant m determines the rate of work hardening, and the rate sensitivity of the material is controlled by the constant n. Decreasing n also lowers the overall level of the stress-strain curve in the plastic range.

Although these models have provided a basis for predicting the complex monotonic behavior of MMCs, their development was not grounded in actual deformation mechanisms. It was assumed that plastic deformation of the matrix governs the overall composite response. However, rarely do composites behave similarly to the matrix material only. More often, a composite may show a combination of damage and plasticity, making interpretation of overall response more complex and prediction even more challenging. Micromechanics computational models based on unit cells that represent or capture the damage and plasticity condition in the composite provides a practical approach to address many of the MMC constitutive modeling difficulties. These models have been quite successful [12] and are rational alternatives to the plasticity-based models discussed earlier.

Ahmad-Nicholas Model

Consider composites in which inelastic deformation characteristics of the constituents are significantly different. For example, in a metal matrix composite the matrix material is generally much more prone to inelastic deformation than the fiber material and (if present) the interfacial zone material. Then, under stress, inelastic deformation of the composite will be largely due to inelastic deformation response of only one of the components, which, in the case of MMCs, is the matrix material. Thus, inelastic deformation of the composite will occur when the stress in the matrix is sufficient to cause inelastic deformation of the matrix material. Inelastic deformation of the matrix material can be described by well-established theories. However, a composite's deformation cannot always be attributed only to elastic and inelastic deformation of its constituents. Micromechanical damage, such as interfacial separation and sliding, matrix cracking, and discrete fiber fractures can contribute to global composite deformation. The contributions of these and other possible forms of damage to the global composite deformation are caused by mutually distinct phenomena. Therefore, a mechanistic model for estimating global deformation response should include representations of (at least the dominant) damage mechanisms. For example, if fiber-matrix separation is a plausible damage mechanism in a certain composite, then conditions for its occurrence and growth should be part of a mechanistic model.

With the above as prelude, the authors propose the following yield function [13]:

$$J_2^A = \sum_{i=1}^{3} \frac{1}{3}(N_{ii}\sigma_{ii} + R_{ii})^2 + (N_{12}\sigma_{12} + R_{12})^2 + (N_{23}\sigma_{23} + R_{23})^2 + (N_{31}\sigma_{31} + R_{31})^2$$

$$- \frac{1}{3}[(N_{11}\sigma_{11} + R_{11})(N_{22}\sigma_{22} + R_{22}) + (N_{22}\sigma_{22} + R_{22})(N_{33}\sigma_{33} + R_{33})$$

$$+ (N_{33}\sigma_{33} + R_{33})(N_{11}\sigma_{11} + R_{11})] \qquad (6)$$

where N_{ij} are non-negative dimensionless functions of stress (σ_{ij}) and inelastic strain (ε_{ij}^I) in the composite. R_{ij} are components of initial stress, and J_2^A is analogous to the second stress invariant (J_2) for homogeneous to the Von Mises yield criterion, we assume that the composite yields when:

$$J_2^A = \sigma_0^2/3 \qquad (7)$$

where σ_0 is a reference stress. Also, Ahmad and Nicholas [13] assume that damage in the form of fiber-matrix separation (debonding) and fiber-matrix sliding occurs when:

$$\kappa_{22}\sigma_{22} = I^r \qquad (8)$$

$$\xi_{ij}|\sigma_{ij}| = I_{ij}^s, \; i \neq j \qquad (9)$$

where I^r and I_{ij}^s are normal and shear interface bond "strengths," and κ_{22} and ξ_{ij} are dimensionless functions of σ_{ij} and ε_{ij}^I. We define an effective stress for the composite as follows:

$$\sigma_e = \sqrt{3J_2^A} \qquad (10)$$

We assume that the inelastic strain-stress relation is that given by the following equation that is analogous to the Prandl-Reuss equations:

$$d\varepsilon_{ij}^I = d\varepsilon_{ij}^D + \frac{\partial J_2^A}{\partial \sigma_{ij}} d\lambda_{ij} \qquad (11)$$

in which $d\varepsilon_{ij}^I$ are components of inelastic strain increment and $d\lambda_{ij}$ are non-negative constants that may vary throughout the loading history. The term $d\varepsilon_{ij}^D$ represents strain induced by micromechanical damage. Thus:

$$d\varepsilon_{ij}^I = d\varepsilon_{ij}^D + d\lambda_{ij}\left(N_{ij} + \sigma_{ij}\frac{\partial N_{ij}}{\partial \sigma_{ij}}\right)(S_{ij}^A + \mathscr{R}_{ij}) \qquad (12)$$

in which S_{ij}^A is analogous to deviatoric stress tensor (S_{ij}) in homogeneous materials and $\mathscr{R}_{ij}$ is the deviatoric stress corresponding to initial stress R_{ij}. S_{ij}^A and $\mathscr{R}_{ij}$ are defined as follows:

$$S_{ij}^A = N_{ii}\sigma_{ii} - \frac{1}{3}\sum_{i=1}^3 N_{ii}\sigma_{ii} \qquad (13)$$

$$\mathscr{R}_{ii} = R_{ii} - \frac{1}{3}\sum_{i=1}^3 R_{ii} \qquad (14)$$

for $i \neq j$:

$$S_{ij}^A = 2N_{ii}\sigma_{ii} \qquad (15)$$

$$\mathscr{R}_{ii} = 2R_{ii} \qquad (16)$$

The factor 2 in Equations (15) and (16) has been included so that shear strain components $(d\varepsilon_{ij}^I, i \neq j)$ conform to the engineering definition of shear strains. An expression for $d\varepsilon_{ij}^D$ should depend on the nature of the active damage mechanism(s). In the case of fiber-matrix debond damage, the following simple expression may be appropriate:

$$d\varepsilon_{ij}^D = \frac{N_{ij}}{E_{ij}}d\sigma_{ij}, \text{ except } d\varepsilon_{11}^D = 0. \qquad (17)$$

THERMAL RESIDUAL STRESSES IN UNIDIRECTIONAL MMCS

Thermal residual stresses can be estimated for MMCs using either closed-form elasticity solution or by conducting finite element stress analysis. Although the closed-form solutions may not be the most accurate, they can be very useful to conceive and understand the over-all effects of the role of fiber and matrix in developing the residual stress in the composite.

Thermal residual stresses are developed in the reinforcement and matrix if both the following conditions occur: the coefficient of thermal expansion of the fiber, α_f, and the matrix, α_m, are different, and the composite experiences a temperature change after solidification of the matrix during fabrication.

For a concentric fiber embedded in matrix (Figure 4), the stresses in matrix for plane strain condition are given by [14]:

$$\sigma^r = \frac{\Delta\alpha\Delta T\,[(\eta^2 / \rho^2) - 1]}{(c + e)\eta^2 + d - c} \tag{18}$$

$$\sigma_\theta = \frac{-\Delta\alpha\Delta T\,[(\eta^2 / \rho^2) - 1]}{(c + e)\eta^2 + d - c} \tag{19}$$

$$\sigma_z = \nu_m(\sigma_r + \sigma_\theta) \tag{20}$$

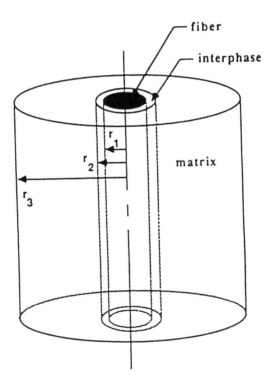

Figure 4. Concentric cylinder model.

and

$$\tau_{r\theta} = 0 \qquad (21)$$

where

$$\eta = \frac{R_m}{R_f}$$

$$\rho = \frac{r}{R_f}$$

$$c = \frac{1 - v_f}{E_f}$$

$$d = \frac{1 - v_m}{E_m}$$

$$e = \frac{1 + v_m}{E_m}$$

$$\Delta\alpha = \alpha_m - \alpha_f$$

v_m, v_f is Poisson's ratio of matrix and fiber; α_m, α_f is thermal expansion coefficient of matrix and fiber; ΔT is temperature range

At various locations close to the fiber, e.g., at the fiber-matrix interface, there is a likelihood of yielding if the combined radial and circumferential stresses are high. For the purpose of analysis, we will assume that the likelihood for initiation of microcracks is highest at a location when yielding occurs.

RESIDUAL STRESSES DUE TO COOLDOWN FROM PROCESSING TEMPERATURE

The fiber and the matrix have the following typical room temperature properties:

$$\alpha_m = 9.9 \times 10^{-6}/°C$$

$$\alpha_f = 4.86 \times 10^{-6}/°C$$

$$E_m = 99.3 \text{ GPa}$$

$$E_f = 400 \text{ GPa}$$

$$v_m = 0.35$$

$$v_f = 0.25$$

Based on these values and Equations (18 to 20), for a processing temperature of 815°C and cooled down to 23°C, the local stresses at interface are $\sigma_r = -190$ MPa, $\sigma_\theta = 378$ MPa, and $\sigma_z = 65$ and the corresponding equivalent stress is 492 MPa. Note that the radial stress is compressive at the interface. All of the calculations are based on a volume fraction of 0.4. If fibers are locally taken to be closely spaced for a fiber volume fraction of 0.7, the equivalent stress will be higher: 500 MPa. However, yielding still would not occur because at room temperature, matrix yield stress is about 800 MPa. Based on these results, it is anticipated that cracks will not occur at the fiber-matrix interface as a result of matrix yielding. However, it is noted that the residual stresses are substantial in magnitude. Nimmer et al. [15] used finite element stress analysis to investigate residual stresses thoroughly for MMCs.

MECHANICAL BEHAVIOR

Tensile Loading

Mechanical behavior of unidirectional MMCs can be understood well in the context of the loading and unloading response because this type of loading can provide valuable information on the relative contributions of plasticity and damage in the material for monotonic loading. Mechanical characterization of MMCs has received considerable attention, and researchers have addressed many aspects of the MMC behavior [16–22].

Loading and Unloading Response

Figure 5 shows the stress-strain response for a 0° specimen. Results include three tested specimens from Battelle's data and one from NASA-Lewis [23]. The reproducibility is excellent with nonlinear de-

Table 1. Temperature-dependent material properties.

Temperature (°C)	Fiber		Matrix	
	E_f (GPa)	υ_f	E_m (GPa)	υ_m
23	400	0.30	99.3	0.32
204	400	0.30	92.4	0.32
426	400	0.30	84.8	0.32
648	400	0.30	84.8	0.32
815	400	0.30	34.5	0.32

formation starting at approximately 0.55% strain. In Figure 5, specimen 0-9 was partially unloaded from about 0.55% strain (point B) and then reloaded to failure. Specimen 0-1 was completely unloaded from a higher value of total strain (point D, 0.8%). Calculation of the slopes of loading and unloading lines indicates that the stiffness of the specimens was essentially constant, with an effective elastic modulus of 176.5 GPa (25.6 Msi). In addition, there was significant strain offset at zero load when specimens were unloaded from the inelastic regimen of deformation. Thus, these results appear to indicate that the primary deformation mode was plasticity.

Figure 6 illustrates the stress-strain behavior for four 90° specimens. Once again, the figure illustrates excellent reproducibility of the stress-strain response. The 90° specimen shows a characteristic three-stage behavior. The initial elastic modulus was 111 GPa (16.1 Msi). The 90-20 specimen shows a bilinear unloading response in stage II when unloaded from strain level of 0.5% strain. The unloading slope is approximately 43% less than the loading stiffness, implying a significant contribution of damage. Although not shown here, additional experiments with the 90% lamina showed that the reloading curves were almost identical with the unloading curves, illustrating the bilinear response of the stress-strain response.

Specimen 90-16 in Figure 6 was unloaded from stage III of the stress-strain curve. The unloading stiffness for this specimen was ap-

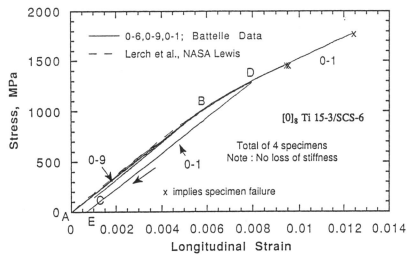

Figure 5. Stress versus longitudinal strain for 0-degree Ti 15-3/SCS6 composites.

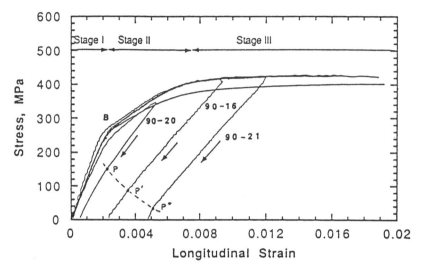

Figure 6. Stress versus the longitudinal strain for 90-degree Ti 15-3/SCS6 specimens. Unloading was done for different specimens at different strain levels.

proximately 56% less than the initial loading stiffness, which is a further reduction of 13% from the stage II initial loading modulus. Also, the offset of 0.23% strain on unloading is significant. These results appear to indicate that the deformation in stage III is controlled by both damage and plasticity. Obviously, the interpretation of the 90° unloading response is much more difficult than the 0° loading response. We will explore the fine points of the loading and unloading behavior of the 90° composite in the following paragraphs.

In Figure 7(a)–7(c), the idealized response of a solid is illustrated for three different deformation mechanisms. If the onset of nonlinearity is due to damage (microcracks and debonding) only [Figure 7(a)] with absence of plasticity and if complete closure takes place on unloading, then the unloading curve should return to zero position as shown. For only plastic deformation, i.e., yielding in a classical sense [Figure 7(b)], the unloading curve will have the same slope as the initial loading slope. When the deformation characteristics include both plasticity and microcracking [Figure 7(c)], then the unloading line will have a lower slope than the initial loading line. Note that Figure 7(a) and 7(b) represent extreme cases in inelastic material deformation and Figure 7(c) represents an intermediate behavior.

Based on the explanation in the previous paragraph, we can attempt to rationalize the unloading behavior in stage II in Figure 6. The first portion of the bilinear unloading curve has a significantly

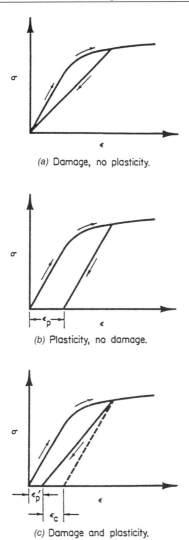

(a) Damage, no plasticity.

(b) Plasticity, no damage.

(c) Damage and plasticity.

Figure 7. Loading-unloading response due to: (a) damage, no plasticity; (b) plasticity, no damage; and (c) damage and plasticity.

lower slope than the initial loading slope, and the unloading line almost passes through the origin: these imply that deformation due to damage is dominant in stage II. The second portion of the bilinear curve has a slope that is larger than the initial unloading slope, resulting in a small strain offset at zero load. Because the strain corresponding to the transition point P (~ 0.002) is approximately the same as the second portion of unloading, slope is dominated by crack closure effects rather than small scale plasticity, although the latter was observed microscopically.

Point P, which is the inflection point, appears to be the point at which partial fiber-matrix debonds are closed on unloading. It appears that residual compressive stress (from prior processing) is still active or unrelaxed in stage II. Fiber-matrix interface above point P is under effective compression from residual compressive stresses. Thus, the difference in slopes of the two portions of the unloading curve represents the differences in the deformation response of the composite under effective tension at the fiber-matrix interface and effective compression at the interface. The unloading in stage III (Figure 6) also contains characteristics of the bilinear unloading response (the inflection point, P' is lowered) with significant debonding damage related to strain, of about 0.0025 and a plastic offset strain of about 0.0025. Thus, at this point of contribution from damage and plasticity are similar. The lowering of inflection point P' on unloading can be attributed to lowering of stresses that occur at the fiber-matrix interface, because of extensive debonding, microcracking at reaction zone, and plastic deformation (which influences stress redistribution) in the matrix during stage III deformation. Additional straining into stage III results in a greater drop of inflection point, P''. A curve is drawn joining these points to illustrate the shift in the inflection point as a function of loading (Figure 6). These inflection points can best be observed by eyeing Figure 6 from its edge in the plane of the figure. The effect of compressive residual stress on raising the yield stress of structural materials is discussed by Calladine [24].

Poisson's Ratio

Poisson's ratio in structural materials may provide valuable information on the state of damage and plasticity. Plastic deformation is associated with an increase in Poisson's ratio with fully plastic conditions occurring at Poisson's ratio of 0.5. If the damage occurs in a material due to microcracks, Poisson's ratio will decrease. A continuum damage mechanics approach was offered by Talreja [25], showing the decrease of elastic coefficients, including Poisson's ratio due to damage in fiber-reinforced elastic composites. That work provides the basis for understanding the effect of damage on elastic properties.

Plots of Poisson's ratio (Figures 8 and 9) for 0 and 90° specimens show that Poisson's ratio slightly increases for the 0° and decreases almost to zero for the 90° specimen. These confirm that the defamation in the 0° specimen is primarily due to plasticity and that of the

90° specimen is dominated by damage in the form of fiber-matrix debonding. Note that in the 90° lamina, even elastic-plastic calculations do indicate that Poisson's ratio should decrease with inelasticity. However, the decreases that were observed were substantially larger than any that would be predicted without incorporating damage.

Compression Loading

Monotonic behavior of MMCs under compression can be significantly different from that observed for monotonic tension. Therefore, no generalization can be made from the tensile response to predict the compressive response. The roles of the fiber, matrix, and interface are altered as local stresses acting on these constituents can change from one loading mode to the other. Furthermore, the failure mechanisms can be quite different.

The typical stress-strain response including unloading for a [0]₈ specimen is shown in Figure 5. The compression test result without buckling guides is also shown in Figure 5. The elastic moduli in tension and compression were similar, ranging between 175 and 179 GPa. Also,

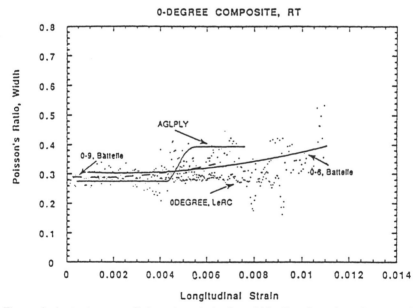

Figure 8. *Instantaneous Poisson's ratio in the width direction plotted versus the longitudinal strain for 0-degree specimens; the lines correspond to least-squares best-fit polynomials to the data. The results of the AGLPLY analysis is also shown.*

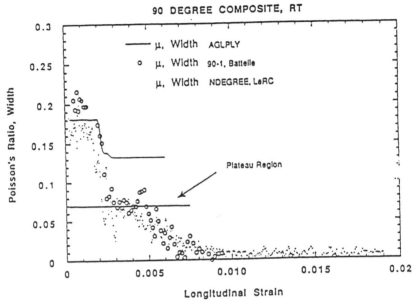

Figure 9. *Instantaneous Poisson's ratio in the width direction plotted versus the longitudinal strain for 90-degree specimens. The results of the AGLPLY analysis is also shown by the inverted S-shaped curve. The arrow points to the plateau region during Stage II of stress-strain response.*

the corresponding tension data are presented in the same plot. It may be noted that the compression capability of the $[0]_8$ lamina is significantly higher than the tension case both in terms of the onset of non-linearity and the higher ultimate strength (when premature buckling is avoided). The compression specimen was not loaded to failure. However, experimental efforts suggest that the compression strength will be quite high. In the absence of restraint for buckling, it is clear that premature buckling takes place, resulting in no useful information in the nonlinear deformation regimen. The unloading line is parallel to the elastic loading line for the $[0]_8$ composite and indicates that the deformation response is controlled primarily by plastic deformation.

The onset of nonlinear deformation at room temperature (RT) in compression (~2200 MPa) was significantly higher than in tension (~980 MPa). This difference is due to a residual tensile stress in the matrix; calculations show that for this system the axial residual stresses at RT are approximately 412 MPa in the matrix at the fiber-matrix interface and 1000 MPa in the fiber. Assuming an elastic one-dimensional response, the residual axial stress can be calculated using the formula $S_r = [(S_c - S_t)/2]*E_m/E_c$, where S_c and S_t are the measured proportional limits in compression and tension, and E_m and E_c are matrix and composite moduli. Using this formula, the tensile axial residual stress in the matrix is calculated to be 335 MPa.

Figure 10 shows the typical stress-strain response of the $[90]_8$ lamina for the three different temperatures, namely, room temperature, 538°C (1000°F), and 650°C (1200°F). In each case, the specimens were unloaded in the nonlinear regimen. There is a progressive loss in elastic modulus and proportional limit as a function of temperature as given in Table 2. The distinct knee observed in tension for the onset of nonlinearity in the $[90]_8$ lamina (Figure 11) is clearly absent in the case of compression. The onset of nonlinearity in the tension stress-strain curve was attributed to fiber-matrix debonding—a dominant damage mechanism for the $[90]_8$ MMC lamina. Note the large work-hardening period for the MMC, extending from approximately 0.5 to approximately 1% strain. This period was much longer than for the matrix material.

The room temperature and 538°C compression stress-strain results are compared in Figure 11 with those of the tension case to accentuate the difference in the response of the same SCS-6/Ti-15-3 composites under different loading conditions. This plot also contains the pure matrix stress-strain response in tension at room temperature

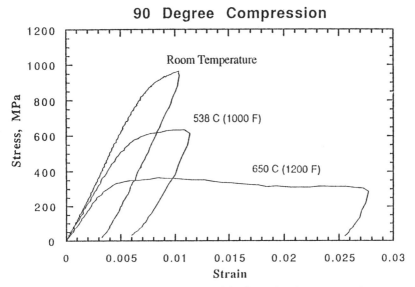

Figure 10. Mechanical response of $[90]_8$ lamina in compression.

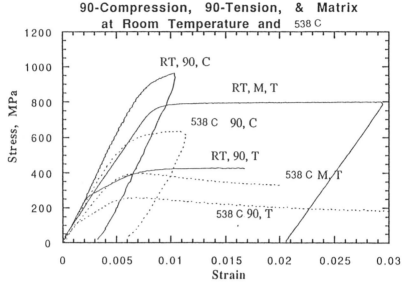

T = Tension, C=Compression, M = Matrix

Figure 11. Comparison of compression and tension response of [90]₈ lamina. Matrix behavior in tension is shown too.

and at 538°C. There are two important observations that can be made from this comparison: (1) the initial elastic or tangent moduli under tension and compression are similar, and (2) the proportional limits in compression are higher than for their corresponding tension counterparts. The higher proportional limit can be explained by the presence of a beneficial tensile residual stress that must be overcome in compression before a compressive proportional limit is reached. A matrix average radial residual compressive stress of approximately 190 MPa at room temperature, estimated from the proportional limits in tension and compression, compares well with a prediction of approximately 200 MPa using finite element analysis (FEA).

Table 2. [90]₈ SCS-6/Ti 15-3 properties in compression.

| Temperature | Young's Modulus (GPa) | Proportional Limit | |
		Stress (MPa)	Strain
RT	125	697	0.006
538 C	108	428	0.004
650 C	92	276	0.003

The unloading characteristics for the $[90]_8$ specimens under compression are quite different from those observed in tension. The initial unloading line is parallel to the initial loading line in compression signifying plastic deformation at work. The offset strains in each case for all temperatures is significantly large (Figure 11). Further unloading results in a somewhat more compliant behavior as zero load is approached.

A considerable amount of information on Timetal composite response is available in the literature. The following summary is provided from the work of Newaz [26]. The typical compressive stress-strain responses for the longitudinal and transverse laminate with SCS-6 and Sigma fibers are shown in Figures 12 and 13. The strain to failure of the SCS-6/TIMETAL®21S composite is slightly higher than the Sigma/TIMETAL®21S composite. It may be noted that the laminates were processed at different times. The exact processing conditions were not known. Furthermore, the fibers have different diameters; for the same volume fraction, there will be more Sigma fibers in the MMC, which may influence local residual stresses and strains.

The longitudinal compression strengths (over 4000 MPa) for the MMC are more than double the tensile strength of the composites (~ 2000 MPa or less). It should be noted that the stress-strain response shown for longitudinal MMCs in Figure 12 represents behavior beyond nonlinearity and not to failure.

Unloading characteristics for the longitudinal composite for the Sigma/TIMETAL®21S composite points out that the longitudinal re-

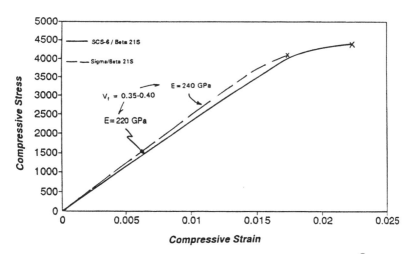

Figure 12. Monotonic compressive response of $[0]_8$ SCS-6/TIMETAL®21S composite at room temperature.

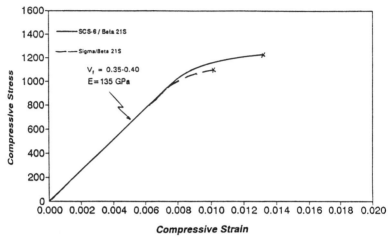

Figure 13. Monotonic compressive response of [90]₈ SCS-6/TIMETAL-21S composite at room temperature.

sponse is primarily dominated by matrix plasticity (loading and unloading compliances are about the same as shown in Figure 14). Poisson's ratio obtained from transverse and longitudinal extensometers point out that it increases slightly (Figure 15). This response further confirms that matrix plasticity dominates the longitudinal response of the composite.

For the transverse lamina, the stress-strain curve flattens out after yielding is reached and follows the nearly perfect plastic behavior exhibited by the TIMETAL®21S matrix. The stress-strain response in Figure 13 represents the behavior to specimen failure.

The unload response shows initially elastic unloading, followed by compliance change at low stresses and continues until zero load is reached (Figure 16). Poisson's ratio shown in Figure 17 reflects that initially the deformation in the transverse lamina is primarily controlled by plasticity up to 0.0035 strain (Possion's ratio increases up to this strain level). This is followed by debonding or damage of the fiber-matrix interface that subsequently lowers the Possion's ratio.

The effect on the mechanical properties, such as modulus and yield strength, are presented in Figures 18 and 19 for the two different composites. Results for MMCs with Sigma fibers were obtained only at RT and at 538°C. The rest of the test data represents SCS-6/TIMETAL®21S MMC. The bars in the plots refer to the standard deviation for three specimens tested at each condition. It is noted that the effect of temperature is greater in lowering the MMC yield strength than in the case of the modulus.

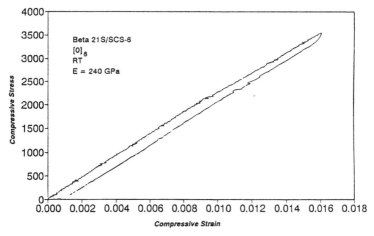

Figure 14. Compressive loading-unloading response of a [0]₈ SCS-6/TIMETAL-21S composite at room temperature showing plasticity dominated response at higher loads.

DAMAGE MECHANISMS AND FAILURE MODES FOR LONGITUDINAL AND TRANSVERSE LAMINAE

Longitudinal Lamina

Optical photomicrographs of replicas taken from polished specimens show the presence of extensive matrix plasticity in the lamina

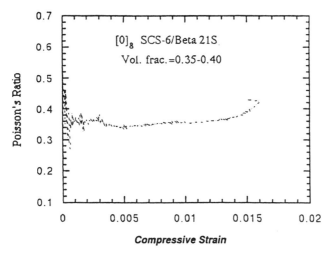

Figure 15. Compressive Poisson's ratio for a [0]₈ SCS-6/TIMETAL-21S composite.

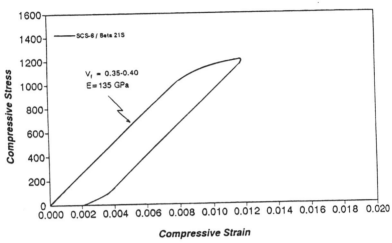

Figure 16. Compressive loading-unloading response of a [90]₈ SCS-6/TIMETAL-21S lamina showing combined effects of plasticity and damage.

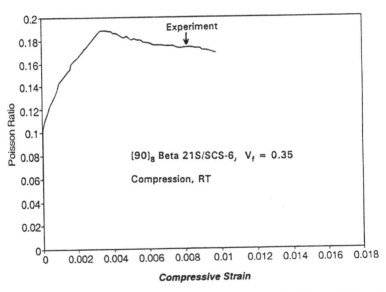

Figure 17. Compressive Poisson's ratio of a [90]₈ SCS-6/TIMETAL-21S lamina at room temperature.

46

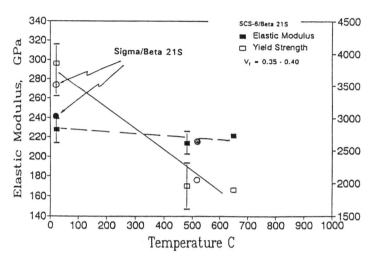

Figure 18. Effect of temperature on compressive elastic modulus and yield strength in [0]$_8$ TIMETAL-21S laminae with Sigma and SCS-6 fibers.

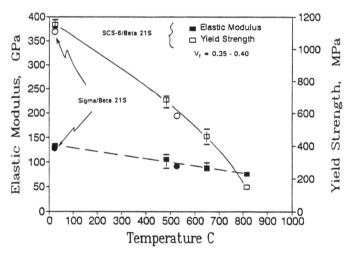

Figure 19. Effect of temperature on compressive elastic modulus and yield strength in [90]$_8$ TIMETAL-21S laminae with Sigma and SCS-6 fibers.

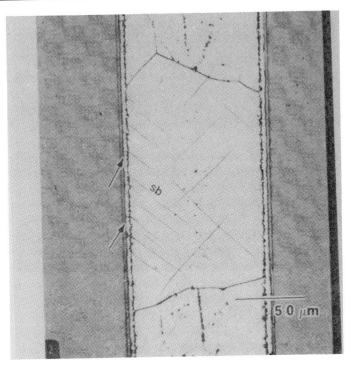

Figure 20. Extensive slip band (sb) formation in matrix (r = 0.008) in 0° specimen. Arrows indicate reaction zone cracks.

under tension (Figure 20). Fiber-matrix debonding is also evident at higher strain levels within the ply.

Transverse Lamina

Debonding is a major failure mode in transverse tension (Figure 21). The differences in tension and compression are depicted in Figure 22. The slip lines in compression are considerably more prominent compared with the similar slip lines observed in the tension case at similar locations [19,20]. In the case of tension, the initial microplasticity is followed by fiber-matrix debonding at the fiber poles at the onset of departure from the linear elastic response (knee). Debonding, as observed in the tension case at the two poles of a fiber, is absent in the compression case, because the matrix is in compression at the poles of the fiber (top and bottom points on the fibers in Figure 22).

Careful observation also reveals a very interesting damage condition in the composite under compression. There are radial cracks within the

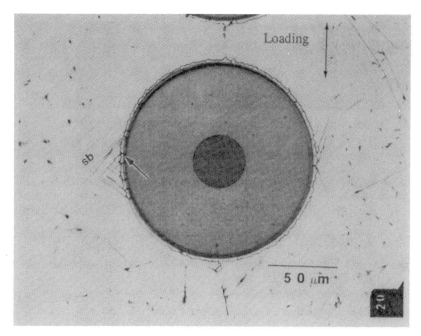

Figure 21. Slip bands associated with reaction zone microcracks in 90° specimen (r = 0.006).

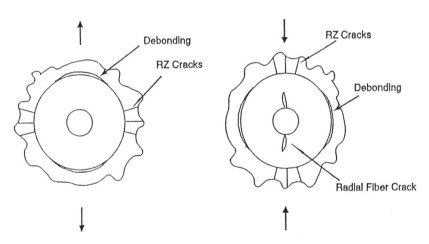

Figure 22. Orientation of damage (cracking) in tension and compression.

49

fibers oriented primarily along the compression loading axis, which form after the initial slip lines have developed in the matrix as mentioned earlier. Reaction zone cracks (RZC) are also observed. Radial fiber cracking is indicative of the low radial fiber strength that may be related to the soft carbon core in the SCS-6 fiber and the columnar grain structure of the CVD deposited SiC with grain boundary facets oriented radially. These cracks are not observed in as-received materials and were not observed in the tension-loading mode.

In the nonlinear regime, further compressive deformation of the composite results in the development of bulk plasticity in the matrix with extensive networks of shear slip bands [8]. Qualitatively, the debonding in compression can be attributed to high local tensile transverse strains at the fiber-matrix interface that develop during compressive deformation of the specimen. Shear strains at the fiber-matrix interface are also expected. However, precise computational analysis will be required to determine the contribution of normal and shear strains to the debonding phenomena observed.

Although the radial cracking of the fiber in compression is a concern, it may be noted that the most debilitating damage, i.e., one that occurs at a fairly low strain level, is fiber-matrix debonding under tension. Radial fiber cracking is observed in compression at a strain level of 0.007, whereas debonding under tension initiates around a strain level of 0.003.

The sequence and location of damage development under tension and compression can explain the different stress-strain responses of the composite. Reaction zone cracks and debonding, which are perpendicular to load axis as is the case in tension, are anticipated to have a larger effect on compliance.

OFF-AXIS LOADING

Newaz and Zhang [27] investigated off-axis response of MMC lamina. Some important results are discussed below.

Analysis of Combined Inplane Stress State

A tension test of a ply with its fibers oriented θ away from the monotonic far field loading (σ_x) is shown in Figure 23. The stresses in the material coordinate system are:

$$\sigma_L = \sigma_x \cos^2 \theta \tag{22}$$

$$\sigma_T = \sigma_x \sin^2 \theta \tag{23}$$

$$\tau_{LT} = -\sigma_x \cos\theta \sin\theta \qquad (24)$$

For $\theta = 10°$, we have:

$$\sigma_L = 0.970\sigma_x \qquad (25)$$

$$\sigma_T = 0.30\sigma_x \qquad (26)$$

$$\tau_{LT} = -0.171\sigma_x \qquad (27)$$

The strain in the loading axes are given by:

$$\varepsilon_x = \sigma_x \left[\frac{\cos^4\theta}{E_L} + \frac{\sin^4\theta}{E_T} + \frac{1}{4}\left(\frac{1}{G_{LT}} - \frac{2\nu_{LT}}{E_L}\right) \sin^2 2\theta \right] \qquad (28)$$

$$\varepsilon_y = -\sigma_x \left[\frac{\nu_{LT}}{E_L} - \frac{1}{4}\left(\frac{1}{E_L} + \frac{2\nu_{LT}}{E_L} + \frac{1}{E_T} - \frac{1}{G_{LT}}\right) \sin^2 2\theta \right] \qquad (29)$$

$$Y_{xy} = -\sigma_x \sin 2\theta \left[\frac{\nu_{LT}}{E_L} + \frac{1}{E_T} - \frac{1}{2G_{LT}} - \cos^2\theta\left(\frac{1}{E_L} + \frac{2\nu_{LT}}{E_L} + \frac{1}{E_T} - \frac{1}{G_{LT}}\right) \right] \qquad (30)$$

where E_L is the longitudinal elastic modulus, E_T is the transverse elastic modulus, G_{LT} is the inplane shear modulus, and ν_{LT} is the inplane Poisson's ratio.

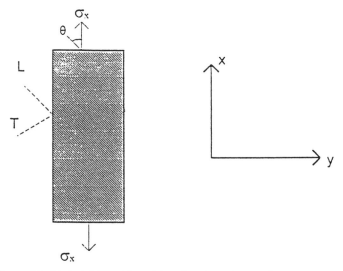

Figure 23. Material (T – L) and loading (x – y) axes for a tension test.

The transformed strains in the material axis are given by:

$$\varepsilon_L = \varepsilon_x \cos^2 \theta + \varepsilon_y \sin^2 \theta + Y_{xy} \cos \theta \sin \theta \qquad (31)$$

$$\varepsilon_T = \varepsilon_x \sin^2 \theta + \varepsilon_y \cos^2 \theta - Y_{xy} \cos \theta \sin \theta \qquad (32)$$

$$Y_{LT} = (\varepsilon_y - \varepsilon_x) \sin 2\theta + Y_{xy} \cos 2\theta \qquad (33)$$

We plotted variation of normalized material-axis strains versus loading angle θ (0~90°) with varied combination of stiffness constants in Figures 24 to 26. We can see that the maximum normalized material axis shear strain ($\varepsilon_{LT}/\varepsilon_x$) appears at different orientation angle, θ due to the difference of stiffness constants. Given the fact that at all fiber orientations, shear strain does not necessarily dominate the MMC lamina response, obtaining pure shear deformation response may not be possible as in the case of 10° off-axis PMC lamina.

Off-axis tests were made on 8-ply unidirectional Ti-15-3/SCS-6 composite, approximately 2-mm thick, with a fiber volume fraction of approximately 0.34. The SCS-6 (SiC) fiber diameter is approximately 140 μm and it contains alternating outer layers of C and Si, which protect the fiber from damage during handling. The Ti-15-3 alloy is a metastable body-centered cubic (bcc) β-Ti-alloy, the bcc phase being

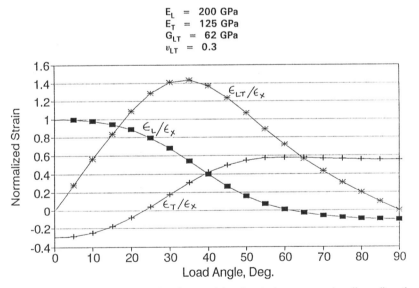

Figure 24. Variation of normalized material-axis strains versus loading direction for SiC/Ti 15-3 unidirectional MMC ply with $G_{LT} = 62$ GPa.

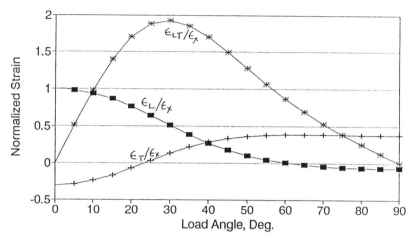

Figure 25. Variation of normalized material-axis strains versus loading direction for SiC/Ti 15-3 unidirectional MMC ply with G_{LT} = 34 GPa.

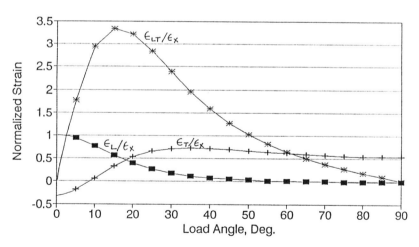

Figure 26. Variation of normalized material-axis strains versus loading direction for graphite/epoxy unidirectional PMC ply results.

stabilized by V. The material is metastable because hexagonal close
packed (hcp) α-phase, which is the stable room temperature phase of
pure Ti, precipitates when the material is held at temperatures as low
as 400°C for long periods of time.

Uniaxial test specimens were machined from the unidirectional
panel using an electric discharge machining (EDM) technique, using
specimen dimensions as shown in Figure 27. The specimens were mechanically polished after EDM machining to remove any damage associated with the machining. Specimens were prepared with 10° orientation, with fibers 10° away from the specimen axis. Monotonic
loading was applied at room temperature. All specimens were tested
in the as-fabricated condition, i.e., no heat treatment was performed
prior to the testing.

Specimens were gripped using friction grips and loaded on a servo-
hydraulic testing machine at a strain rate of approximately 0.002/sec.
The longitudinal and width strains were measured using both strain
gauges and extensometers. Following mechanical testing, specimens
were sectioned slowly using a diamond wafering blade; debonding and
fiber cracking can be significant during machining if adequate care is
not taken during sectioning. Specimens were metallographically polished, with the polished surfaces corresponding to the faces of the test
specimen. Faces were polished at surfaces and up to the first set of

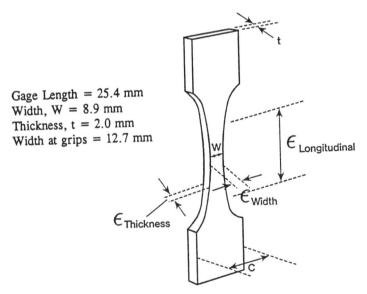

Gage Length = 25.4 mm
Width, W = 8.9 mm
Thickness, t = 2.0 mm
Width at grips = 12.7 mm

$\epsilon_{Longitudinal}$

ϵ_{Width}

$\epsilon_{Thickness}$

*Figure 27. Specimen dimensions and showing definitions of the longitudinal
(loading direction), width, and thickness strains.*

fibers. A fairly complex procedure has been developed for evaluating MMCs to preserve the characteristics of deformation. Repeated polishing was performed on a few tested and untested samples to confirm that any cracking, debonding, or slip bands was due to deformation alone and not associated with the polishing procedure. Specimens were etched using Kroll's reagent, which was effective in revealing slip bands. Metallographic specimens were examined optically and also using a scanning electron microscope (SEM).

Effect of Fiber Orientation and Matrix Shear Modulus

Figures 24 and 25 are for two MMCs materials with the same E_L (200 GPa), E_T (125 GPa), and v_{LT} (0.3), but shear modulus G_{LT} was changed from 62 to 34 GPa to study its influence. The maximum normalized material-axis shear strain changes from 1.44 at 35° to 1.8 at 30°. We can see within 10 to 15° the normalized material-axis shear strain and normalized longitudinal axis strain have similar weight of effect (1:1) with G_{LT} = 34 GPa and different effect (0.6:1) with G_{LT} = 62 GPa. Observed from Figure 26 for PMC material, the maximum normalized material-axis shear strain occurs around θ = 15°, with its value obviously larger than that of MMCs, and also within 10° to 15°, the primary effect is due to shear strain only. The interaction of normal strain between MMC and PMC are different, and severity of interaction occurs at different θ.

COMPARISON OF MECHANICAL RESPONSE OF LONGITUDINAL AND TRANSVERSE AND OFF-AXIS LAMINAE

Figure 28 shows the stress-strain response for 0, 10, and 90° specimens. Results include typical stress-strain curves for two 10° specimens and 0° and 90° specimens from the work of the first author and work from other researchers. Table 3 gives data on stress-strain at the onset of inelastic deformation for 0, 10, and 90° specimens, respectively. We can observe that the elastic portion (before onset of the inelastic response) shrank significantly (onset of nonlinearity was lowered significantly) when loading angle θ was changed from 0 to 10°, though there was no significant change in elastic modulus. This means that with a 10° loading from the fiber direction we will lose almost 40% of the yield stress capacity (stress: 1000 MPa →600 MPa; strain: 0.006→0.003) without any noticeable change in elastic modulus (E). The variation in modulus of 0 and 10° laminates may be due to fiber volume fraction between different researchers. Our experience

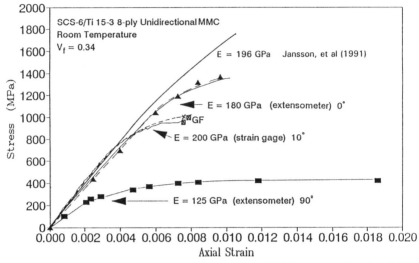

Figure 28. Stress-strain responses for two 8-ply SiC/Ti 15-3 undirectional MMC specimens with 10° loading and comparison with those of 0° and 90° specimens.

has been that a fiber volume fraction variation of 30 to 35% has been common for many Ti-15-3 MMC laminates.

Referring to References [28,29,30] the 0° response curve was divided into elastic stage I and plasticity-dominated inelastic stage II. The 90° response curve was divided into linear stage I, both plasticity and damage dominated inelastic stage III (portion of the response curve parallel to strain axis), and damage dominated inelastic stage II (transient portion between stages I and III). For 10° off-axis lamina, we anticipate that the response behavior and relative damage contribution to inelastic deformation should fall between the behavior observed for 0 and 90° laminates.

Figure 28 also shows the elastic modulus for 0 and 10° were close. It implies that the elastic responses were almost the same for both cases before the onset of inelastic deformation. The elastic modulus of 90° was 37.5% lower than that of 10° specimen. This means matrix and fiber-matrix bonds instead of fibers were taking most of the load

Table 3. Onset of inelastic deformation in Ti 15-3/SCS-6 laminates.

Ply Orientation	Stress (MPa)	Strain
0°	1000	0.006
10°	600	0.003
90°	250	0.002

in its elastic range. We can see that the yield stress capacity loss (stress: 600 MPa →250 MPa; strain: 0.003→0.002) from 10 to 90° off-axis loading was relatively small compared with the loss in which the loading orientation changed from 0 to 10°. The significance is that the yield stress capacity drops sharply when loading is off-axis for MMCs.

OFF-AXIS DAMAGE MECHANISMS AND FAILURE MODES

Figure 29 shows the appearance of a tested specimen just after failure. We can see the matrix cracking along the fiber direction. More detailed inspection shows there were a few fiber fractures along with matrix failure and fiber-matrix debonding. In situ examination of specimens using the replication technique and microscopy of specimens unloaded from different strains provided information on the evolution of damage and plasticity.

Figure 30(a), (b), (c), and (d) shows SEM of replica for a sequence of 10° specimen deformation stages. Figure 31(a) shows there were fiber-

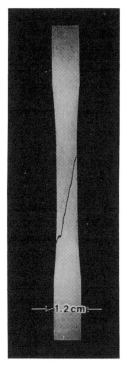

Figure 29. An 8-ply SiC/Ti 15-3 unidirectional MMC specimens with 10° loading just after failure.

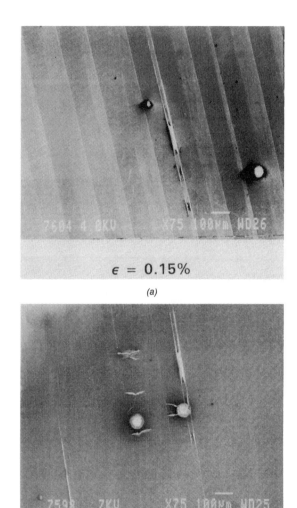

$\epsilon = 0.15\%$

(a)

$\epsilon = 0.37\%$

(b)

Figure 30. *SEM of replica for a sequence of deformation stages [(a) ε = 0.15%, (b) ε = 0.37%, (c) ε = 0.6%, (d) ε = 0.92%] of 8-ply SiC/Ti 15-3 unidirectional MMC specimens with 10° loading.*

58

$\epsilon = 0.6\%$

(c)

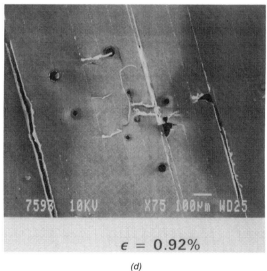

$\epsilon = 0.92\%$

(d)

Figure 30. (cont.) SEM of replica for a sequence of deformation stages (a) ε = 0.15%, (b) ε = 0.37%, (c) ε = 0.6%, (d) ε = 0.92% of 8-ply SiC/Ti 15-3 unidirectional MMC specimens with 10° loading.

59

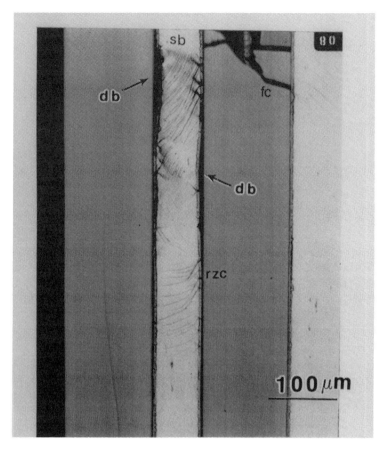

Figure 31. Fiber fractures (fc), fiber matrix debonding (db), shear bands (sb) and slip bands (sb) induced by reaction zones (rz) in a 10° off-axis specimen.

matrix debonding as early as $\varepsilon = 0.15\%$. However, obvious fiber failure and more fiber-matrix debonding appear in Figure 30(b) ($\varepsilon = 0.37\%$), which is after inelastic deformation onset strain ($\varepsilon = 0.3\%$). Figure 30(c) and (d) provides us the image of how damage developed through high level of inelastic deformation before ultimate failure of the composite occurs.

Figures 31 and 32 show the inelastic deformation mechanisms at failure. From Figure 31 we can see fracture of fibers, serious fiber-matrix debonding, reaction zone cracks, obvious shear bands, and extensive slip bands induced by reaction zone cracks. In Figure 32, re-

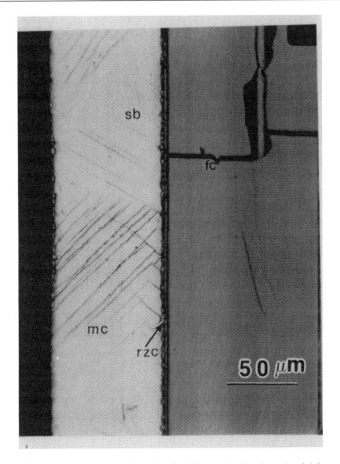

Figure 32. Reaction zone cracks (rzc) induced slip bands (sb) eventually evolved into microcracks (mc) in matrix. Fiber fracture is also seen.

action zone crack-induced slip bands are shown, which were eventually found to evolve into microcracks in the matrix.

Off-axis tension tests were performed by Sun et al. [31] on SCS-6/Ti-6-4 metal matrix composite. A one-parameter plasticity model was used to characterize the elastic-plastic properties. In addition, a micromechanical model was developed assuming elastic fiber and elastic and plastic matrix properties. This model was employed to relate the apparent yielding with the fiber-matrix separation in the MMC. From the micromechanical model, the fiber-matrix interface bond strength was estimated, and with the aid of a damage model,

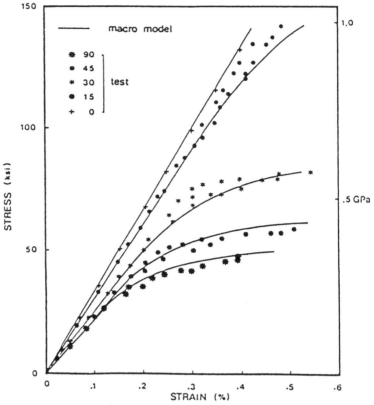

Figure 33. Comparison of the predicted and experimental off-axis stress-strain curves.

the nonlinear off-axis stress-strain curves were accurately predicted [31] as in Figure 33.

COMBINED AND SHEAR LOADING

In part I [32], experimental results from cyclic multiaxial tests on unidirectional and angle-ply silicon carbide and titanium (SiC-Ti) tubular specimens were compared with theoretical prediction from a micromechanical model that included imperfect bonding between the fiber and matrix. Fiber-matrix interfacial debonding had deleterious effects on the stress-strain response under some multiaxial loadings. Part II [33] presented experimental results indicating that [±45°]$_s$ SiC-Ti-accumulated axial strain had both time-dependent and time-independent contributions. The time-independent accumulated strain was associated with shear coupling, and the time-dependent accumulated strain was due to matrix creep. In part III [34], postfailure microstructural evaluation was used to assess the actual damage state after testing. It was not economically feasible to conduct interrupted tests due to the high cost of the tubular specimens. The results show that damage, which accrued during multiaxial loading of the tubular specimens, is concentrated around the fiber-matrix interface. Metallographic observations were compared with the stress-strain response measured by strain gauge rosettes and predictions from a theoretical model.

On numerous occasions shear loading of MMCs can be expected for a number of MMC structural components. For MMCs, this area has not received much attention, primarily due to the difficulty in developing reliable data through characterization tests. A number of test methods attempt to generate appropriate data for MMCs, which include the $[\pm45]_{2s}$ laminate. Some limited results from Reference [35] are presented in this section.

$[\pm45]_{2s}$ Laminate Response

The stress-strain response of the $[\pm45]_{2s}$ laminate at room temperature is shown in Figure 34. The $[\pm45]_{2s}$ composite certainly shows the three-stage behavior as observed in the case of $[90]_8$ specimen.

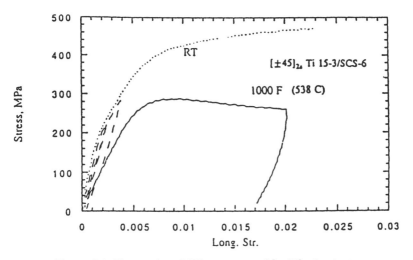

Figure 34. Elevated and RT response of $[\pm45]_{2s}$ laminate.

Unloading in stage II shows that there is some permanent offset. However, much of the strain is recovered after unloading. This is similar to what we have observed for the $[90]_8$ specimen. It appears that in this stage, the inelastic deformation of the $[\pm 45]_{2s}$ laminate is primarily due to damage, possibly due to fiber-matrix debonding. At 538°C, unloading of a $[\pm 45]_{2s}$ laminate in stage III as shown in Figure 34 exhibits significant permanent offset in strain, implying that at this stage, inelastic deformation in this laminate is controlled by plasticity.

The results presented here suggest that the monotonic response of $[0]_8$ and $[90]_8$ laminae and $[\pm 45]_{2s}$ laminate retain similar characteristics at elevated temperature compared with what is observed at room temperature. The influence of elevated temperature can be best understood by recognizing two critical factors: (1) influence on flow characteristics of the matrix material and (2) relaxation of residual stress in the composites.

The $[\pm 45]_{2s}$ laminate behavior at room temperature shows the three-stage behavior as in the case of $[90]_8$ lamina, although not as pronounced. At 538°C, stage I is lost. Again, both matrix flow stress and residual stresses are altered to influence the overall response and inelastic deformation characteristics of the $[\pm 45]_{2s}$ laminate.

SUMMARY AND CONCLUSIONS

Generalizations on the constitutive behavior of MMCs under monotonic loading cannot be made because lamina orientation and imposed loading determine the relative degrees of interaction of plastic deformation and damage (microcracking) in the composite. Deformation and failure mechanisms can be quite different for each of the lamina and laminate orientations and applied loading. In general, it is best to accept that both microcracking and plasticity will affect the overall nonlinear response. The compressive characteristics of unidirectional MMCs are significantly better than their tensile characteristics, which is contrary to the conventional notion for filamentary composites in which fiber buckling is always a concern in compression. For complex damage modes involving matrix plastic deformation and microcracking, computational constitutive models can provide more realistic predictions of the behavior of the composite. However, this modeling approach requires readjustment of the unit cell with another state of damage at different strain levels, thereby requiring more computational time to predict the overall response of the composite.

Table 4. Deformation mechanisms in tension and compression.

[90]$_8$ SCS-6/Ti 15-3

Inelastic Deformation Mechanisms	Tension		Compression	
	Strain	Stress (MPa)	Strain	Stress (MPa)
Plastic slip band	0.002	225	0.002[a]	225[a]
Debonding	0.003-0.005	280-340	0.006-0.008	700-900
Reaction zone cracks	0.005	350	0.006	700
Radial fiber cracking	—	—	0.006-0.008	700-900
Extensive matrix plasticity	>0.006	400	>0.01	900

[a]Data not retrievable at earlier strain/stress levels because of loss of replica.

For design, basic properties of MMCs have both potential and concerns. Although good strength and stiffness are likely for unidirectional lamina, the poor transverse strain of 0.003 for the onset of nonlinearity representing fiber-matrix debonding is a major concern. Residual stress-induced mechanical bonding and lack of some level of chemical bonding are other concerns because relaxation of residual stresses may result in poor performance of the transverse lamina. The key deformation characteristics of unidirectional 90° SCS-6/Ti-15-3 MMCs for tension and compression are compared in Table 4.

REFERENCES

1. Dvorak, G. J., and Bahei-El-Din, Y. A., "Elastic–Plastic Behavior of Fibrous Composites," *Journal of the Mechanics and Physics of Solids,* Vol. 27, 1997, pp. 51–72.

2. Bahei-El-Din, Y. A., Dvorak, G. J., and Utku, S., "Finite Elements Analysis of Elastic-Plastic Fibrous Composite Structures," *Computers and Structures,* Vol. 13, No. 1–3, June 1981, pp. 321–330.

3. Bahei-El-Din, Y. A., Dvorak, G. J., "Plasticity Analysis of Laminated Composites Plates," *Journal of Applied Mechanics,* Vol. 49, 1982, pp. 740–746.

4. Dvorak, G. J., and Bahei-El-Din, Y. A., "Plasticity Analysis of Fibrous Composites," *Journal of Applied Mechanics,* Vol. 49, 1982, pp. 327–335.

5. Aboudi, J., "Elastoplasticity Theory of Composite Materials," *Solid Mechanics Archives,* Vol. 11, 1986, pp. 141–183.

6. Aboudi, J., "Closed Form Constructive Equations for Metal Matrix Composites," *International Journal of Engineering Science,* Vol. 25, 1987, pp. 1229–1240.

7. Newaz, G. M., and Majumdar, B. S., "Deformation and Failure Mechanisms in MMC," *ASME Winter Annual Meeting,* AD-22, Atlanta, GA, 1991, pp. 55–66.

8. Newaz, G. M., and Majumdar, B. S., "Inelastic Deformation Mechanisms in a Transverse MMC Lamina under Compression," AD-Vol. 27, Fracture and Damage, *ASME Winter Annual Meeting,* Anaheim, CA, 1992, pp. 77–84.

9. Newaz, G. M., and Majumdar, B. S., "Failure Modes in Transverse MMC Lamina Under Compression," *Journal of Materials Science & Letters,* December 1992, pp. 551–552.

10. Bearden, K. L., and Mall, S., "Mechanical Behavior and Failure Mechanisms of SCS-9/β21S Metal Matrix Composite at Room Temperature," *Proceeding of American Society for Composite,* Cleveland, OH, 1993.

11. Bodner, S. R., "Review of Unified Elastic-Viscoplastic Theory," *Unified Constitutive Equations for Plastic Deformation and Creep of Engineering Alloys* (ed. A. K. Miller), Elsevier Applied Science Publishers, 1987.

12. Brust, F. W., Majumdar, B. S., and Newaz, G. M., "Constitutive and Damage Response of Ti 15-3/SCS-6 MMC," presented at the *ASTM Conference of Fatigue and Fracture*, Pittsburgh, PA (May 4–5, 1992).

13. Ahmad, J., and Nicholas, T., "Modelling of Inelastic Metal Matrix Composite Response under Multiaxial Loading," *ASME Symposium on Failure Mechanisms and Mechanism Based on Modelling in High Temperature Composites*, ASME Winter Annual Meeting, Atlanta, GA, November 1996.

14. Corten, H. T., "Micomechanics and Fracture Behavior of Composites," in *Modern Composite Materials* (ed. L. J. Broutman and R. H. Krock) Addison-Wesley Publishing Co., New York, April 1967.

15. Nimmer, R. P., Bankert, R. J., Russell, E. S., Smith, G. A. and Wright, K. P., "Micromechanical Modeling of Fiber/Matrix Interface Effects in Transversely Loaded SiC/Ti-6-4 Metal Matrix Composites," *Journal of Composites Technology and Research*, Vol. 13, No. 1, 1991, pp. 3–13.

16. Johnson, W. S., Lubowinski, S. J., Highsmith, A. L., Brewer, W. D., and Hoogstraten, C. A., "Mechanical Characterization of SCS6/Ti 15-3 Metal Matrix Composites at Room Temperature," NASA Technical Memorandum 1014, NASA TM-1000628, NASA Langley Research Center, 1988.

17. Sun, C. T., "Modeling Continuous Metal Matrix Composite as an Orthotropic Elastic-Plastic Material," *ASTM STP 1032*, 1989, pp. 148–160.

18. Lerch, B. A., and Saltsman, J. F., "Tensile Deformation Damage in SiC Reinforced Ti-15V-3Cr-3Al-3Sn," NASA Technical Memorandum 103620, April 1991.

19. Majumdar, B. S., and Newaz, G. M., "Thermomechanical Fatigue of a Quasi-Isotropic Metal Matrix Composites," *ASTM STP 1110*, 1991, pp. 732–752.

20. Newaz, G. M., Majumdar, B. S., and Brust, F. W., "Thermal Cycling Response of Quasi-Isotropic Metal Matrix Composites," *ASME Journal of Engineering Materials and Technology*, April 1992, pp. 156–161.

21. Majumdar, B. S., and Newaz, G. M., "Inelastic Deformation Mechanisms in MMC: Compression and Fatigue," HITEMP Review, NASA Conference Publication 10104, 1992, pp. 49–1 through 49–8.

22. Gunawardena, S. R., Jansson, S., and Leckie, F. A., "Transverse Ductilities of Metal Matrix Composites," presented at the *ASME Winter Annual Meeting*, Atlanta, GA, December, Vol. AD-22, 1991, pp. 23–30.

23. Lerch, B. A., NASA-Lewis, Private Communication, December 1990.

24. Calladine, C. R., *Plasticity of Engineers,* John Wiley & Sons, New York, 1985.

25. Talreja, R., *Fatigue of Composite Materials,* Dept. of Solid Mechanics, The Technical University of Denmark, 1985.

26. Newaz, G. M., "Constitutive Response and Deformation Mechanisms in Unidirectional Metal Matrix Composites under Compression," *Composite Materials: Testing and Design (Twelfth Volume), ASTM STP 1247* (ed. R. B. Deo and C. R. Saff), American Society for Testing and Materials, 1996, pp. 278–291.

27. Newaz, G. M., and Zhang, K., "Inelastic Response of Off-Axis MMC Lamina," accepted for publication in the *ASME Journal of Materials and Technology,* 1997.

28. Newaz, G. M., and Majumdar, B. S., "Deformation and Failure Mechanism in Metal Matrix Composite," The *Winter Annual Meeting of the America Society of Mechanical Engineers,* Atlanta, GA, 1991, pp. 1–6.

29. Majumdar, B. S., and Newaz, G. M., "Inelastic Deformation of Metal Matrix Composite: Plasticity and Damage Mechanisms," *Philosophical Magazine A,* Vol. 66, No. 2, 1992, pp. 187–212.

30. Newaz, G. M., and Majumdar, B. S., "Failure Modes in Transverse MMC Lamina under Compression," *Journal of Materials Science and Letters,* Vol. 12, 1993, pp. 551–552.

31. Sun, C. T., Chen, J. L., Sha, G. T., and Koop, W. E., "Mechanical Characterization of SCS-6/Ti-6-4 Metal Matrix Composite," *Journal of Composite Materials,* Vol. 24, 1990, pp. 1029–1059.

32. Lissenden, C. J., Herakovich, C. T., and Pindera, M.J., "Response of SiC/Ti under Combined Loading—Part I: Theory and Experiment for Imperfect Bonding," *Journal of Composite Materials,* Vol. 29, 1995, pp. 130–155.

33. Lissenden, C. J., Herakovich, C. T., and Pindera, M-J., "Response of SiC/Ti under Combined Loading—Part II: Room Temperature Effects," *Journal of Composite Materials,* Vol. 29, 1995, pp. 1404–1417.

34. Lissenden, C. J., Lerch, B. A., and Herakovich, C. T., "Response of SiC/Ti under Combined Loading—Part III: Microstructure Evaluation," *Journal of Composite Materials,* Vol. 30, 1996, pp. 84–108.

35. Newaz, G. M., and Majumdar, B. S., "A Comparison of Mechanical Response of MMC at Room and Elevated Temperatures," *Journal of Composites Science & Technology,* Vol. 50, 1994, pp. 85–90.

————————— 3 —————————

Micromechanical Theories

GEORGE J. DVORAK
Rensselaer Polytechnic Institute
Troy, NY 12181

YEHIA A. BAHEI-EL-DIN
Cairo University
Cairo, Egypt

JOSEPH R. ZUIKER
Materials Directorate
Wright Laboratory
Wright-Patterson AFB, OH 45433-7817

INTRODUCTION

Motivation

TITANIUM MATRIX COMPOSITES (Tmcs) have been considered for a wide variety of aerospace applications, including structural components of the National AeroSpace Plane; actuators in the F-22 [1,2]; and a variety of turbine engine applications, including fan and compressor rings, engine shafts, hollow fan blades, and weight and stiffness critical static structures [3,4]. Fabrication, processing, and in-service thermal and mechanical loading of these components often cause inelastic deformations in the material. In many actual systems, the elastic-strain range of the matrix is much smaller than the failure strain of the elastic-brittle fiber. Similarly, the temperature changes that may cause yielding in a stress-free composite are often smaller than those encountered in service. However, the total strains seen in fibrous systems are also small, seldom exceeding the failure strain of the fiber, which is usually found in the range of 0.01 to 0.02. Therefore, in contrast to metals, inelastic deformation of TMCs is usually

69

limited to small strains, but it may affect much of the useful load range of structural parts designed to utilize the high strength of the fibers. Under some loading conditions, local stress concentration caused in the presence of fibers may cause plastic flow of the matrix at overall stress levels that are smaller than the yield stress of the unreinforced matrix. This is especially true in the case of TMCs, which typically have weakly bonded fiber-matrix interfaces that may separate under transverse loading conditions, thereby inducing high-stress concentrations in the nearby matrix material.

The purpose of this chapter is to give a brief review of the micro-mechanical methods for predicting the inelastic response of TMCs. First, we shall discuss some of the features of TMCs that must be considered to develop accurate micromechanical models. Then, we consider general features of thermoelastic behavior, such as evaluation of overall thermomechanical properties and local thermomechanical fields; micromechanical techniques for estimating phase concentration factors, thermal concentration factors, and transformation stress and strain concentration factors; and the use of transformation fields to simulate a variety of behaviors. Next, the inelastic behavior of heterogeneous materials is reviewed. The focus is on recent efforts in the areas of transformation field analysis and their application to metal matrix composites with variable constitutive response.

The notation used is similar to that introduced by Hill [5]. Vectors are denoted by lowercase boldface letters, e.g., a, b; matrices are denoted by uppercase boldface letters, e.g., L, M. In the contracted notation used, those will typically be (6×1) vectors and (6×6) matrices. L^{-1} denotes the inverse of L, defined if it exists to satisfy $LL^{-1} = I = L^{-1}L$, where I is the unit matrix.

Unique Features of TMCs

TMCs possess several unique features that must be accounted for in developing accurate methods of mechanical analysis and life prediction. Any analysis of such material systems is limited in accuracy if it does not account for each of these features in a sufficiently accurate manner. A concern in the implementation of TMCs is the large mismatch in the coefficients of thermal expansion (CTEs) between fiber and matrix materials. Current fibers of interest include SiC with a coefficient of thermal expansion on the order of $5 \times 10^{-6}/°C$ over the temperature range from room temperature to 700°C [6]. Potential matrix materials include TIMETAL®21S (or β21S) with a CTE on the order of $10 \times 10^{-6}/C°$ over a similar temperature range (see, e.g., [7]). One manufacturing technique for such composites involves hot isostatic pressing (HIPing) of alternating layers of titanium alloy foil and mats of aligned fibers at temperatures in excess of 700°C. Thus, upon cooling, large residual stress fields may be induced in the composite due to the CTE mismatch. In addition, depending on the intended application, TMC components can be expected to see temperature changes of several hundred degrees (C) over very short periods of time, often in combination with either in-phase or out-of-phase mechanical loading.

Over the range of anticipated service temperatures, it is generally sufficient to consider the SiC fiber's mechanical behavior to be linearly elastic. The matrix behavior, however, is much more complicated. Potential matrix materials exhibit inelastic deformations at stress levels well within the range expected in service. Furthermore, at higher temperatures this inelastic behavior is highly dependent on the rate of loading [7,8]. Thus, accurate estimates of the matrix stress and strain state, and therefore the overall composite response, over a wide range of conditions require the use of unified constitutive models. Several reviews are available on such models (see, e.g., [9–11]. Recent work on TIMETAL®21S using the Bodner-Partom viscoplastic constitutive model with directional hardening [12] indicates that existing models may not be sufficient to accurately characterize the material response over ranges of conditions found in a single component test, especially at low sustained stress levels [13]. Extension of these models is required for accurate correlation over the temperature-strain rate range of interest [8].

MICROMECHANICAL MODELS OF ELASTIC MEDIA

Local Fields and Overall Response

We start with an outline of a general procedure for the evaluation of overall thermomechanical properties of multiphase composite media in terms of thermoelastic constants and volume fractions of the phases. The elastic response contributes to the total strain during inelastic loading, and it is the sole source of this strain in any elastic unloading step. Consider a fibrous composite material consisting of one or more continuous phases of cylindrical shape, which are aligned parallel to the x-axis of a Cartesian coordinate system and are embedded in a continuous matrix phase. The phases remain bonded and are free of voids and cracks during deformation. A representative volume V with surface S is chosen so that, under certain boundary conditions, it represents the macroscopic response of the composite.

Within V, each phase $r = 1, 2, \ldots$ occupies a volume V_r, and the volume fractions $c_r = V_r/V$ satisfy $c_1 + c_2 + \ldots + c_n = 1$.

The volume V is subjected to certain uniform overall stresses σ, or strains ε, and to a uniform temperature change θ. Suppose that the constitutive relations of the phases are known, e.g., from experiments on neat matrix samples and on the fibers, and are written in the form:

$$\sigma(x) = L_r \varepsilon(x) + l_r \theta, \qquad \varepsilon(x) = M_r \sigma(x) + m_r \theta \qquad (1)$$

where L_r and M_r are elastic stiffness and compliance matrices, which have full diagonal symmetry; $L_r = L_r^T$; $M_r = M_r^T$; $LM = I$; l_r and m_r are the thermal stress and strain vectors, such that $l_r = -L_r m_r$. As long as the phase remains elastic, the coefficients of these matrices are constant in V_r. Note that the components of m_r are the linear coefficients of thermal expansion of the phase, which are not affected by deformation and are assumed, for now, to be independent of temperature.

The local fields in Equation (1) are generally not uniform. Therefore, it is often convenient to work with volume averages of the nonuniform fields defined by the integral:

$$\{.\}v' = \frac{1}{V'} \int_{v'} (.)dV \qquad (2)$$

Phase volume averages of local fields follow from Equation (2) if one takes $V' = V_r$ and integrates both sides in Equation (1):

$$\sigma_r = L_r \varepsilon_r + l_r \theta, \qquad \varepsilon_r = M_r \sigma_r + m_r \theta \qquad (3)$$

where

$$\sigma_r = \{\sigma(x)\}_{v_r}, \qquad \varepsilon_r = \{\varepsilon(x)\}_{v_r} \qquad (4)$$

Because L_r and M_r are constant in V_r, Equation (3) are exact analogs of Equation (1) for phase volume averages. One can also obtain the overall stresses and strains as averages of the respective local quantities over the representative volume V and write the overall constitutive relations as:

$$\bar{\sigma} = L\varepsilon + l\theta, \qquad \bar{\varepsilon} = M\sigma + m\theta \qquad (5)$$

where

$$\bar{\sigma} = \{\sigma(x)\}_v, \qquad \bar{\varepsilon} = \{\varepsilon(x)\}_v \qquad (6)$$

The implication is that the representative volume V of the composite aggregate is regarded as a macroscopically homogeneous medium and that, under uniform overall stress or strain, the L, M are the elastic overall stiffness and compliance matrices, and m, $l = -Lm$ are the overall thermal strain and stress vectors of this composite medium.

To determine L, M, l, and m of an elastic composite, one can proceed as follows. Suppose that for the composite system considered, one could evaluate the actual local fields [Equation (1)] and write them in the form:

$$\sigma(x) = B_r(x)\bar{\sigma} + b_r(x)\theta, \qquad \varepsilon(x) = A_r(x)\bar{\varepsilon} + a_r(x)\theta \qquad (7)$$

Of course, in many practical situations, one cannot find the actual fields, but it is usually possible to evaluate an estimate of average local fields in the two phases under the prescribed load increment. The result is the integral Equation (2) of Equation (7), taken over V_r:

$$\sigma_r = B_r\bar{\sigma} + b_r\theta, \qquad \varepsilon_r = A_r\bar{\varepsilon} + a_r\theta \qquad (8)$$

where A_r, B_r, a_r, and b_r are certain mechanical and thermal strain and stress concentration factors. If those are known, one can utilize Equation (2) to write:

$$\bar{\sigma} = \sum_{r=1}^{n} c_r\sigma_r, \qquad \bar{\varepsilon} = \sum_{r=1}^{n} c_r\varepsilon_r \qquad (9)$$

where n is the total number of phases in the body. Then, for $\theta = 0$, Equations (3), (8), and (9) give the overall mechanical properties:

$$L = \sum_{r=1}^{n} c_r L_r A_r, \qquad M = \sum_{r=1}^{n} c_r M_r B_r \qquad (10)$$

together with the results:

$$\sum_{r=1}^{n} c_r A_r = \sum_{r=1}^{n} c_r B_r = I$$

$$A_r M = M_r B_r, \qquad B_r L = L_r A_r \qquad (11)$$

This sequence, first outlined by Hill [5,14], enables evaluation of the overall instantaneous L and M in terms of $n-1$ mechanical concentration factors. The overall thermal strain and stress vectors m and l can be evaluated from known overall mechanical moduli or from local properties and concentration factors as (Reference [15]:)

$$m = \sum_{r=1}^{n} c_r B_r^T m_r, \qquad l = \sum_{r=1}^{n} c_r A_r^T l_r \qquad (12)$$

Averaging Methods

INCLUSION PROBLEMS

In the previous section, the averages of local fields and the overall elastic properties were found in terms of stress or strain concentration factors. These factors can be evaluated in several different ways. First, we review an approach based on the solution of an inclusion problem, which will be useful in some of the micromechanical models discussed in the following. A more detailed accounting is presented in Reference [16]. In what follows, an inclusion is defined as a region in a homogeneous body that is subjected to a prescribed eigenstrain distribution, and an inhomogeneity is defined as a region within an otherwise homogeneous body with different elastic properties from the remainder of the body. These properties may be different in magnitude or, in the case of anisotropic bodies, orientation.

Eshelby [17] established the important result that if an ellipsoidal inclusion in an isotropic body is subjected to a prescribed constant eigenstrain, then the strain field developed inside the inclusion is also constant and can be related to the applied eigenstrain as:

$$\varepsilon_p = S\mu \tag{13}$$

where S is the matrix form of the Eshelby tensor, which is constant inside the inclusion and relates the constant applied eigenstrain μ to the resulting strain field ε_p in the inclusion. ε_p can be thought of as the perturbation to the overall strain field in the inclusion caused by the presence of the eigenstrain, μ, and the constraint of the surrounding material. Kinoshita and Mura [18] later extended this method to allow anisotropic bodies and found solutions of similar form.

Consider the problem of a single inhomogeneity with properties defined by stiffness matrix L_2 and embedded in a large volume of surrounding material, denoted as the matrix material, of stiffness L_1. Far from the inhomogeneity, the matrix is subjected to displacement boundary conditions that are derived from a uniform strain field ε_o. In the absence of the inhomogeneity, this is the strain everywhere in the body. The presence of the inhomogeneity perturbs this strain field by an amount $\varepsilon_p(x)$ such that the total field at any point is:

$$\varepsilon(x) = \varepsilon_o + \varepsilon_p(x) \tag{14}$$

At any point inside the inhomogeneity the stress field is:

$$\sigma_2(x) = L_2\varepsilon_2(x) = L_2[\varepsilon_o + \varepsilon_p(x)] \tag{15}$$

Alternately, the same stress field can be generated inside and outside the inhomogeneity by replacing the inhomogeneity with an equivalent inclusion of identical size and shape. That is, the inhomogeneity is replaced with matrix material of stiffness L_1 and a prescribed eigenstrain μ, denoted as the equivalent eigenstrain, is applied such that:

$$\sigma_2(x) = L_2\varepsilon_2(x) = L_1(\varepsilon_2(x) - \mu). \tag{16}$$

Since $\varepsilon_2(x)$ is uniform under these conditions in any ellipsoidal inhomogeneity [17,18], the strain distribution in the inhomogeneity and equivalent inclusion may be written as:

$$\varepsilon_2(x) = \varepsilon_o + \varepsilon_p = \varepsilon_2 \tag{17}$$

Now, substitute Equation (13) into Equation (17), and the result into Equation (16) to obtain:

$$\varepsilon_2 = [I - SL_1^{-1}(L_1 - L_2)]^{-1}\varepsilon_o \tag{18}$$

Thus, the strain field in the ellipsoidal inhomogeneity may be obtained in terms of the applied strain field, the properties of the inhomogeneity (L_2) and the matrix (L_1), and the shape and orientation of the inhomogeneity (S).

Equation (18) may be written as:

$$\varepsilon_2 = A_2\varepsilon_o \tag{19}$$

where A_2 is a matrix of mechanical strain concentration factors relating the strain in the ellipsoidal inhomogeneity to the applied strain field. A_2 has been derived under the assumption of a single inclusion in an infinite matrix, or equivalently, that the number of inhomogeneities is small and they are distributed sufficiently far apart that interaction between them is negligible. This is the dilute approximation of the strain concentration factors (see, e.g., [19]). Finally, A_2 may be substituted into the third part of Equation (11) to obtain the solution for the stress concentration factor as:

$$B_2 = L_2[I - SL_1^{-1}(L_1 - L_2)]^{-1}M \tag{20}$$

where M is the effective compliance of the overall composite.

THE SELF-CONSISTENT METHOD

So far, we have considered only the problem of a single inclusion, embedded in an elastic medium, as a stepping stone to the more important problem of finding the stresses in the constituents of a composite medium. The latter problem can be solved in several different ways. One approach is based on the self-consistent approximation [5,20], which assumes that the stress and strain field averages in the fiber are equal to those found in the single inclusion problem above, provided that the fiber is embedded in a homogeneous medium that has the properties of the composite. Thus, the solution is exactly the same as that shown in Equation (18) except that the stiffness of the matrix is replaced with the stiffness of the effective medium, L. The strain concentration factor matrix is found as:

$$A_2 = [I - SL^{-1}(L - L_2)]^{-1} \tag{21}$$

Note that the effective stiffness of the medium is not known a priori. Thus, Equation (21) must be solved iteratively. Using the first part of Equation (10) we may rewrite Equation (21) in terms of only the concentration factors as:

$$A_r = \left[I - S\left(\sum_{s=1}^{n} c_s L_s A_s\right)^{-1}\left(\sum_{s=1}^{n} c_s L_s A_s - L_r\right)\right]^{-1} \tag{22}$$

The effective stiffness is then found from the first part of Equation (10). Hill also gives a proof that the estimates of L obtained via Equation (22) lie between the Hashin-Shtrikman bounds, as discussed below.

THE MORI-TANAKA METHOD

Another, and somewhat simpler, estimate of phase concentration factors of composite media was proposed by Mori and Tanaka [21] and restated recently in a more tractable form by Benveniste [22]. To take into account the interaction between nearby inhomogeneities, the Mori-Tanaka method (MTM) approximates the strain and stress distributions in an inhomogeneity embedded in a matrix reinforced by many inhomogeneities and loaded by a remote strain field as that which occurs when a single inhomogeneity embedded in an infinite matrix is subjected to boundary conditions derived from a remote strain field equal to the average matrix strain rather than the remotely applied strain. The unknown matrix strain can be calculated in closed form.

The problem under consideration, then, is a single ellipsoidal inhomogeneity of stiffness L_r ($r \neq 1$) embedded in a body of stiffness L_1.

Far from the inhomogeneity, the body is subjected to traction or displacement boundary conditions derived from a uniform stress or strain field, respectively. Given the applied strain field, the solution to this problem can be obtained in the form:

$$\varepsilon_r = T_r \varepsilon_r \tag{23}$$

where T_r is a partial concentration factor matrix defined by:

$$T_r = [I - SL_1^{-1}(L_1 - L_r)]^{-1}, \quad r = 2, 3, \ldots, n, \qquad T_r = I, \quad r = 1 \tag{24}$$

and $r = 1$ denotes the matrix phase. To relate matrix phase strains to remote strains, the relations between local and remote strains are recalled from Equation (9). From Equations (9), (23), and (24), we compare with Equation (19) and obtain:

$$A_1 = \left[c_1 I + \sum_{r=1}^{n} c_r T_r \right]^{-1}, \qquad A_r = T_r \left[c_1 I + \sum_{r=1}^{n} c_r T_r \right]^{-1}, r = 2, 3, \ldots, n \tag{25}$$

Note that in contrast to Equation (22), these are explicit algebraic equations for the overall properties. Proofs of self-consistency $LM = I$ and full diagonal symmetry of L, M for binary fibrous media are available; also, the estimates of L derived via Equation (25) lie within rigorous bounds on overall moduli [22–24]. Explicit expressions of the Mori-Tanaka estimates of the transversely isotropic effective moduli of a continuous matrix containing many aligned transversely isotropic fibers are given by Chen et al. [25]. In terms of Hill's moduli [26] k, l, m, n, and p, the effective properties are found as:

$$k = \frac{k_1 k_2 + m_2(c_1 k_1 + c_2 k_2)}{c_1 k_2 + c_2 k_1 + m_2}$$

$$l = \frac{c_1 l_1(k_2 + m_2) + c_2 l_2(k_1 + m_2)}{c_1(k_2 + m_2) + c_2(k_1 + m_2)}$$

$$n = c_1 n_1 + c_2 n_2 + (l - c_1 l_1 - c_2 l_2)\frac{l_1 - l_2}{k_1 - k_2} \tag{26}$$

$$m = \frac{m_1 m_2(k_2 + 2m_2) + k_2 m_2(c_1 m_1 + c_2 m_2)}{k_2 m_2 + (k_2 + 2m_2)(c_1 m_2 + c_2 m_1)}$$

$$p = \frac{2c_1 p_1 p_2 + c_2(p_1 p_2 + p_2 p_2)}{2c_1 p_2 + c_2(p_1 + p_2)}$$

THE DIFFERENTIAL SCHEME

An alternative model is the differential scheme [27,28] that begins by embedding a dilute concentration of inhomogeneities in the matrix phase and calculating the effective properties by the dilute method. An additional dilute concentration of inhomogeneities is then embedded in the effective medium calculated in the first pass, and new effective properties are calculated. The process is repeated until the desired concentration of inhomogeneities has been obtained. Comparison with the generalized self-consistent method [29] has shown that the resulting effective properties are dependent on the steps in dilute concentrations followed to obtain the final properties and can be significantly in error for high concentrations.

Unit Cell Models

Methods of determining effective properties noted previously make few assumptions on the distribution of reinforcing fibers within the composite. In many cases, particularly those involving TMCs with relatively large fibers (SCS-6 fibers, e.g., are 140 μm in diameter), it is possible to control the spacing of the fibers such that they are almost periodically distributed throughout the matrix, generally in square or hexagonal arrays. In such cases, a variety of additional techniques may be used to estimate the effective properties. It is generally assumed that, because the fiber matrix architecture varies periodically, far from free surfaces and applied uniform loads, the solution stress and strain fields will also vary periodically. The stress, strain, and displacement fields in a repeating unit cell of the material are then obtained by determining solutions that satisfy equilibrium and compatibility within the unit cell and periodicity of the resulting displacements and tractions on the unit cell boundaries. Solutions can be obtained analytically with boundary conditions satisfied at each point [30–32] or on average [33]. Alternately, finite element or other numerical techniques may be used to solve the periodic boundary value problem [31,34,35]. The homogenization method has evolved over the years to solve such problems using perturbation methods. A brief review has been presented by Hollister and Kikuchi [36]. More comprehensive reviews are presented elsewhere [37–39].

Bounds on Overall Moduli

The methods discussed in the previous two sections lead to simple estimates of local fields and overall properties, but they are heuristic in nature and thus do not provide an assurance of accuracy of the results. The legitimacy of the self-consistent. Mori-Tanaka, and similar estimates is derived from proofs that show that, in certain cases of practical interest, the estimates are bracketed by rigorous bounds on overall moduli. Early efforts along these lines were made by Voigt [40] who assumed a constant strain distribution throughout the RVE, and Reuss [41], who assumed a constant stress distribution. Hill [42] proved that the Voigt and Reuss estimates provide rigorous bounds on the effective properties of perfectly bonded composite materials. The Voigt and Reuss bounds are insensitive to microstructural geometry and thus are not particularly sharp. Often, these bounds are too far apart to serve as useful estimates of the effective properties. These are referred to as one-point bounds (see, e.g., [43]).

More restrictive bounds, which reflect essential features of the geometry, were originally derived by Hashin and Shtrikman [44–46], and for fibrous composites by Hashin and Rosen [47] and Walpole [48,49]. These are often referred to as two-point bounds. The admissibility of new micromechanical methods is often confirmed by showing that predictions lie within these bounds. Further details on one- and two-point bounds are discussed in Bahei-El-Din and Dvorak [50] and Dvorak [16]. Significant effort has gone to developing improved bounds. Kröner [51] has presented a concise introduction to the subject, and Torquato [52] has presented an extensive review of the literature. Although very tight theoretical bounds can be obtained by these methods, evaluation of specific systems requires the measurement of n-point ($n > 2$) probability densities that are often difficult or impossible to obtain experimentally.

INELASTIC RESPONSE

Solution Strategies

During elastic deformation of the composite medium, the constituent phases are homogeneous, and their properties are known constants. This simplifies modeling to a great extent because the local stress and strain fields and their averages are linearly related by the constant stiffness and compliance tensors. Overall properties can be evaluated from the phase field averages that can be found from solutions of certain inclusion problems in a representative volume of the aggregate. Continuum modeling of the inelastic behavior is much more difficult. When at least one of the phases deforms inelastically, its local properties depend on the deformation history, and the phase is no longer homogeneous. This creates a major obstacle in evaluation of local fields that now depend on the deformation history, as does the relationship between the stress and strain averages.

In addition to complicating the evaluation of local fields, the presence of significant inelasticity presents certain numerical difficulties. Numerical analysis becomes nonlinear and must generally be solved in an incremental manner, which can significantly increase the computational effort. A variety of techniques have been invoked to deal with these issues. The simplest, and computationally most efficient, is to treat the inelastic composite as an effective homogeneous medium that follows an anisotropic inelastic constitutive law (see, e.g., [53,54]. Material constants for the equivalent material are obtained from tests on the composite system, with the number of constants in the constitutive equation determining the number of tests required to characterize the composite system.

The principal drawback of a macromechanical approach is the neglect of the micromechanical inhomogeneity of the system. Stress and strain predictions are a smeared average of the fiber and matrix states and do not reflect the actual distribution in either phase. Also, changes in the microstructural makeup of the composite cannot be easily accommodated. A change in fiber volume fraction, e.g., would require a new set of experiments to characterize the model. Because many life prediction techniques for TMCs rely on knowledge of fiber and matrix stress or strain [55], such methods are of particularly limited use when life prediction estimates are desired.

The alternative is to proceed with an analysis that leads to estimates of the stress and strain states in both the fiber and the matrix, often following strategies similar to those outlined previously in the Micromechanical Models section. The simplest of these circumvent the difficulty of estimating stress, strain, and effective property distributions throughout the inelastic phase by assuming that the local fields are uniform [56–59]. Other models approximate in various ways the actual nonuniform fields [33,60–62], but at least one such approach also provides upper and lower bounds on certain instantaneous stiffness and compliance coefficients [31]. In general, there is a continuous trade-off between computational expense and computational accuracy. The optimal choice is, of course, dependent on the required degree of accuracy, which can vary significantly between applications.

Transformation Fields

Under infinitesimal deformation, the strains introduced at a material point by a mechanical load cycle $\pm\sigma_{ij}$, together with any nonmechanical effects such as temperature variations and moisture absorption, are referred to as transformation or eigenstrains. These strains include inelastic deformations caused by mechanical loads that exceed the proportional or elastic limit of the material. Under a fully constrained domain, the eigenstrains cause internal stresses, which are referred to as transformation or eigenstresses.

Considering an elastic body, consisting of a single or a multiphase material, we assume that the total strains and stresses at a material point with coordinates x in a selected cartesian system associated with the body can be additively decomposed as:

$$\varepsilon(x, t) = \varepsilon^e(x, t) + \mu(x, t), \qquad \sigma(x, t) = \sigma^e(x, t) + \lambda(x, t) \qquad (27)$$

where t denotes the current time. The ε^e and μ in the first part of Equation (27) denote, respectively, the elastic strain and an eigenstrain at the material point. Similarly, the σ^e and λ in the second part of Equation (27) denote the elastic stress and eigenstress.

As already mentioned, the eigenstrain and eigenstress fields, henceforth referred to jointly as transformation fields, may consist of contributions of distinct physical origin and thus may be decomposed further. For example, if only thermal and inelastic effects are considered:

$$\mu(x, t) = m\, \theta(t) + \varepsilon^{in}(x, t) + \ldots$$
$$\lambda(x, t) = \ell\, \theta(t) + \sigma^{re}(x, t) + \ldots \qquad (28)$$

where m and ℓ are the thermal strain and stress tensors. The coefficients of m represent the linear thermal expansion coefficients, the ε^{in} denotes an inelastic strain, and σ^{re} a relaxation stress. Contributions due to other transformation effects can be added.

With these definitions, Equations (27) become:

$$\varepsilon(x, t) = M\, \sigma(x, t) + m\, \theta(t) + \varepsilon^{in}(x, t)$$
$$\sigma(x, t) = L\, \varepsilon(x, t) + \ell\, \theta(t) + \sigma^{re}(x, t) \qquad (29)$$

together with the interrelations:

$$m = -M\, \ell, \qquad \varepsilon^{in}(x, t) = -M\, \sigma^{re}(x, t)$$
$$\ell = -L\, m, \qquad \sigma^{re}(x, t) = -L\, \varepsilon^{in}(x, t) \qquad (30)$$

where L and $M = L^{-1}$ are the elastic stiffness and compliance tensors of the material located at point x, assumed to be diagonally symmetric, positive definite, and for now, independent of temperature.

In describing the overall response of the heterogeneous medium under overall uniform stress or strain and uniform temperature variation, the elastic body under consideration is selected as a representative volume V of the material as described previously. In heterogeneous media with randomly distributed and oriented phases, the representative volume is selected as a sufficiently large sample that contains many phases and reflects typical macroscopic properties [14]. On the other hand, heterogeneous materials in which the phases are uniformly dispersed can be represented by a suitably selected unit cell of a (usually) periodic model of the actual material geometry. In either case, macroscopically homogeneous response and the implied existence of certain overall or effective properties are assumed under macroscopically uniform overall stress σ (t) or uniform overall strain ε (t), prescribed through surface tractions or displacements specified on the surface S of V. In unit cell models, the uniform overall quantities must be reduced to certain periodic boundary conditions for the representative cell.

With these definitions, the overall and local fields are connected by [cf. Equations (2), (4), and (6)]:

$$\varepsilon(t) = \frac{1}{V} \int_V \varepsilon(\boldsymbol{x}, t) \, dV, \qquad \sigma(t) = \frac{1}{V} \int_V \sigma(\boldsymbol{x}, t) \, dV \qquad (31)$$

When phase eigenstrains are present, the above definition of the representative volume needs to be expanded to incorporate overall response of the heterogeneous solid to local eigenstrains in phases or subvolumes of such phases. In particular, $\mu(\boldsymbol{x}, t)$ represent an eigenstrain field defined in V such that, if the surface S of V is traction free, this field causes surface displacements on S that are consistent with a macroscopically uniform overall strain μ (t). Similarly, if the volume V is constrained such that no displacements are permitted on the surface S, then $\lambda(\boldsymbol{x}, t)$ represent an eigenstress field defined in V such that it causes surface tractions on S that are consistent with a macroscopically uniform overall stress $\lambda(t)$. In analogy with Equation (28), the μ (t) and $\lambda(t)$ will be referred to as the overall transformation fields.

Recall now that the local and overall transformation fields are connected by the generalized Levin [15] formula (Reference [63], Equations 7 and 8)

$$\lambda(t) = \frac{1}{V} \int_V \boldsymbol{A}^T(\boldsymbol{x}) \, \lambda(\boldsymbol{x}, t) \, dV, \qquad \mu(t) = \frac{1}{V} \int_V \boldsymbol{B}^T(\boldsymbol{x}) \, \mu(\boldsymbol{x}, t) \, dV \qquad (32)$$

where the $\boldsymbol{B}(\boldsymbol{x})$ and $\boldsymbol{A}(\boldsymbol{x})$ are the mechanical stress and strain influence functions [Equation (7)]. If the transformation fields in Equation

(28) are caused only by a uniform change in temperature, then the mechanical and thermal influence functions define the local fields as:

$$\sigma(\boldsymbol{x}, t) = \boldsymbol{B}(\boldsymbol{x})\, \sigma(t) + \boldsymbol{b}(\boldsymbol{x})\, \theta(t)$$
$$\varepsilon(\boldsymbol{x}, t) = \boldsymbol{A}(\boldsymbol{x})\, \varepsilon(t) + \boldsymbol{a}(\boldsymbol{x})\, \theta(t)$$
(33)

where the $\boldsymbol{b}(\boldsymbol{x})$ and $\boldsymbol{a}(\boldsymbol{x})$ are the thermoelastic influence functions.

Under purely thermoelastic deformation of the heterogeneous aggregate, the local fields are provided by Equation (33), as a superposition of the contributions due to the overall mechanical stresses and strains, and the thermally induced local transformations in Equation (28). In the same spirit, one can superimpose the effect of any other local transformation field $\mu(\boldsymbol{x}, t)$ or $\lambda(\boldsymbol{x}, t)$ by writing the local fields in the form [64]:

$$\varepsilon(\boldsymbol{x}, t) = \boldsymbol{A}(\boldsymbol{x})\, \varepsilon\,(t) + \boldsymbol{D}(\boldsymbol{x}, \boldsymbol{x}')\, \mu(\boldsymbol{x}', t),$$
$$\sigma(\boldsymbol{x}, t) = \boldsymbol{B}(\boldsymbol{x})\, \sigma(t) + \boldsymbol{F}(\boldsymbol{x}, \boldsymbol{x}')\, \lambda(\boldsymbol{x}', t).$$
(34)

Noting that Equations (28) to (30) imply:

$$\mu(\boldsymbol{x}', t) = -\boldsymbol{M}(\boldsymbol{x}')\, \lambda(\boldsymbol{x}', t) \qquad \lambda(\boldsymbol{x}', t) = -\boldsymbol{L}(\boldsymbol{x}')\, \mu(\boldsymbol{x}', t),$$
(35)

we observe that $\boldsymbol{F}(\boldsymbol{x}, \boldsymbol{x}')$ and $\boldsymbol{D}(\boldsymbol{x}, \boldsymbol{x}')$ are eigenstress and eigenstrain influence functions which evaluate the effect at $\boldsymbol{x}$ induced by a transformation at $\boldsymbol{x}'$ under overall uniform applied stress $\sigma(t)$ or strain $\varepsilon(t)$. For a stress-free homogeneous medium with a transformed ellipsoidal inclusion, $\boldsymbol{D}(\boldsymbol{x}', \boldsymbol{x}') = \boldsymbol{S}$, the Eshelby tensor.

LOCAL FIELDS AND OVERALL RESPONSE

In actual solutions, the continuous fields are usually replaced by piecewise uniform approximations in the phases or in subvolumes Ω_ρ, $\rho = 1, 2, \ldots M$, of a discretized unit cell. Then, Equations (31) and (32) are reduced to:

$$\varepsilon(t) = \sum_{\rho=1}^{M} c_\rho \varepsilon_\rho(t), \qquad \sigma(t) = \sum_{\rho=1}^{M} c_\rho \sigma_\rho(t),$$
(36)

$$\lambda(t) = \sum_{\rho=1}^{M} c_\rho\, \boldsymbol{A}_\rho^T\, \lambda_\rho(t), \qquad \mu(t) = \sum_{\rho=1}^{M} c_\rho \boldsymbol{B}_\rho^T\, \mu_\rho(t),$$
(37)

where the concentration factor tensors $\boldsymbol{A}_\rho$ and $\boldsymbol{B}_\rho$ represent the volume averages of the respective influence functions over Ω_ρ.

With regard to Equations (31), (32), and (34), the relation between the overall total and transformation stresses and strains at any time t may be written in terms of the overall elastic stiffness L and compliance M:

$$\varepsilon(t) = M\,\sigma(t) + \mu(t), \qquad \sigma(t) = L\,\varepsilon(t) + \lambda(t) \tag{38}$$

where $M = L^{-1}$, $\lambda(t) = -L\,\mu(t)$, $\mu(t) = -M\,\lambda(t)$.

If the decomposition [Equation (28)] is applied to the overall quantities, one recovers:

$$\varepsilon(t) = M\,\sigma(t) + m\,\theta\,(t) + \varepsilon^{in}(t)$$
$$\sigma(t) = L\,\varepsilon(t) + \ell\,\theta\,(t) + \sigma^{re}(t) \tag{39}$$

and the relations [Equations (36) and (33)] provide the well-known connections [14,65]:

$$L = \sum_{\rho=1}^{M} c_\rho L_\rho A_\rho, \qquad M = \sum_{\rho=1}^{M} c_\rho M_\rho B_\rho \tag{40}$$

$$\ell = \sum_{\rho=1}^{M} c_\rho (\ell_\rho + L_\rho\,a_\rho), \qquad m = \sum_{\rho=1}^{M} c_\rho (m_\rho + M_\rho\,b_\rho) \tag{41}$$

where L and $M = L^{-1}$ are the overall elastic stiffness and compliance tensors, and ℓ, $m = -M\,\ell$ are the overall thermal stress and strain tensors.

The local elastic and transformation fields, the respective influence functions in Equation (37), and all components of the local fields are also sought in terms of averages and piecewise uniform approximations, respectively, within phase volumes V_r, or subvolumes or subelements Ω_ρ of discretized phases. In particular, the representative volume V, or the unit cell volume Ω, are subdivided into subelements ρ, $\eta = 1, 2, \ldots M$ of volume Ω_ρ, $\Omega_\eta \in V_r$, where $M \geq N$, so that each subelement resides in only one phase r. Conversely, each phase may contain one or more subelements. The stresses and strains are approximated as uniform in each Ω_ρ, Ω_η, and, if all local transformations [Equation (28)] are superimposed, Equation (34) is replaced by:

$$\varepsilon_\rho(t) = A_\rho\,\varepsilon(t) + \sum_{\eta=1}^{M} D_{\rho\eta}\,[m_\eta\,\theta(t) + \varepsilon_\eta^{in}(t)] \tag{42}$$

$$\sigma_\rho(t) = B_\rho\,\sigma(t) + \sum_{\eta=1}^{M} F_{\rho\eta}\,[\ell_\eta\,\theta(t) + \sigma_\eta^{re}(t)] \tag{43}$$

This form describes the response of the elastic composite to certain uniform overall mechanical fields and piecewise uniform local transformation fields. The A_ρ and B_ρ are the mechanical concentration factor tensors. Under overall strain $\varepsilon(t) = 0$, the $D_{\rho\eta}$ gives the strain caused in Ω_ρ by a unit uniform eigenstrain located in Ω_η. Under overall stress $\sigma(t) = 0$, the $F_{\rho\eta}$ defines the stress in Ω_ρ due to a unit eigenstress in Ω_η. Any additional eigenstrains or eigenstresses of interest can be incorporated. In what follows, the $D_{\rho\eta}$ and $F_{\rho\eta}$ will be referred to as the transformation concentration factor tensors.

INFLUENCE FUNCTIONS FOR BINARY MATERIALS

Two-Phase Averaging Models

A particularly simple evaluation of these functions is possible in two-phase media, with phases denoted as $r = \alpha$, β. The mechanical strain and stress concentration factors are given by Reference [14]:

$$A_r = (L_r - L_s)^{-1} (L - L_s)/c_r$$

$$r, s = f, m \tag{44}$$

$$B_r = (M_r - M_s)^{-1} (M - M_s)/c_r \tag{46}$$

The transformation strain and stress influence functions are evaluated from Reference [64]:

$$D_{r\alpha} = (I - A_r) (L_\alpha - L_\beta)^{-1} L_\alpha, \qquad D_{r\beta} = -(I - A_r) (L_\alpha - L_\beta)^{-1} L_\beta \tag{46}$$

$$F_{r\alpha} = (I - B_r) (M_\alpha - M_\beta)^{-1} M_\alpha, \qquad F_{r\beta} = -(I - B_r) (M_\alpha - M_\beta)^{-1} M_\beta \tag{47}$$

It is of interest to record here the expressions for the concentration factor tensors found by Dvorak and Benveniste [63] for multiphase solids in terms of estimates derived with the self-consistent or Mori-Tanaka methods. For both methods, the results are:

$$D_{rs} = (I - A_r)(L_r - L)^{-1}(\delta_{rs} I - c_s A_s^T) L_s$$

$$r, s = 1, 2 \ldots N \tag{48}$$

$$F_{rs} = (I - B_r)(M_r - M)^{-1}(\delta_{rs} I - c_s B_s^T) M_s \tag{49}$$

with the connection:

$$F_{\rho\eta} = L_\rho[\delta_{\rho\eta} I - c_\eta A_\rho B_\eta^T - D_{\rho\eta}] M_\eta \tag{50}$$

Here, the L and M are the respective estimates of the overall elastic stiffness and compliance tensors, whereas A_r and B_r are the related estimates of the elastic mechanical concentration factor tensors; L_r, M_r are the phase elastic properties. The δ_{rs} is the Kronecker symbol, but no summation is indicated by repeated indices. It can be verified that in two-phase media, Equations (48) and (49) reduce to Equations (46) and (47).

Unit Cell Models

The results of Equations (44) to (47) provide simple estimates of the transformation concentration factor tensors of individual phases, but their accuracy may prove inadequate when the transformation fields exhibit large variations, as they do, e.g., during inelastic deformation. Subdivision of the phases is then indicated and is usually accomplished in the context of a unit cell model of an idealized composite material. If a periodic arrangement of the microstructure is selected, then the unit cell is subjected to appropriate periodic boundary conditions, and the local fields are evaluated by the finite element method. Many such models have been described in the literature; in the examples that follow, we will utilize the periodic hexagonal array (PHA) model of a fibrous composite [31,34].

To illustrate the finite element evaluation of the piecewise uniform approximations in Equations (42) and (43) of the influence functions in Equation (34), we subdivide the domain under consideration by constant strain 3-D elements Ω_η, $\eta = 1, 2, \ldots M$, interconnected at nodes $i, j, \ldots = 1, 2, \ldots R$. The coefficients of the mechanical strain or stress concentration factor tensors A_η or B_η in Equation (42) or (43) are then found from solutions of six successive elasticity problems. In each solution, the domain, free from any eigenstrains, is subjected to overall strain ε or stress σ which have only one nonzero component of unit magnitude. The (6×1) strain or stress vector found in the element Ω_η is the column of the (6×6) matrix A_η or B_η, corresponding to the selected nonzero component.

The transformation concentration factors can be obtained in a similar way by applying, in turn, each single component of a unit eigenstrain vector μ_ρ in the element Ω_ρ within an otherwise strain-free domain and finding the local strains in all elements. Each such local strain in Ω_η represents one column of the (6×6) matrix $D_{\eta\rho}$ for Ω_η in (42). An analogous sequence involving substitution of the $D_{\eta\rho}$ matrices in (50) yields the columns of $F_{\eta\rho}$. If the evaluation of the $D_{\eta\rho}$ and $F_{\eta\rho}$ matrices is made by using a standard finite element code, the unit local eigenstrains may be produced as thermal strains by appropriate selections of nonzero thermal expansion coefficients.

A much more efficient evaluation of these matrices using the stiffness matrix of the unit cell domain and load vectors generated from each of the eigenstrain components present in each Ω_η, $\eta = 1, 2, \ldots$ M, elements is described by Dvorak et al. [66]. In their approach, the equilibrium equations at the nodal points where the elements are interconnected are written as:

$$K \, \alpha = f, \qquad K_{ij} = \sum_{\eta=1}^{M} K_{ij}^{\eta}, \qquad f_i = -\sum_{\eta=1}^{M} f_i^{\eta} \tag{51}$$

where K is the overall stiffness matrix, f is the overall load vector, assembled from the elemental matrices K_{ij}^{η} and f_i^{η}, $\alpha = [\alpha_1 \, \alpha_2 \, \ldots \, \alpha_R]^T$ lists the as yet unknown nodal displacements, and R is the total number of nodes in the unit cell.

If constant strain elements are selected, the elemental matrices are written as:

$$K_{ij}^{\eta} = \Lambda_i^{\eta T} L_\eta \Lambda_j^{\eta} \Omega_\eta \tag{52}$$

$$f_i^{\eta} = -\Lambda_i^{\eta T} L_\eta \mu_\eta \Omega_\eta \tag{53}$$

where the Λ_i^{η} matrices relate the small strains in element Ω_η to the nodal displacements α_i^{η} by $\varepsilon_\eta = \sum_{i=1}^{p} \Lambda_i^{\eta} \alpha_i^{\eta}$, and P is the total number of element nodes. Matrix L_η represents the elastic stiffness of the element, and μ_η is an eigenstrain present in the element that is otherwise free of stress. The procedure for evaluation of the strain transformation factors $D_{\rho\eta}$, ρ, $\eta = 1, 2, \ldots M$, is completed by solving Equations (51) to (53) for 6M independent f vectors; each corresponds to an eigenstrain $\mu_\eta = i_k$, $k = 1, 2, \ldots 6$; $\eta = 1, 2, \ldots M$, where i_k is the kth column of the (6×6) identity matrix.

Because the overall stiffness matrix K of the unit cell depends on the cell geometry, the finite element type and displacement shape functions and on material elastic properties, it is a constant matrix, and one may write the first part of Equation (51) as:

$$K [\alpha^{(1)} \, \alpha^{(2)} \, \ldots \, \alpha^{(6M)}]^T = [f^{(1)} \, f^{(2)} \, \ldots f^{(6M)}]^T \tag{54}$$

With a constant coefficient matrix K, the solution of Equation (54) can be evaluated efficiently using standard solvers of algebraic equations with constant coefficients. The element nodal displacements α_η, $\eta = 1$, $2, \ldots M$, are then extracted from the solution for $\alpha^{(i)}$, $i = 1, 2, \ldots$ $6M$, and used to obtain the strains ε_η, which represent the columns of $D_{\rho\eta}$, ρ, $\eta = 1, 2, \ldots M$. Because most unit cell models replace the actual material geometry with a certain periodic approximation, the unit cells themselves may possess internal symmetries that can facilitate evaluation of the transformation influence factors [66].

Whether obtained for two-phase averaging models or unit cell models, the mechanical and transformation influence functions must satisfy certain general properties, derived by Dvorak and Benveniste [63]. We list the principal results here and refer the reader to the above reference for specific proofs. Two exact relations exist for the actual transformation influence functions that evaluate the local fields in Ω_η caused by uniform transformations in Ω_ρ [cf. Equations (34), (42), and (43)]:

$$\sum_{\rho=1}^{M} D_{\eta\rho}(x) = I - A_\eta(x) \qquad \sum_{\rho=1}^{M} F_{\eta\rho}(x) = I - B_\eta(x) \tag{55}$$

$$\sum_{\rho=1}^{M} D_{\eta\rho}(x) M_\rho = 0 \qquad \sum_{\rho=1}^{M} F_{\eta\rho}(x) L_\rho = 0 \tag{56}$$

In addition, the concentration factor tensors in Equations (42) and (43) must satisfy:

$$c_\eta D_{\eta\rho} M_\rho = c_\rho M_\eta D_{\rho\eta}^T \qquad c_\eta F_{\eta\rho} L_\rho = c_\rho L_\eta F_{\rho\eta}^T \tag{57}$$

$$\sum_{\rho=1}^{M} c_\rho D_{\rho\eta} = 0 \qquad \sum_{\rho=1}^{M} c_\rho F_{\rho\eta} = 0 \tag{58}$$

where η, ρ = 1, 2, . . . M, the number of elements, and $c_\eta = \Omega_\eta/V$. These connections are exact but not independent; note that Equations (56) and (57) give Equation (58). Actual solutions then show that only Equations (55) and (56) or (58) are independent. This provides $(2 \times M)$ independent relations for the $(M \times M)$ unknown transformation concentration factor tensors.

Analysis of Transformation Fields

CLOSED FORM RATE EQUATIONS

Once the transformation concentration factors for a selected micromechanical model are known, rate equations for evaluation of the inelastic local fields can be formulated in closed form [67]. A specific constitutive relation must be adopted in Ω_η to connect the current values or rates of $\varepsilon_\eta^{in}(t)$ or $\sigma_\eta^{re}(t)$ to the history of $\sigma_\eta(t - \tau)$ or $\varepsilon_\eta(t - \tau)$, and θ $(t - \tau)$, respectively. Examples of such constitutive equations for elastic-plastic, viscoplastic, and viscoplastic materials for metal matrix composites are given by Bahei-El-Din [68], Dvorak [16], Shah [69], and

Wafa [70]. Assuming piecewise uniform fields in the local volumes, one can formally write such constitutive relations in the general form:

$$\sigma_\eta^{re}(t) = g\,(\varepsilon_\eta(t - \tau),\, \theta\,(t - \tau)), \qquad \varepsilon_\eta^{in}(t) = f\,(\sigma_\eta(t - \tau),\, \theta\,(t - \tau)) \quad (59)$$

When substituted into Equations (42) and (43), this provides the governing equations for the total local fields:

$$\varepsilon_\rho(t) + \sum_{\eta=1}^{M} D_{\rho\eta}\, M_\eta\, g\,(\varepsilon_\eta(t - \tau),\, \theta(t - \tau)) = A_\rho\, \varepsilon(t) + a_\rho\, \theta(t) \quad (60)$$

$$\sigma_\rho(t) + \sum_{\eta=1}^{M} F_{\rho\eta}\, L_\eta\, f\,(\sigma_\eta(t - \tau),\, \theta(t - \tau)) = B_\rho\, \sigma(t) + b_\rho\, \theta(t) \quad (61)$$

where the thermal concentration factor tensors are evaluated from Reference [16]:

$$a_\rho = \sum_{\eta=1}^{M} D_{\rho\eta} m_\eta = -\sum_{\eta=1}^{M} D_{\rho\eta} M_\eta\, \ell_\eta, \qquad b_\rho = \sum_{\eta=1}^{M} F_{\rho\eta}\, \ell_\eta = -\sum_{\eta=1}^{M} F_{\eta\rho}\, L_\eta\, m_\eta \quad (62)$$

The mechanical and transformation concentration factors depend only on elastic moduli and on local geometry. If those remain constant, the governing equations can be differentiated and used for evaluation of stress and strain increments or their time rates. In this manner, the inelastic deformation problem for any heterogeneous medium is reduced to evaluation of the various concentration factor tensors or matrices and to solution of either Equation (60) or (61) along the prescribed overall stress or deformation path. The final form of the governing equations, however, depends on the constitutive law assumed for the phases.

Elastic-Plastic Phases

Considering elastic-plastic behavior of the phases, we write the phase response under the locally uniform fields in V_r or Ω_η in the form:

$$d\sigma_r = \mathscr{L}_r(\varepsilon_r - \beta_r)\, d\varepsilon_r + l_r(\varepsilon_r - \beta_r)\, d\theta, \qquad d\varepsilon_r = \mathscr{M}_r(\sigma_r - \alpha_r) + m_r(\sigma_r - \alpha_r)\, d\theta \quad (63)$$

where $\mathscr{L}_r$ and $\mathscr{M}_r$ are the instantaneous stiffness and compliance tensors, and l_r, m_r are the thermal stress and strain tensors that typically describe the consequence of a variation of yield stress with temperature. The α_r and β_r represent back stress and back strain that define the current centers of the yield and relaxation surfaces. All these tensors depend in some given form on the past deformation history; hence, in an actual material the instantaneous magnitudes of their coefficients will vary within each local volume V_r or Ω_η. To prevent large errors in evaluation of the local response, it is advisable to choose material models that permit refined subdivisions of the representative volume.

Rewriting Equation (59) as:

$$d\sigma_r^{re} = \mathscr{L}_r^p \, d\varepsilon_r + l_r^p \, d\theta, \qquad d\varepsilon_r^{in} = \mathscr{M}_r^p \, d\sigma_r + m_r^p \, d\theta \qquad (64)$$

with

$$\mathscr{L}_r^p = \mathscr{L}_r - L_r, \qquad l_r^p = (l_r - \ell_r) \qquad (65)$$
$$\mathscr{M}_r^p = \mathscr{M}_r - M_r, \qquad m_r^p = (m_r - m_r)$$

and substituting for the eigenstrain terms into Equations (60) and (61), one finds the following two systems of equations for local fields in elastic-plastic heterogeneous media:

$$d\varepsilon_\rho + \sum_{\eta=1}^{M} D_{\rho\eta} \, M_\eta \, \mathscr{L}_\eta^p \, d\varepsilon_\eta = A_\rho \, d\varepsilon + (a_\rho - \sum_{\eta=1}^{M} D_{\rho\eta} \, M_\eta \, \ell_\eta^p) \, d\theta \qquad (66)$$

$$d\sigma_\rho + \sum_{\eta=1}^{M} F_{\rho\eta} \, L_\eta \, \mathscr{M}_\eta^p \, d\sigma_\eta = B_\rho \, d\sigma + (b_\rho - \sum_{\eta=1}^{M} F_{\rho\eta} \, L_\eta \, m_\eta^p) \, d\theta \qquad (67)$$

In actual numerical solutions, one may reduce these to the matrix forms:

$$\{d\varepsilon_\rho\} = [diag \, (I) + [D_{\rho\eta} \, M_\eta \, \mathscr{L}_\eta^p]]^{-1} \, \{[A_\rho] \, d\varepsilon - \{[D_{\rho\eta} \, M_\eta]\{l_\eta + l_\eta^p\} \, d\theta\} \quad (68)$$

$$\{d\sigma_\rho\} = [diag \, (I) + [F_{\rho\eta} \, L_\eta \, \mathscr{M}_\eta^p]]^{-1} \, \{[B_\rho] \, d\sigma - \{[F_{\rho\eta} \, L_\eta]\{m_\eta + m_\eta^p\} \, d\theta\} \quad (69)$$

If one of these is solved for the local fields, then the overall response of the representative volume V follows from Equations (31), (32), or (36) to (38). Integration of Equation (68) or (69) along the thermomechanical loading path specified by the history of overall strain $\varepsilon^0(t)$ or stress $\sigma^0(t)$ and temperature $\theta^0(t)$ provides the piecewise uniform local strains or stresses. This can be achieved efficiently with the Runge-Kutta formula [66].

Of course, for rather coarse subdivisions of the microstructure, the governing equations can be solved in closed form. For example, in a two-phase system $r = \alpha, \beta$, or $M = 2$ in Equations (66) and (67), one can find the solution of these equations as Reference [16]:

$$d\varepsilon_r = \mathscr{A}_r \, d\varepsilon + a_r \, d\theta, \qquad d\sigma_r = \mathscr{B}_r \, d\sigma + b_r \, d\theta, \qquad (70)$$

where, for $r = \alpha$,

$$\mathcal{A}_\alpha = [I + D_{\alpha\alpha}M_\alpha \mathcal{L}^p_\alpha - (c_\alpha/c_\beta)\,D_{\alpha\beta}M_\beta\,\mathcal{L}^p_\beta]^{-1}[A_\alpha - (1/c_\beta)\,D_{\alpha\beta}M_\beta\,\mathcal{L}^p_\beta] \quad (71)$$

$$a_\alpha = [I + D_{\alpha\alpha}M_\alpha \mathcal{L}^p_\alpha - (c_\alpha/c_\beta)\,D_{\alpha\beta}M_\beta\,\mathcal{L}^p_\beta]^{-1}$$
$$[a_\alpha - D_{\alpha\alpha}M_\alpha \ell^p_\beta - D_{\alpha\beta}M_\beta \ell^p_\beta]$$

$$\mathcal{B}_\alpha = [I + F_{\alpha\alpha}L_\alpha \mathcal{M}^p_\alpha - (c_\alpha/c_\beta)\,F_{\alpha\beta}L_\beta\,\mathcal{M}^p_\beta]^{-1}[B_\alpha - (1/c_\beta)\,F_{\alpha\beta}L_\beta\,\mathcal{M}^p_\beta] \quad (72)$$

$$b_\alpha = [I + F_{\alpha\alpha}L_\alpha \mathcal{M}^p_\alpha - (c_\alpha/c_\beta)\,F_{\alpha\beta}L_\beta\,\mathcal{M}^p_\beta]^{-1}$$
$$[b_\alpha - F_{\alpha\alpha}L_\alpha m^p_\alpha - F_{\alpha\beta}L_\beta m^p_\beta]$$

$\mathcal{A}_\beta$, a_β, etc., follow from an exchange of subscripts.

Once the local fields are known, the overall instantaneous response is described in analogy with Equations (39) to (41) as:

$$d\sigma = \mathcal{L}d\varepsilon + l\,d\theta, \qquad d\varepsilon = \mathcal{M}d\sigma + m\,d\theta \quad (73)$$

$$\mathcal{L} = \sum_{r=1}^{M} c_r \mathcal{L}_r \mathcal{A}_r, \qquad l = \sum_{r=1}^{M} c_r(\mathcal{L}_r\,a_r + l_r) \quad (74)$$

$$\mathcal{M} = \sum_{r=1}^{M} c_r \mathcal{M}_r \mathcal{B}_r, \qquad m = \sum_{r=1}^{M} c_r(\mathcal{M}_r\,b_r + m_r) \quad (75)$$

The instantaneous overall properties [Equation (73)] of multiphase aggregates or unit cells subdivided into many elements can be evaluated in a similar way as described by Dvorak et al. [66]. The overall instantaneous stiffness matrix $\mathcal{L}$, compliance matrix $\mathcal{M}$, thermal stress matrix l, and thermal strain matrix m for an aggregate of Ω_ρ, ρ, $\eta = 1, 2, \ldots M$, subvolumes are given by:

$$[\mathcal{L}] = [c_\rho \mathcal{L}_\rho]\,[diag\,(I) + [D_{\rho\eta}\,M_\eta \mathcal{L}^p_\eta]]^{-1}[A_\rho] \quad (76)$$

$$[\mathcal{M}] = [c_\rho \mathcal{M}_\rho]\,[diag\,(I) + [F_{\rho\eta}\,L_\eta \mathcal{M}^p_\eta]]^{-1}[B_\rho] \quad (77)$$

$$\{\ell\} = [c_\rho \mathcal{L}_\rho]\,[diag\,(I) + [D_{\rho\eta}\,M_\eta \mathcal{L}^p_\eta]]^{-1}\{\{a_\rho\} - [D_{\rho\eta}\,M_\eta]\{l^p_\eta\}\} + \left\{\sum_{\eta=1}^{M} c_\eta l_\eta\right\} \quad (78)$$

$$\{m\} = [c_\rho \mathcal{M}_\rho]\,[diag\,(I) + [F_{\rho\eta}\,L_\eta \mathcal{M}^p_\eta]]^{-1}\{\{b_\rho\} - [F_{\rho\eta}\,L_\eta]\{m^p_\eta\}\} + \left\{\sum_{\eta=1}^{M} c_\eta m_\eta\right\} \quad (79)$$

These expressions provide the overall instantaneous moduli and compliances that may be used in analysis of elastic-plastic composite structures, such as laminated plates and shells.

Viscoplastic Phases

Next, consider the governing equations for the local stresses in composite materials with viscoplastic phases. In this case, the local constitutive equations may be specified with a certain unified theory that connects the local inelastic strain rate $\dot{\varepsilon}_\eta^{in}(t)$ to the local stress history or the local relaxation stress rate $\dot{\sigma}_\eta^{re}(t)$ to the local strain history. For example, the local inelastic strain rate may be specified by a power law of an internal stress variable R_η:

$$\dot{\varepsilon}_\eta^{in}(t) = \kappa_\eta(\theta)\, R_\eta^{p_\eta(\theta)}\, \boldsymbol{n}_\eta \tag{80}$$

where $\kappa_\eta(\theta)$, $p_\eta(\theta)$ are material parameters for the element volume Ω_η, and $\boldsymbol{n}_\eta$ specifies the direction of the inelastic strain rate in the local stress space. Evolution laws for the internal stress variable R_η have been postulated in several publications (see, e.g., [7–9,69,71]). When substituted into Equation (43), with $\dot{\sigma}_\eta^{re}(t) = -\boldsymbol{L}_\eta\, \dot{\varepsilon}_\eta^{in}(t)$, Equation (80) provides the following differential equation for the local stresses:

$$\dot{\sigma}_\rho(t) = \boldsymbol{B}_\rho\, \dot{\sigma}^0(t) - \sum_{\eta=1}^{M} \boldsymbol{F}_{\rho\eta}\, \boldsymbol{L}_\eta\{\boldsymbol{m}_\eta\, \dot{\theta}^0(t) + \kappa_\eta(\theta)\, R_\eta^{p_\eta(\theta)}\, \boldsymbol{n}_\eta\} \tag{81}$$

where $\sigma^0(t)$ is the overall stress history and $\theta^0(t)$ is the temperature history. The power law [Equation (80)] leads to a stiff differential Equation (81), which requires integration by an implicit multistep method, such as the Adams method, with Newton iteration to solve the resulting nonlinear system of equations, and backward differentiation to evaluate the functional derivatives found in the Newton's method.

Application of the transformation field analysis to simple fibrous composite structures, such as symmetric laminates, have been achieved, leading to closed form rate equations for the local stresses in all plies [70,72].

FINITE ELEMENT METHOD

An alternate to the closed form, transformation field analysis (TFA), described above for evaluation of the local fields and overall response in composite materials, is the finite element method. In problems where a refined local field is sought, the finite element method is expected to be more efficient than the TFA method when the number of subvolumes selected in the unit cell of the composite aggregate is large [66]. Moreover, the finite element method can be applied to composite structures, taking into consideration the underlying microgeometry of the material as described by Bahei-El-Din [73] and briefly outlined here.

The problem of an aggregate of viscoplastic finite volumes, V_m, m = 1, 2, . . . , M, interconnected at N nodes and subjected to a given load and temperature history can be obtained numerically with the initial strain formulation given by Zienkiewicz and Cormeau [74]. The method has been used extensively in finite element routines for viscoplastic homogeneous materials. Assuming that the strain can be additively decomposed into a recoverable elastic mechanical strain component, ε_m^e, and a transformation or eigenstrain μ_m, [Equation (27)], the elemental stiffness and load matrices are given by:

$$K_{ij}{}^m = \int_{V_m} [\Lambda_i^m]^T L_m \Lambda_j^m \, dV_m \tag{82}$$

$$\dot{f}_i^m = -\int_{V_m} [\Lambda_i^m]^T L_m \dot{\mu}_m \, dV_m \tag{83}$$

where K_{ij}^m is the ij partition of the element stiffness matrix, and f_i^m represents the forces at node i caused by the uniform eigenstrain μ_m. The L_m denotes the elastic stiffness matrix, and Λ_i^m represents the shape functions for the finite elements, such that if $\dot{\boldsymbol{\epsilon}}_m$ is the strain vector in element m and $\dot{\boldsymbol{a}}_m$ is the nodal displacements for the same element, then $\dot{\varepsilon}_m = \sum_{i=1}^{p} \Lambda_i^m \dot{\boldsymbol{a}}_i^m$, where P is the total number of element nodes. The governing equations for the nodal displacements $\dot{\alpha}$ are given by:

$$K \dot{\alpha} = \dot{f}, \quad K_{ij} = \sum_{m=1}^{M} K_{ij}{}^m, \quad \dot{f}_i = \dot{p}_i - \sum_{m=1}^{M} \dot{f}_i^m \tag{84}$$

where K is the standard overall stiffness matrix, $\dot{f}$ is the vector of overall load rates, and $\dot{\alpha} = [\dot{\alpha}_1 \ \dot{\alpha}_2 \ . \ . \ \dot{\alpha}_N]^T$ lists the rates of the nodal displacements. The $\dot{p}_i$ vector represents the rate of applied nodal loads.

Except when the elastic moduli L_m vary with temperature, the stiffness matrix K_{ij}^m is constant. On the other hand, the load vector f_i^m and the applied load p_i vary with time, and their time rate functions must be updated. The frequency of updating the load vector is dependent on the ordinary differential equation solver used in evaluating the nodal displacements α_i, i = 1, 2, . . ., N, and the stress σ_m and strain ε_m in the elements V_m, m = 1, 2, . . . , M, from Equations (29) and (30).

When this procedure is applied to multiphase materials, the overall elastic stiffness matrix L_m and the transformation strain μ_m are to be specified for a selected representative volume of the aggregate. Although the overall elastic properties can be measured experimentally, bounded or estimated for some heterogeneous materials, the overall transformation strain μ_m for multiphase media needs to be evaluated for the assumed microgeometry using the transformation field equations described previously. The interface between the transformation field analysis and the finite element scheme for viscoplastic materials is described by Bahei-El-Din [73]. The result is a system of first-order differential equations (ODE) that can be written in the form:

$$\dot{y}_j(t) = g_j(t, y_1, y_2, \ldots, y_R) \tag{85}$$

The unknown functions g_j, $j = 1, 2, \ldots, R$, are identified with the nodal displacements α_i, $i = 1, 2, \ldots, N$, the element stress σ_m and strain ε_m, $m = 1, 2, \ldots, M$, the phase stress σ_ρ and strain ε_ρ, $\rho = 1, 2, \ldots, Q$, and any internal variables required to define the eigenstrain rate $\dot{\mu}_\rho$, e.g., the scalar function R in Equation (80).

The algorithm for computing the rate functions g_j, in Equation (85) for an assemblage of M finite elements interconnected at N nodes, where each element consists of Q phases, can be outlined for a given nodal load history, $p_i(t)$, $i = 1, 2, \ldots, N$, and temperature history $\theta_m(t)$, $m = 1, 2, \ldots, M$, as follows [73].

1. Compute the overall stiffness matrix K at current temperature:

$$K_{ij} = \sum_{m=1}^{M} - \int_{V_m} [\Lambda_i^m]^T L_m(\theta_m) \Lambda_j^m \, dV_m$$

2. For elements $m = 1, 2, \ldots, M$:

   (a) Compute the transformation strain rate at temperature θ_m for phases $\rho = 1, 2, \ldots, Q$:

$$\dot{\mu}_{ij}^{(\rho m)} = [\alpha_{ij}^{(\rho m)} + (\partial M_{ijkl}^{(\rho m)}/\partial\theta) \sigma_{kl}^{(\rho m)}] \dot{\theta}_m + (k(\theta) \phi^{p(\theta)})_{\rho m} v_{ij}^{(\rho m)}$$

$$\dot{\mu}_{\rho m} = [\mu_{11} \mu_{22} \mu_{33} 2\mu_{23} 2\mu_{31} 2\mu_{12}]_{\rho m}$$

   (b) Compute the transformation strain rate at time t for element m:

$$\dot{\mu}_m = \sum_{\rho=1}^{Q} c_\rho B_\rho^T \dot{\mu}_{\rho m}$$

   (c) Compute contribution of element m to nodal load rates:

$$\dot{f}_i^m = -\int_{V_m} [\Lambda_i^m]^T L_m(\theta_m) \dot{\mu}_m \, dV_m$$

3. Assemble current rates of nodal loads:

$$\dot{f}_i = \dot{p}_i - \sum_{m=1}^{M} \dot{f}_i^m$$

4. Compute rates of nodal deflections at time t:

$$\dot{\alpha} = K^{-1} \dot{f}$$

5. For elements $m = 1, 2, \ldots, M;$
   (a) Compute overall strain rate: $\quad \dot{\varepsilon}_m = \Lambda_m \dot{\alpha}_m$
   (b) Compute overall stress rate: $\quad \dot{\sigma}_m = L_m(\theta_m)(\dot{\varepsilon}_m - \dot{\mu}_m)$
   (c) For phases $\rho = 1, 2, \ldots, Q$ of element m:

   Compute local strain rate: $\quad \dot{\varepsilon}_\rho = A_\rho(\theta_m)\dot{\varepsilon} + \sum_{\eta=1}^{Q} D_{\rho\eta}(\theta_m)\dot{\mu}_\eta$

   Compute local stress rate: $\quad \dot{\sigma}_\rho = L_\rho(\theta_m)(\dot{\varepsilon}_\rho - \dot{\mu}_\rho)$

Assuming elastic response of the phases in the initial state, the rate equations can be integrated for a specified time period using an ODE solver that is appropriate for stiff differential equations (e.g., the Gear [75] method) that are normally encountered in viscoplastic response modeled with the powerlaw assumed in Equation (80).

Modeling of Interface Damage

Fiber debonding and/or sliding are usually the initial damage states in fibrous composite plies subjected to off-axis loading of sufficient magnitude. The interface strengths are typically quite small. For example, Clyne and Withers [76] give interface shear strength magnitudes in the range of 50 to 100 MPa; normal strength values do not appear to be available but are expected to be lower. Cyclic shear sliding tends to reduce the number of adhering asperities and thus the apparent interface shear strength. Therefore, transverse tension and shear, and longitudinal shear stresses in plies can initiate interface damage. Of course, processing and heat treatment may affect the actual magnitudes, e.g., growth of brittle interfacial reaction layers may promote early debonding and matrix crack growth, whereas treatments that promote favorable residual stress states may increase the apparent strength.

However, as long as some adhering asperities are left at the interface after debonding, they should provide for shear stress transfer that may prevent fiber pullout. For example, the average surface shear (τ) and normal (σ) stresses in the fiber are connected by the force equilibrium relation $2\pi r L \tau = \pi r^2 \sigma$ over a length L of a fiber with radius r, i.e., $\sigma/\tau = 2L/r$. If $L = 25$ mm is selected as a transfer length and the radius of the Sigma fiber $r = 50$ μm, then one finds that $\sigma = 1000 \, \tau$. This suggests that large normal stress may be transferred into a fiber even if the interface shear stress is very low. Therefore, apart from fiber breaks, fiber pullout is not expected, and the integrity of the fibrous plies in the axial direction should not be impaired in the presence of fiber sliding. Guided by results of our related studies, which suggest that large-diameter fibers release a substantial amount of energy during debonding and thus experience nearly total, instantaneous separation, we represented the interface failure under normal or shear stresses by reducing certain fiber stress components to zero through abrupt changes in the fiber elastic moduli. In this part, the fiber was regarded as transversely isotropic, with moduli equal to isotropic material magnitudes. If the interface normal stress component reached the selected strength magnitude, then the transverse shear and Young's moduli G_T^f and E_T^f were reduced to zero. Similarly, if the interface longitudinal shear stress exceeded the selected shear strength, the longitudinal shear modulus G_L^f was reduced to zero. This approach has been implemented in the transformation analysis and finite element method described above. After fiber moduli reductions, the ply stiffnesses and compliances and the concentration factors were reevaluated and used in the governing equations [73].

IMPLEMENTATION AND RESULTS

Both the transformation field analysis (TFA) and the finite element (FE) analysis described above have been implemented and used to examine several important problems for titanium matrix composites. The TFA/FE solution scheme described above was implemented for fibrous composites, including interface damage, and encoded in the VISCOPAC program by Bahei-El-Din [77]. Solution of the TFA closed form rate equations given previously for unidirectionally reinforced composites is provided by Dvorak et al. [66]. The solution for laminates is provided by Wafa [70] and Dvorak and Bahei-El-Din [72]. The problems described herein were examined by the VISCOPAC program.

First, the local stresses caused by fabrication of a titanium matrix (Ti-15-3) reinforced by silicon-carbide fibers (SCS-6) are evaluated. In particular, the typical pressure-temperature profile used by the manufacturer in processing of the SCS-6/Ti-15-3 composite by hot isostatic pressing (HIP) was applied. The actual profile is shown by the solid curves in Figure 1, whereas the dashed curves indicate the modeled profile applied in the numerical simulation. Assuming the composite to be fully consolidated either before or within the initial stage of the HIP cycle where the temperature is very high and the matrix

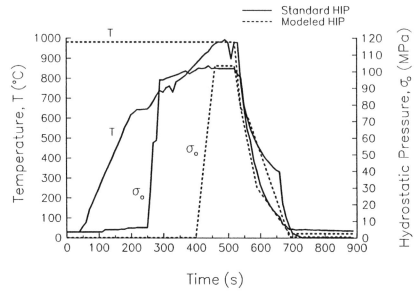

Figure 1. *Standard and modeled HIP profiles for SCS6/Ti-15-3 composite material.*

yield stress is very low, the modeled profile retained those parts of the actual profile that influence the residual fields.

At the processing temperature of 980°C (1800°F), the hydrostatic pressure was applied at the rate of $\dot{\sigma}_o$ = 1.725 MPa/sec for 60 sec, until σ_o = 103.5 MPa (15 ksi). After 60 sec at this pressure level, both the pressure and temperature were reduced along the path indicated in Figure 1. Under these axisymmetric loads, we chose the composite cylinder model of Figure 2, with fiber volume fraction c_f = 0.325, under hydrostatic pressure σ_o, and a uniform change in temperature. The unit cell is then selected as the small wedge segment "abcdef" of the cylinder, under generalized plane strain in the direction of the x_3-axis, and axisymmetric boundary conditions on the planes "abed" and "acfd." The central angle of the wedge segment was selected as 1^0, and the matrix domain was divided into 10 elements in the transverse plane. The SCS-6 fiber is assumed to remain elastic during deformation, whereas the titanium matrix follows the thermo-viscoplastic constitutive equations described by Bahei-El-Din et al. [71], together with the relevant property data derived from experimental observations of neat matrix response [78]. The isotropic phases are assumed to have temperature-dependent elastic moduli and coefficient of thermal expansion.

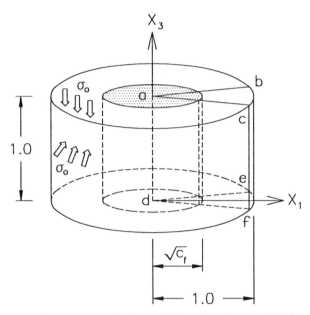

Figure 2. Composite cylinder model under axisymmetric loads.

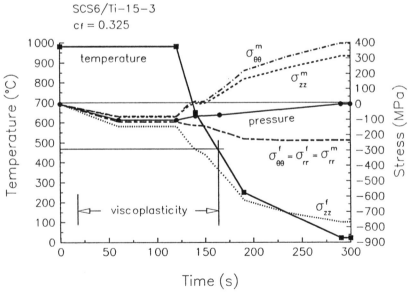

Figure 3. Evolution of interface stresses during standard HIP of a SCS6/Ti-15-3 composite.

98

Analysis of the unit cell using the VISCOPAC program provided the local stress distribution under the HIP profile of Figure 1. Figure 3 shows the stress components evaluated at selected time intervals at the fiber-matrix interface. Here, the $\sigma_{\theta\theta}^m$ is the hoop stress in the matrix, $\sigma_{rr}^m = \sigma_{rr}^f$ is the radial stress, also equal to the fiber hoop stress $\sigma_{\theta\theta}^f$, and σ_{zz}^m, σ_{zz}^f are the axial normal stresses. For reference, the temperature and pressure histories applied during hot isostatic pressing are also shown in Figure 3. As expected, the stresses remain low during viscoplastic deformation of the matrix at the fabrication temperature. The stress buildup starts at the onset of cooling, even in the presence of pressure release. At the end of the cycle at room temperature, the fiber supports a rather high axial compressive stress. The radial stress is also compressive so that the material remains well bonded, but both the hoop and axial normal stresses in the matrix are high and positive (tensile), even at a distance from the interface. This is illustrated in Figure 4 that shows the stress distribution in a radial section of the composite cylinder. With this detailed evaluation of the local stresses, it is possible to examine how they are affected by various fabrication parameters. For example, the cooling rate after HIP, and the magnitude of the hydrostatic pressure σ_o applied during the HIP process are expected to affect the local stress distribution. A detailed evaluation of these effects in the unidirectional SCS-6/Ti-15-3 composite has been provided by Bahei-El-Din and Dvorak [79].

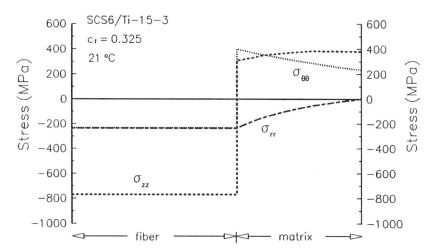

Figure 4. Distribution of local stresses found at room temperature following standard HIP of a SCS6/Ti-15-3 composite.

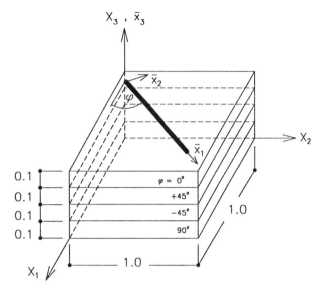

Figure 5. *Finite element model for a (0/±45/90)ₛ laminate.*

We next evaluated the local stresses and overall response of a $(0/\pm45/90)_s$ Sigma and TIMETAL®21S laminate under isothermal cyclic tension applied at 650°C. The silicon-carbide Sigma fiber was regarded as elastic, and the TIMETAL®21S titanium matrix as thermo-viscoplastic. Elastic moduli of the Sigma fiber were assumed to be the same as those of the SCS-6 fiber [78]. The inelastic response of the matrix is again derived from a unified viscoplasticity theory described by Bahei-El-Din et al. [71]. The finite element model for the laminate is shown in Figure 5. The overall elastic moduli of the plies and the phase mechanical and transformation concentration factors were derived from the Mori-Tanaka model. Internal damage by debonding or sliding of the fibers was considered in the present analysis. Interface failure by fiber separation takes place when the interface normal stress component reaches the normal bond strength magnitude of 50 MPa and by fiber sliding when the interface longitudinal shear stress exceeds the shear bond strength of 80 MPa. Because in two-phase averaging models, such as the Mori-Tanaka model used here, the fiber stress is uniform, then the interface normal and longitudinal stresses equal to the corresponding fiber stresses. Thermomechanical loading of the laminate consists of hot isostatic pressing at 899°C (1650°F) and 103.5 MPa (15 ksi) hydrostatic pressure for 2 hours, subsequent cooling to room temperature, aging at 621°C for 8

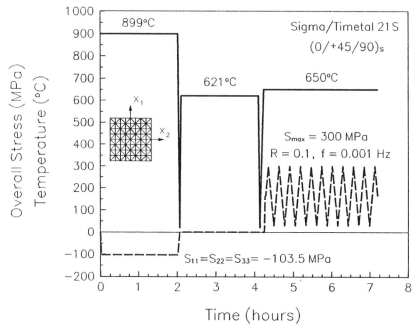

Time (hours)

Figure 6. *Thermomechanical loading history applied to the (0/±45/90)ₛ laminate.*

hours, reheating to 650°C, and cyclic axial tensile stress applied at that temperature. The cyclic loading conditions are $R = S_{min}/S_{max} = 0.1$, $S_{max} = 300$ MPa, at 0.001 Hz. Figure 6 shows the temperature and load history applied during fabrication and testing of the laminate, with 11 load cycles at 0.001 Hz that were sustained by an actual test sample prior to failure [80]. We note that the aging period of 8 hours, during which the titanium matrix remained elastic, has been reduced in Figure 6 to 2 hours.

Under the selected values for the interface normal and shear bond strength, failure of the fiber interface took place in the various plies of the laminate during mechanical loading to the mean stress at 1.375 MPa/sec. The overall stress levels at the onset of interface failure and the corresponding local stress averages in the individual plies are indicated in Table 1. The local stress averages in the fiber and matrix are also listed in Table 1. The stresses in the ply, fiber, and matrix are defined in the coordinate system of each ply such that σ_{11} is the normal stress parallel to the fiber longitudinal axis, σ_{22} and σ_{33} are the normal stresses in the transverse plane, and σ_{12} is the longitudinal

Table 1. Local stress averages (MPa) in plies of a Sigma/Timetal 21S (0/±45/90), laminate caused by hot isostatic pressing and subsequent static loading at 650°C.

Ply	Local Stress[a]	HIP Residual Stress @21°C	HIP Residual Stress @650°C	Under Overall Axial Stress (MPa) 18[b]	26[c]	59[d]	65[e]
0°	σ_{11}^p	−157	−55	−5	39	210	217
	σ_{22}^p	157	55	54	59	85	40
	σ_{11}^f	−1095	−293	−195	−109	256	320
	σ_{22}^f	−52	15	15	20	50	0
	σ_{33}^f	−232	−53	−53	−55	−65	0
	σ_{11}^m	492	110	127	141	177	145
	σ_{22}^m	302	82	82	86	110	67
	σ_{33}^m	161	36	37	38	45	0
45°	σ_{11}^p	−157	−55	−33	−10	53	77
	σ_{22}^p	157	55	69	84	47	50
	σ_{12}^p	0	0	15	26	54	54
	σ_{11}^f	−1095	−293	−255	−214	−30	21
	σ_{22}^f	−52	15	32	50	0	0
	σ_{33}^f	−232	−53	−54	−56	0	0
	σ_{12}^f	0	0	20	34	78	80
	σ_{11}^m	492	110	121	131	110	115
	σ_{22}^m	302	82	95	108	80	85
	σ_{33}^m	161	36	37	39	0	0
	σ_{12}^m	0	0	11	19	37	37
90°	σ_{11}^p	−157	−55	−61	−82	−77	−58
	σ_{22}^p	157	55	84	43	61	65
	σ_{11}^f	−1095	−293	−314	−325	−319	−281
	σ_{22}^f	−52	15	50	0	0	0
	σ_{33}^f	−232	−53	−54	0	0	0
	σ_{11}^m	492	110	115	85	90	97
	σ_{22}^m	302	82	107	72	103	110
	σ_{33}^m	161	36	38	0	0	0

[a] σ_{ij}^p = ply stress, σ_{ij}^f = fiber stress, σ_{ij}^m = matrix stress; x_1 = axial direction.
[b] Corresponds to onset of fiber debonding in 90° ply at σ_b = 50 MPa.
[c] Corresponds to onset of fiber debonding in ±45° plies at σ_b = 50 MPa.
[d] Corresponds to onset of fiber debonding in 0° ply at σ_b = 50 MPa.
[e] Corresponds to onset of fiber sliding in ±45° plies at τ_b = 80 MPa.

shear stress (Figure 5). Also shown in Table 1 are the residual stresses caused by fabrication at room temperature and at the test temperature (650°C). As expected, in the quasi-isotropic laminate the residual stresses after HIP are independent of the fiber orientation. Moreover, it is evident that the residual stresses caused by processing do contribute to the subsequent stress magnitudes. This contribution need not be detrimental; in the present case, the residual axial fiber stress σ_{11} is compressive and thus helps to reduce the tensile stress magnitude under subsequent tensile mechanical loading. However, a tensile residual stress, σ_{22}, of 15 MPa is present in the fiber at 650°C, which is likely to promote debonding under subsequent axial tension loading.

Evolution of the local stresses during processing and subsequent isothermal cyclic load history has been evaluated numerically. A sample is shown in Figures 7 and 8 for the average stresses generated in the fiber and matrix of the 0°-ply. These are defined in the coordinate system of each ply as defined before. It is seen that the fiber stress is enhanced under the applied tension-tension load cycles. This is caused by stress relaxation in the viscoplastic matrix stresses, which, under the slow load cycles applied at 0.001 Hz, reach a nearly steady state after only few cycles. In contrast, the stresses computed for a fast cycle at 0.1 Hz (not shown here) reached a steady state after 90 cycles. In this case, actual test specimen sustained 207 cycles [79]. The maximum fiber stress found at failure, however, appears to be in-

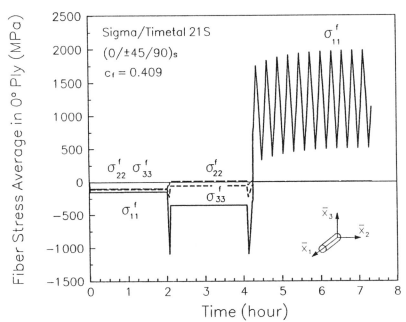

Figure 7. *Fiber stress average computed in the 0° ply of a Sigma/Timetal 21S (0/±45/90)ₛ laminate subjected to HIP and isothermal cyclic loading.*

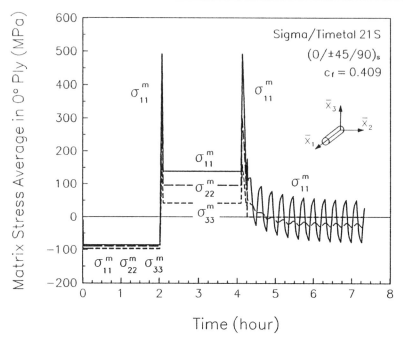

Figure 8. Matrix stress average computed in the 0° ply of a Sigma/Timetal 21S (0/±45/90)$_s$ laminate subjected to HIP and isothermal cyclic loading.

dependent of the loading frequency. Specifically, the axial stress computed in the 0° fiber at the failure cycle is 1981 MPa under 0.001 Hz and 1988 MPa under 0.1 Hz. This suggests that failure of the laminate is controlled by the strength of the fiber in the 0° ply and that the fatigue life is dependent on the rate at which the fiber tensile strength is reached.

Analysis of laminates under in-phase and out-of-phase thermomechanical load cycles or any complex thermomechanical loading path can be also performed. The computed overall response and internal stresses in the Sigma and TIMETAL®21S (0/±45/90)$_s$ laminate under in-phase thermomechanical load cycles in the temperature range of 150 to 650°C have been examined by Bahei-El-Din and Dvorak [81].

CLOSURE

This brief summary of analytical and numerical methods for solving problems arising in thermomechanical loading of structural parts made of TMC and other metal matrix systems provides an introduction to the subject. Several methods are available for evaluation of estimates or bounds on overall elastic constants and local fields; together, they give a range of values that are likely to correspond to those observed experimentally on undamaged systems.

When inelastic response of the matrix exerts a significant influence on the overall behavior, the instantaneous matrix properties are no longer constant; instead, they depend on the deformation history of each material point. Accordingly, analysis of the local fields becomes much more complex. The two-phase approximations, Equations (71) or (72), obtained with the transformation analysis are appealing but are likely to be of value only in simple monotonic or cyclic loading situations. If more complex loading histories are applied, then unit cell models and the transformation field analysis appear to be advantageous when the microstructure is subdivided into relatively few subvolumes. Several comparisons with ABAQUS-generated plasticity solutions [66] showed that the finite element method may be more efficient in problems involving more that 100 elements.

The principal advantage of the transformation analysis is derived from the fact that the concentrations factors in Equations (42) and (43), which reflect entirely the effect of local geometry, depend only on elastic constants. They can be obtained either with the averaging methods using Equations (44) to (50) or evaluated with an elastic finite element program. No inelastic inclusion problems need to be solved. The governing equations for the local fields [Equations (60) or (61)] can then be completed with appropriate substitutions of the phase constitutive relations and solved with a suitable solver. On the other hand, special material routines and solution procedures are required in inelastic finite element programs.

Although substantial progress has been made in modeling of inelastic composite systems, several unresolved problems deserve further attention. In particular, realistic description of the constitutive relations of the inelastic phase poses a challenge, because the fiber-constrained deformation of small matrix volumes may not be accurately predicted by tests on much larger, uniformly deforming samples of neat matrix. Extraction of in situ matrix properties is desirable, but so far it can be accomplished only by phenomenological comparisons of model predictions with selected experiments. The role of local geometry and fiber arrangements appears to be significant. Therefore, models should approximate the actual geometry as closely as possible, and use of unrealistic shapes, such as fibers of rectangular cross sections, should be avoided.

The presence of damage by fiber-matrix decohesion and matrix cracking complicates the analysis further. Although the relatively simple treatment used in elastic-plastic phases appears adequate under simple cyclic loading, unit cell models with properly designed interface elements should be employed for applications involving complex loading.

In summary, the analysis of inelastic fibrous composites is now well understood, both in terms of available analytical and numerical approaches and as yet unresolved problem areas that deserve special attention in future research.

ACKNOWLEDGMENT

The work surveyed in this paper was supported by grants from the Air Force Office of Scientific Research and by the Mechanics Division of the Office of Naval Research. Drs. George Haritos, Walter Jones, and Yapa Rajapakse served as program monitors.

REFERENCES

1. Larsen, J. M., Revelos, W. C., and Gambone, M. L. 1992, An overview of potential titanium aluminide composites in aerospace applications. In: *Intermetallic Composites II, MRS Proceedings Vol. 273*. MRS, Pittsburgh, PA.

2. Wilson, T. M. 1994, Applications of titanium matrix composite to large airframe structure. In: *Characterization of Fibre Reinforced Titanium Metal Matrix Composites, AGARD-R-796*. Specialized Printing Services, Ltd., Loughton, Essex, UK.

3. Larsen, J. M., Russ, S. M. and Jones, J. W. 1994, Possibilities and pitfalls in aerospace applications of titanium matrix composites. In: *Characterization of Fibre Reinforced Titanium Metal Matrix Composites, AGARD-R-796*. Specialized Printing Services, ltd., Loughton, Essex, UK.

4. Larsen, J. M., Russ, S. M., and Jones, J. W. 1996, An evaluation of titanium matrix composites for advanced high temperature structural applications. *Met. and Mat. Trans.*, in press.

5. Hill, R. 1965, A self-consistent mechanics of composite materials. *J. Mech. Phys. Sol., 13*, pp. 213–222.

6. Hillmer, N. J. 1991, Thermal expansion of high-modulus fibers. *Int. J. Thermophysics, 12*, pp. 741–750.

7. Neu, R. W. 1993, Nonisothermal material parameters for the Bodner-Partom model. In: *Material Parameter Estimation for Modern Constitutive Equations, MD-43*, pp. 211–226, American Society of Mechanical Engineers, New York, NY.

8. Neu, R. W. and Bodner, S. R. 1995, *Contributive Research and Development Volume 6, Determination of the Material Constants of Timetal21S for a Constitutive Model.* Systran Corporation, Dayton, OH.

9. Krempl, E. 1987, Models of viscoplasticity some comments on equilibrium (back) stress and drag stress. *Acta Mechanica, 69,* pp. 25–42.

10. Miller, A. K., ed. 1987, *Unified Constitutive Equations for Creep and Plasticity,* Elsevier Applied Science, Amsterdam.

11. Chaboche, J. L. 1989, Time-independent constitutive theories for cyclic plasticity. *Int. J. Plasticity, 2,* pp. 149–188.

12. Chan, K. S., Bodner, S. R. and Lindholm, U. S. 1988, Phenomenological modeling of hardening and thermal recovery in metals. *ASME J. Eng. Materials Tech., 110,* pp. 1–8.

13. Zuiker, J. R. and Sanders, B. P. 1995, Analysis of pinned joints in viscoplastic monolithic plates. In: *Numerical Implementation and Application of Constitutive Models in the Finite Element Method, AMD-213/MD-63,* American Society of Mechanical Engineers, New York, NY, pp. 37–48.

14. Hill, R. 1963, Elastic properties of reinforced solids: Some theoretical principles. *J. Mech. Phys. Solids,* pp. 357–372.

15. Levin, V. M. 1967, Thermal expansion coefficients of heterogeneous materials. *Mekhanika Tverdogo Tela, 2,* pp. 88–94; English translation: *Mech. of Solids, 11,* pp. 58–61.

16. Dvorak, G. J. 1991, Plasticity theories for fibrous composite materials. In: *Metal Matrix Composites, Mechanisms and Properties* (ed. R. K. Everett and R. J. Arsenault), vol. 2, pp. 1–77. Academic Press, Boston, MA.

17. Eshelby, J. D. 1957, The determination of the elastic field of an ellipsoidal inclusion, and related problems. *Proc. Roy. Soc. Lond., A241,* pp. 376–396.

18. Kinoshita, N. and Mura, T. 1971, Elastic fields of inclusions in anisotropic media. *Phys. Stat. Sol. A, 5,* pp. 759–768.

19. Christensen, R. M. 1979, *Mechanics of Composite Materials,* John Wiley & Sons, New York, NY.

20. Budiansky, B. 1965, On the elastic moduli of some heterogeneous materials. *J. Mech. Phys. Sol., 13,* pp. 223–227.

21. Mori, T. and Tanaka, K. 1973, Average stress in matrix and average elastic energy of materials with misfitting inclusions. *Acta Metallurgica, 21,* pp. 571–574.

22. Benveniste, Y. 1987, A new approach to the application of Mori-Tanaka's Theory in composite materials. *Mech. Materials, 6,* pp. 147–157.

23. Weng, G. J. 1984, Some elastic properties of reinforced solids, with special reference to isotropic ones containing spherical inclusions. *Int. J. Eng. Sci., 22,* pp. 845–856.

24. Benveniste, Y., Dvorak, G. J. and Chen, T. 1991, On diagonal and elastic symmetry of the approximate stiffness tensor of heterogeneous media. *J. Mech. Phys. Sol., 39,* pp. 927–946.

25. Chen, T., Dvorak, G. J. and Benveniste, Y. 1992, Mori-Tanaka estimates of the overall elastic moduli of certain composite materials. *ASME J. Appl. Mechanics, 59,* pp. 539–546.

26. Hill, R. 1964, Theory of mechanical properties of fibre-strengthened materials: I. Elastic behavior. *J. Mech. Phys. Sol., 12,* pp. 199–212.

27. McLaughlin, R. 1977, A study of the differential scheme for composite materials. *Int. J. Eng. Sci., 15,* pp. 237–244.

28. Norris, A. N. 1985, A differential scheme for the effective moduli of composites. *Mechanics of Materials, 4,* pp. 1–16.

29. Christensen, R. M. 1990, A critical evaluation for a class of micromechanics models. *J. Mech. Phys. Sol., 38,* pp. 379–404.

30. Iwakuma, T. and Nemat-Nasser, S. 1983, Composites with periodic microstructure. *Composites Struct., 16,* pp. 13–19.

31. Teply, J. and Dvorak, G. J. 1988, Bounds on overall instantaneous properties of elastic-plastic composites. *J. Mech. Phys. Sol., 36,* pp. 29–58.

32. Nemat-Nasser, S. and Hori, M. 1993, *Micromechanics: Overall Properties of Heterogeneous Materials,* Elsevier Science Publishers, Amsterdam, Netherlands.

33. Aboudi, J. 1991, *Mechanics of Composite Materials: A Unified Micromechanical Approach.* Elsevier, Amsterdam, Netherlands.

34. Dvorak, G. J. and Teply, J. L. 1985, Periodic hexagonal array models for plasticity analysis of composite materials. In: *Plasticity Today: Modeling, Methods and Applications, W. Olszak Memorial Volume* (ed. A. Sawczuk and V. Bianchi), pp. 623–642, Elsevier Science Publishers, Amsterdam, Netherlands.

35. Suquet, P. M. 1987, Elements of homogenization for inelastic solid mechanics. In: *Homogenization Techniques for Composite Media* (ed. E. Sanchez-Palencia and A. Zaoui) Springer-Verlag, Berlin, Germany, pp. 193–278.

36. Hollister, S. J. and Kikuchi, N. 1992, A comparison of homogenization and standard mechanics analyses for periodic porous composites. *Computational Mechanics, 10,* pp. 73–95.

37. Bensoussan, A. and Lions, J.-L. 1978, *Asymptotic Analysis for Periodic Structures,* North-Holland Publishing Co., Amsterdam, Netherlands.

38. Sanchez-Palencia, E. and Zaoui, A., ed. 1987, *Homogenization Techniques for Composite Media.* Springer-Verlag, Berlin, Germany.

39. Bakhvalov, N. and Panasenko, G. 1989, *Homogenization: Averaging Processes in Periodic Media,* Kluwer Academic Publishers, Dordrecht, Netherlands.

40. Voigt, W. 1887, Theoretische Studien über die Elastizitätsverhältnisse der Krystalle. *Abh. Kgl. Ges. Göttingen Math. Kl., 34,* p. 47.

41. Reuss, A. 1929, Berechnung der Fliessgrenze von Mischkristallen auf Grund der Plastizitätsbedingung für Einkristalle. *Z. Ang. Math. Mech., 9,* p. 49.

42. Hill, R. 1952, The elastic behavior of a crystalline aggregate, *Proc. Roy. Soc. Lond., A65,* p. 349.

43. Beran, M. J. 1968, *Statistical Continuum Theories,* Interscience Publishers, New York, NY.

44. Hashin, Z. and Shtrikman, S. 1962a, On some variational principles in anisotropic and nonhomogeneous elasticity. *J. Mech. Phys. Sol., 10,* pp. 335–342.

45. Hashin, Z. and Shtrikman, S. 1962b, A variational approach to the theory of the elastic behavior of polycrystals. *J. Mech. Phys. Sol.*, *10*, pp. 343–352.

46. Hashin, Z. and Shtrikman, S. 1963, A variational approach to the theory of the elastic behavior of multiphase materials. *J. Mech. Phys. Sol.*, *11*, pp. 127–140.

47. Hashin, Z. and Rosen, B. W. 1964, The elastic moduli of fiber reinforced materials. *J. Appl. Mech.*, *31*, pp. 223–232.

48. Walpole, L. J. 1966a, On bounds for the overall elastic moduli of inhomogeneous systems I. *J. Mech. Phys. Solids*, *14*, pp. 151–162.

49. Walpole, L. J. 1966b, On bounds for the overall elastic moduli of inhomogeneous systems II. *J. Mech. Phys. Solids*, *14*, pp. 289–301.

50. Bahei-El-Din, Y. A. and Dvorak, G. J. 1989, A review of plasticity theory of fibrous composite materials. In: *Metal Matrix Composites: Testing, Analysis, and Failure Modes, ASTM STP 1032*, (ed. W. S. Johnson), American Society for Testing and Materials, Philadelphia, PA, pp. 103–129.

51. Kröner, E. 1986, Statistical modelling. In: *Modelling Small Deformations of Polycrystals*, Elsevier, London, UK.

52. Torquato, S. 1991, Random heterogeneous media: Microstructure and improved bounds on effective properties. *Appl. Mech. Rev.*, *44*, pp. 37–76.

53. Robinson, D. N., Duffy, S. N. and Ellis, J. R. 1986, A Viscoplastic Constitutive Theory for Metal Matrix Composites at High Temperature. NASA CR-179530.

54. Krempl, E. and Hong, B. Z. 1989, A simple laminate theory using the orthotropic viscoplasticity theory based on overstress. Part I. In-plane stress-strain relationships for metal matrix composites. *Comp. Sci. Tech.*, *35*, pp. 53–74.

55. Nicholas, T. 1995, An approach to fatigue life modeling in titanium-matrix composites. *Mat. Sci. Eng.*, *A200*, pp. 29–37.

56. Bahei-El-Din, Y. A. and Dvorak, G. J. 1982, Plasticity analysis of laminated composite plates. *J. Appl. Mech.*, *104*, pp. 740–746.

57. Dvorak, G. J. and Bahei-El-Din, Y. A. 1982, Plasticity analysis of fibrous composites. *J. Appl. Mech.*, *49*, pp. 327–335.

58. Lagoudas, D. C., Gavazzi, A. C. and Nigam, H. 1991, Elastoplastic behavior of metal matrix composites based on incremental plasticity and the Mori-Tanaka averaging scheme. *Computational Mech.*, *8*, p. 193.

59. Zuiker, J. R. 1996, Modeling the creep response of ceramic matrix composites: A tool for inferring in-situ matrix properties. In: *Thermal and Mechanical Tests Methods and Behavior of Continuous-Fiber Ceramic Matrix Composites, ASTM STP 1309*, American Society for Testing and Materials, Philadelphia, PA.

60. Coker, D., Ashbaugh, N. E. and Nicholas, T. 1993, Analysis of thermomechanical cyclic behavior of unidirectional metal matrix composites. In: *Thermomechanical Fatigue Behavior of Materials, ASTM STP 1186*, American Society for Testing and Materials, Philadelphia, PA, pp. 50–69.

61. Kroupa, J. L. and Neu, R. W. 1994, The nonisothermal viscoplastic behavior of a titanium-matrix composite. *Comp. Eng.*, *4*, pp. 965–977.

62. Kroupa, J. L., Neu, R. W., Nicholas, T., Coker, D. et al. 1996, A comparison of analysis tools for predicting the inelastic cyclic response of cross-ply titanium matrix composites. In: *Life Prediction Methodology for Titanium Matrix Composites*. *ASTM STP 1253*, American Society for Testing and Materials, Philadelphia, PA.

63. Dvorak, G. J. and Benveniste, Y. 1992, On transformation strains and uniform fields in multiphase elastic media. *Proc. R. Soc. Lond., A437*, pp. 291–310.

64. Dvorak, G. J. 1990, On uniform fields in heterogeneous media. *Proc. R. Soc. Lond. A431*, pp. 89–110.

65. Laws, N. 1973, On the thermostatics of composite materials. *J. Mech. Phys. Solids, 21*, pp. 9–17.

66. Dvorak, G. J., Bahei-El-Din, Y. A. and Wafa, A. M. 1994, Implementation of the transformation field analysis for inelastic composite materials. *Comput. Mech., 14*, pp. 201–228.

67. Dvorak, G. J. 1992, Transformation field analysis of inelastic composite materials. *Proc. R. Soc. Lond., A 437*, pp. 311–327.

68. Bahei-El-Din, Y. A. 1990, Plasticity analysis of fibrous composite laminates under thermomechanical loads. In: *Thermal and Mechanical Behavior of Metal Matrix and Ceramic Matrix Composites* (ed. J. M. Kennedy, H. H. Moeller and W. S. Johnson), *ASTM STP 1080*, pp. 20–39. American Society for Testing and Materials, Philadelphia, PA.

69. Shah, R. S. 1991, Modeling and analysis of high temperature inelastic deformation in metal matrix composites. Ph.D. thesis, Rensselaer Polytechnic Institute, Troy, NY.

70. Wafa, A. M. 1994, Application of the Transformation Field Analysis to Inelastic Composite Materials and Structures. Ph.D. thesis, Rensselaer Polytechnic Institute, Troy, NY.

71. Bahei-El-Din, Y. A., Shah, R. S. and Dvorak, G. J. 1991, Numerical analysis of the rate-dependent behavior of high temperature fibrous composites. In: *Mechanics of Composites at Elevated and Cryogenic Temperatures* (ed. S. N. Singhal, W. F. Jones, T. Cruse and C. T. Herakovich), *ASME AMD-vol. 118*, pp. 67–78. American Society of Mechanical Engineers, New York, NY.

72. Dvorak, G. J. and Bahei-El-Din, Y. A. 1995, Transformation analysis of inelastic laminates. In: *IUTAM Symposium on Microstructural-Property Interactions in Composite Materials,* Aalborg, Denmark (ed. R. Pyrz), Kluwer Academic Publishers, Netherlands, pp. 89–100.

73. Bahei-El-Din, Y. A. 1996, Finite element analysis of viscoplastic composite materials and structures. *Mechanics of Composite Materials and Structures*, pp. 1–28.

74. Zienkiewicz, O. C. and Cormeau, I. C. 1974, Viscoplasticity-plasticity and creep in elastic solids—A unified numerical solution approach. *Int. J. Numer. Methods Engng., 8*, pp. 821–845.

75. Gear, C. W. 1971, *Numerical Initial Value Problems in Ordinary Differential Equations*. Prentice-Hall, Englewood Cliffs, NJ.

76. Clyne, T. W. and Withers, P. J. 1993, *An Introduction to Metal Matrix Composites.* Cambridge University Press, Cambridge, UK.

77. Bahei-El-Din, Y. A. 1994, *VISCOPAC Finite Element Program for Viscoplastic Analysis of Composites, User's Manual.* Structural Engineering Department, Cairo University, Giza, Egypt.

78. Johnson, W. S., Mirdamadi, M. and Bahei-El-Din, Y. A. 1993, Stress-strain analysis of a [0/90]2s Titanium matrix laminate subjected to a generic hypersonic flight profile. *J. Composites Technology and Research, 15,* pp. 297–303.

79. Bahei-El-Din, Y. A. and Dvorak, G. J. 1995, Mechanics of hot isostatic pressing of a densified unidirectional SiC/Ti composite. *Acta Metallurgica et Materialia,* pp. 2531–2539.

80. Dvorak, G. J., Nigam, H. and Bahei-El-Din, Y. A. 1996, Isothermal fatigue behavior of Sigma/Timetal 21S laminates—I. Experimental results. *Mechanics of Composite Materials and Structures,* in press.

81. Bahei-El-Din, Y. A. and Dvorak, G. J. 1996. Isothermal fatigue behavior of Sigma/Timetal 21S laminates—II. Modeling and numerical analysis. *Mechanics of Composite Materials and Structures,* in print.

Fiber-Matrix Interface

BHASKAR S. MAJUMDAR*

Materials Directorate, Wright Laboratory
Wright-Patterson AFB, OH 45433-7817

IMPORTANCE OF THE FIBER-MATRIX INTERFACE

THE FIBER-MATRIX INTERFACE is a critical constituent of composites because load transfer from the matrix to the high-modulus fiber and vice versa must occur through the interface. It is this load transfer behavior that distinguishes composites from monolithic materials. Figure 1 illustrates a number of practical situations in composites where the role of the interface is important.

To emphasize how interface characteristics can influence composite performance, two special examples are considered initially. Additional details of the effects of interface properties on composite response will be provided later in this chapter.

The first example is on the creep of unidirectional fiber-reinforced metal matrix composites (MMCs) loaded in the transverse (90°) direction. The reason for choosing this example is that the application of MMCs in aerospace engines appears to be currently limited by the transverse creep performance. Finite element computations performed sometime ago by Crossman et al. [1] suggest that the creep response of transversely loaded MMCs can be effectively represented by considering the MMC to be of the matrix material but under a modified stress (σ_{mod}), such that:

$$\sigma_{mod} = \sigma_{applied} \cdot exp \ \{-\eta \cdot v\} \tag{1}$$

where v is the volume fraction of fibers, $\sigma_{applied}$ is the applied stress on the composite, and η is a factor that depends on the normal bond strength between the matrix and fibers, being also slightly dependent

*UES, Inc., 4401 Dayton-Xenia Rd., Dayton, OH 45432-1894.

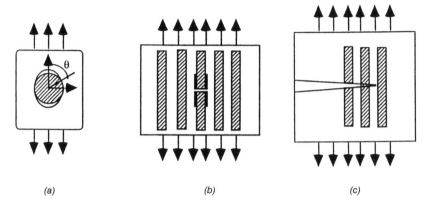

(a)   (b)   (c)

Figure 1. A number of loading conditions where the role of the interface is important. (a) Transverse loading of the MMC, (b) longitudinal loading in the presence of fiber breaks, and (c) fatigue crack growth and crack bridging.

on the packing geometry. For a square array, η is approximately 1.4 for infinite bond strength, and approximately -2.0 for no bond strength.

In the secondary creep regime, the creep rate of most metallic alloys follow:

$$d\varepsilon/dt = A\,\sigma^n \tag{2}$$

Substituting Equation (2) in Equation (1):

$$\frac{\left(\dfrac{d\varepsilon}{dt}\right)_{MMC}}{\left(\dfrac{d\varepsilon}{dt}\right)_{Matrix}} = exp^{-n\eta v} \tag{3}$$

Thus, for a typical case of $n = 5$ and $v = 0.3$, the MMC would be creeping at a rate of about one-tenth of the matrix for an interface with infinite bond strength, whereas it would be creeping about 20 times faster than the matrix for an interface with zero bond strength, a difference by a factor close to 200. These are illustrated in Figure 2. Clearly, a strong bond is desirable for improved transverse creep strength. The applicability of Equation (3) has indeed been observed for a SCS-6/Ti-6A1-4V MMC [2], creep tested at 800°F, wherein a zero bond strength ($\eta = -2.0$) appeared to fit the data reasonably well.

Transverse Creep of a TMC as Influenced by Interface Strength

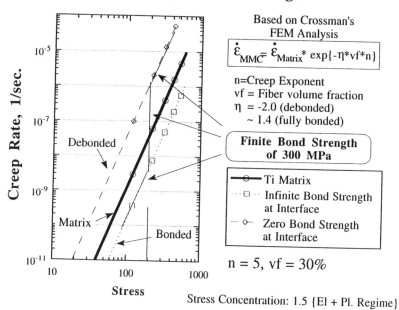

Based on Crossman's
FEM Analysis

$$\dot{\varepsilon}_{MMC} = \dot{\varepsilon}_{Matrix} * \exp\{-\eta * vf * n\}$$

n=Creep Exponent
vf = Fiber volume fraction
η = -2.0 (debonded)
~ 1.4 (fully bonded)

Finite Bond Strength
of 300 MPa

—⊖— Ti Matrix
···□··· Infinite Bond Strength
at Interface
--◇-- Zero Bond Strength
at Interface

n = 5, vf = 30%

Stress Concentration: 1.5 {El + Pl. Regime}

Figure 2. Predicted creep rate as a function of far-field applied stress for a transversely loaded unidirectional titanium matrix composite, compared with that of the matrix. The stepped thin solid line represents the case of a finite nonzero interface bond strength.

No systematic transverse creep data of SiC-Ti-alloy MMCs appear to exist in the case of a finite nonzero value of normal interface strength. However, using the Crossman model as a basis, it is possible to speculate the overall creep characteristics. This is illustrated by the thin solid line (stepped) in Figure 2, wherein a 300 MPa bond strength is assumed, and the normal stress at the fiber-matrix interface along the bond line [θ = 0°, see Figure 1(a)] is estimated to be approximately 1.5 times the far-field applied stress. This stress concentration factor (SCF) of 1.5 is typical of SiC-Ti-alloy MMCs with volume fractions less than 0.4 and results in References [3,4] show that the SCF is not significantly affected by plasticity. Although the behavior suggested by the thin solid line in Figure 2 has not been reported thus far in titanium matrix composites, the SCS-6/Ti-6Al-4V data in Reference [2] do suggest that the MMC creep response could cross the matrix response at very low stresses. Unfortunately, the data were likely influenced by narrow specimen geometries and exposure of the interface at specimen edges to oxidation damage. The creep data on B/Al [1] with different bond strengths also are qualitatively in agreement with such an analysis.

The second example corresponds to the case of longitudinal creep of an SCS-6/Ti-6Al-4V MMC, with the assumption that fiber breaks can occur based on their statistical strength distribution. The Du and McMeeking model [5] is employed, but thermal residual stresses also are taken into account. In the analyses procedure, fiber fractures are assumed to follow Curtin's global load sharing model [6], so that the effective stress-strain curve for the totality of fibers in the composite can be represented through a bundle stress (σ_{bundle}) as:

$$\sigma_{bundle} = \left(1 - \frac{r\,T^{m+1}}{2\tau L_o \sigma_o^m}\right) \cdot T \qquad (4)$$

where $T = \varepsilon \cdot E_f$, ε is the axial strain in the MMC (equal to the strain in the uncracked fibers), E_f is the fiber modulus, τ is the interfacial friction stress, r is the fiber radius, m and σ_o are the Weibull modulus and strength of the fibers, respectively, and L_o is the gauge length on which the Weibull strength of fibers is based.

Figure 3 illustrates the composite's longitudinal creep responses for the loading conditions prescribed in the figure for two different values of the interface frictional stress, τ. The role of τ is to alter the ineffective length of the fiber associated with a fiber break. For the case of a friction stress of 25 MPa, no composite failure was predicted at least up to 10,000 hours. On the other hand, a modest decrease in interface strength to 8 MPa can result in MMC rupture within a very short time. This plot once again emphasizes the importance of interface strength in creep, but this time for a longitudinally loaded MMC.

Both of the above examples suggest that a stronger interface strength is beneficial, particularly for elevated temperature use. On the other hand, weaker interfaces may be desirable for reducing fatigue crack growth (FCG) rates through crack bridging and also for dissipating energy in the form of matrix plasticity ahead of a sharp notch or crack. Although FCG constitutes a rather restricted set of situations in the application of MMCs, it had dominated interface development in the past, particularly the emphasis on very weak interfaces. Perhaps this emphasis was influenced by the ceramic matrix composites (CMC) literature, where the single most important requirement is crack bridging. However, even for CMCs, analyses [7]

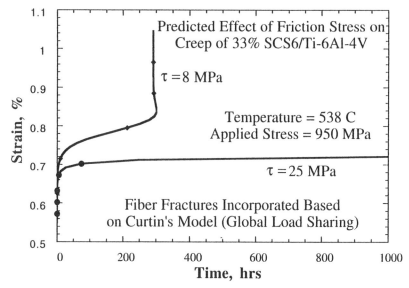

Figure 3. *Predicted creep response of an SCS-6/Ti-6A1-4V, where fiber fractures are taken into account. The difference between the curves illustrate the effect of the interface friction stress, τ.*

show that a higher friction stress is desirable for higher "yield" strength, provided crack bridging is not compromised. Indirect experimental evidence of the benefits of fiber-matrix bonding is also available from some recent work on C-SiC composites, which showed that certain textured pyrolytic carbon coatings (with limited degree of bonding) provided higher CMC strength than when there was no coating at all (the case of weakest bonding) [8]; the strength of the latter was low despite extensive fiber-matrix debonding. As indicated in the previous paragraphs, there is a number of high-temperature applications of MMCs where a higher interface strength may be desirable. Clearly, there are issues of trade-offs and optimization that have to be based on the following: (1) rational understanding of the interface failure process, (2) mechanical characterization and modeling of the interface, (3) interface effects on the bulk composite response, and (4) chemical compatibility and stability of the interface.

In the rest of this chapter, an attempt is made to illustrate some of the important issues, to emphasize its role on MMC response and life prediction, and to provide a background for developing interfaces for optimized MMC performance.

STRENGTH AND TOUGHNESS ISSUES FOR INTERFACES

Strength Issues

The simplest parameters that characterize the mechanical performance of ceramic and metal interfaces are the normal and shear strengths. However, unlike monolithic structures, the measurement of strengths is complicated by the presence of residual stresses that develop as the joint is cooled from the fabrication temperature. Consequently, residual stress calculations become imperative, preferably needing validation by residual stress measurements.

An added level of complexity generally arises from a highly nonuniform stress distribution around the interface, often involving stress singularities at the free surface. The nature of those singularities have been addressed theoretically (see, e.g., the work of Bogy [9]). They pose the problem of extracting an average bond strength in a highly nonuniform stress field, particularly near a free surface. Nevertheless, considering the simplicity of using a strength parameter, a practical approach that has been relied on to some extent in the past and that deserves constant attention, is the use of specimen geometries that reduce gradients of stresses at the bond line, where those stresses arise from thermal mismatch and from the applied load. An example of a geometry where free surface effects are essentially eliminated in interface strength determination is provided later.

In MMCs, the normal interface strength has direct bearing on off-axis (particularly 90°) response, as was indicated in the previous section. It also has bearing on normal interface debonding ahead of a matrix crack tip, thereby establishing conditions for the existence of a bridged matrix crack. A shear strength, on the other hand, appears to have less direct influence on composite performance, because shear failure is a progressive phenomenon. Rather, a more useful parameter is an average shear stress (often considered as frictional) in the shear-failed region of the interface. The sliding stress is important in fatigue crack growth problems and for predicting fiber fragmentation and MMC instability in the absence of matrix cracks.

When interpreting strength, it is important to identify the location of failure, which may not necessarily lie along the interface. For example, in zirconia-steel joints, interface failure was found to oscillate between the ceramic-braze interface and the locally oxygen-depleted region of the ceramic [10,11]. In MMCs, the extent of fiber-matrix reactions can shift the failure from the interface or interphase to the ceramic fiber [12]. By interphase is meant the assemblage of fiber coatings and reaction zones that separate the major load-carrying part of the fiber from the matrix alloy. Because residual stresses differ in the different constituents, a strength can be based on only the identified failure location. Additionally, the failure location can change, depending on the mode of loading. For fiber pull out or push out type of loading in SCS-6/Ti-alloy MMCs, the preferred interphase failure location is at one of the SCS carbon layers [13,14]. Correspondingly, under transverse loading, interface failure in SCS-6/Ti-6A1-4V MMCs often has been observed [12] between the outer carbon layer and the Ti_5Si_3-TiC reaction zone. Although these differences can possibly be rationalized on the basis of how the load is transmitted to the interface, the observations do underscore the importance of determining failure locations for predicting MMC response and for developing improved interfaces.

Toughness Issues

The usefulness of fracture toughness as an interface characterizing parameter is especially appealing when one considers the nonuniform stress distribution at the interface. On the other hand, the toughness parameter is a more difficult one to incorporate for predicting composite response. Current usage of interface toughness in MMC applications is generally limited to interface characterization and to prediction of fiber bridging in the presence of a matrix crack.

The modulus mismatch at ceramic metal interfaces adds the complexity of mode mixity, when considering characterization in terms of the stress intensity factors. The first Dundurs parameter, α, defined as:

$$\alpha = \frac{E_2 - E_1}{E_2 + E_1} \tag{5a}$$

plays a central role in the interface toughness problem, where E_1 and E_2 are the modulus of the two media. In the MMC problem, generally the primary crack is located in medium 1, so that E_1 corresponds to the matrix, and E_2 corresponds to the fiber. The second Dundurs parameter (β) is:

$$\beta = \frac{1}{2} \cdot \frac{\mu_1(1 - 2\nu_2) - \mu_2(1 - 2\nu_1)}{\mu_1(1 - \nu_2) + \mu_2(1 - \nu_1)} \tag{5b}$$

where μ and ν are the shear moduli and Poisson's ratios of the two media, respectively. β is usually small and is neglected from an engineering standpoint. The class of problems encompasses the situation where the crack lies at the interface, to where it is perpendicular to the interface, and all intermediate approach angles (see Figure 4). In almost all these plane problems, there is a combination of mode I and mode II loading, arising from elastic mismatch effects. This causes the mode mixity (Ψ) at the interface crack tip $\{\Psi = \tan^{-1}(K_2/K_1)\}$ to be different from the homogeneous material $\{\tan^{-1}(K_{II}/K_I)\}$; here K_1 and K_2 are the local stress intensity factors, and K_I and K_{II} are the stress intensity factors that would exist in a homogeneous material for the same crack geometry and loading. Hutchinson's [15] and Rice's [16] studies provide a concise description of the issues involved in the bimetallic fracture problem.

One interesting relation that emerges when a thin layer of material 2 (such as a fiber) is sandwiched between two large blocks of material 1 (the matrix) is that the local crack driving force (G) of the interface crack, for a sufficiently long crack, is identical to that for the crack in a homogeneous material of the same material 1, although the local K's are different from the homogeneous material [15]. This, along with the complex behavior of local K_1 and K_2, suggests that G is a more useful parameter for characterizing interfaces, rather than using the stress intensity factors. However, the material parameter G_c at fracture is now a function of Ψ.

G_c is generally an increasing function of Ψ, based on data from a number of ceramics [17,18] and from limited ceramic-metal interface studies [10,11,19,20]. In References [15,21], the G_c versus Ψ relationship was modeled based on the interface roughness and the shielding of K_2 caused by surface asperities. Experimentally, for the case of the zirconia-steel system, G_c increased slowly with Ψ from 0° to about 45° but thereafter increased rapidly such that $G_c(\sim75°)/G_c(0°)$ was approximately 4 [10,11] (see Figure 5). In this system, an active-metal

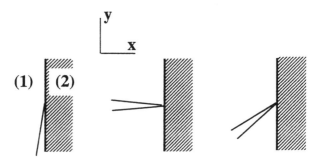

Figure 4. Various crack-interface configurations.

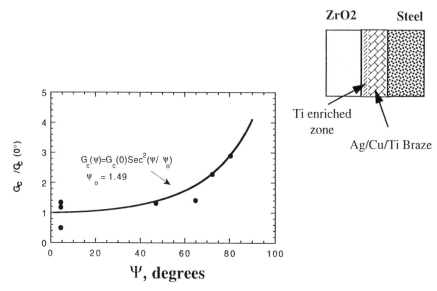

Figure 5. G_c *as a function of* Ψ *(= $tan^{-1}K_2/K_1$) for a zirconia-braze (Ag-Cu-Ti-In)/steel joint, from diametrically cracked disk specimens. During brazing, most of the Ti in the braze deposited as a thin layer (~1 μm) on the zirconia. Failure oscillated between the oxygen-depleted zirconia and the zirconia-braze interface [10,19].*

braze was used, so that the ceramic-metal interface was in effect a zirconia-Ti interface. There does not appear to exist a systematic study of toughness of the SiC-Ti interface for different values of mode mixity, but G_c versus Ψ relationship is not anticipated to differ greatly from other ceramic-metal interfaces, because interface-interphase surface roughness is generally similar; however, quantified data on surface roughness are not available. Toughness values for the fiber-matrix interface in SiC-Ti-alloy MMCs are generally obtained from push out or pull out tests. In these tests, the local mode mixity is estimated to be approximately 75°, based on the far-field mixity (Ψ ~90°), and the phase difference ω (~15°) between the local and far-field mixity, which is a function of α (~0.53 for SiC-Ti-alloy systems), and tabulated in Reference [22].

If it is assumed that the $G_c(\Psi)$ relationship for SiC-Ti is similar to that for zirconia-steel, then for phase angles lying between $0°(K_2 = 0)$ and $45°$ $(K_2 = K_1)$, G_c is estimated to be between 0.25 and 0.33 of that from push out and pull out tests; however, experimental validation is necessary. The additional point to be remembered when interpreting G_c from push out and pull out tests is the existence of significant residual radial compression at the fiber-matrix interface at room temperature (RT). Such a radial compression likely has the effect of increasing measured G_c, near mode II conditions, to values well above those based simply on interface roughness considerations, because of increased asperity rubbing and possible crack instabilities under superimposed compressive stresses. The axial residual stresses also alter the local G and phase angles, as discussed in Reference [23]. Thus, for the pull out problem, the mode I component is increased, and the mode II component is decreased by thermal residual stresses.

Importance of Interface Strength and Toughness in Crack Bridging

Consider a matrix crack approaching a fiber (Figure 6), and pose the question whether interface debonding would occur in preference to fiber fracture. This problem is critical for CMC performance. In the case of MMCs, such situations constitute a limited range of applications and arise when there is a matrix crack, generated either under fatigue loading or as a result of a brittle matrix (actually, past experience with α-2 titanium aluminide and γ-titanium aluminide composites suggest that it may be prudent to avoid brittle matrices in MMC applications).

The reason that debonding is beneficial is that it relieves the stress on the fiber along the crack plane, as is illustrated in Figure 7. A weak ductile matrix that is strongly bonded to the fiber would also provide a similar but less relaxation of stress ahead of the crack tip, as has been shown in Reference [24]. Once the fiber stress is relaxed, it retains integrity, allowing crack bridging and producing a sharp decrease in local stress intensity as the matrix crack approaches the next fiber.

The question of debonding versus fiber fracture has been addressed by a number of researchers. From a mechanism perspective, one of the earliest discussions can be credited to Cook and Gordon [25], who considered the stress components on a plane ahead of a through-thickness elliptical notch with a large ratio of major to minor axis.

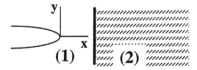

Figure 6. Matrix crack approaching an interface.

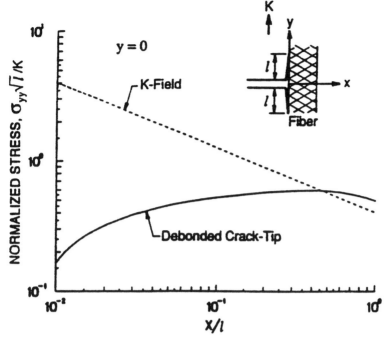

Figure 7. Normal stress ahead of a doubly deflected crack for equal values of E in the two mediums. The stress is normalized by the $K/\sqrt{l}$, where K is the stress intensity factor associated with the main crack. Note the attenuation of the stress over a distance of approximately one delamination length. (Reprinted from Reference [24] with permission from TMS & ASM International.)

The plane (that can be identified as an interface plane) was considered to be perpendicular to the major axis of the ellipse. It was shown that the ratio of the transverse (normal) opening stress on the interface to the stress at the interface along the loading direction (that can be identified with a fiber strength) reached a maximum value of one-fifth at approximately one crack tip radius ahead of the notch tip. Additionally, although the stress along the loading direction had a high gradient, the transverse stress had a flat profile near its peak value. Accordingly, they suggested that so long as the transverse bond strength was less than one-fifth to one-third of the strength of the second medium (the fiber), then debonding would prevail. Furthermore, using data generated by Orowan [26] on mica sheets pulled to failure using different geometries, they indicated that material cracks would always have great difficulty in penetrating crystallographic planes of easy cleavage. Although great sophistication in analyses have been achieved since Cook and Gordon's study, the simple arguments and the numbers derived have retained general validity and will become apparent in the discussions that follow. Additionally, there have been recent attempts at making interfaces in CMCs that employ cleavable planes (such as with beta alumina), which may be considered as direct application of their concepts. Shear failure of the interface ahead of the crack tip and at locations above and below the crack plane were also discussed in Reference [25]. Although this mode of failure was not considered significant for all-brittle systems, it is possible that such interface failures could constitute an important debonding mode during fatigue crack growth in MMCs with weak interfaces.

At a more rigorous scale, the issue of debonding versus fiber fracture has been addressed in two separate ways: (a) toughness considerations and (b) strength considerations. We shall discuss the toughness aspect initially. Erdogan [27] provides a nice description of the issues involved. Referring to Figure 6, the analyses [27] show that the stress intensity factor at the crack tip continually decreases as the crack tip approaches the interface, so long as the modulus of medium 2 (fiber) is higher than medium 1 (matrix); conversely, the stress intensity factor monotonically increases if $E_2 < E_1$. As long as the crack tip does not impinge on the interface, the stress field is describable by a $r^{-1/2}$ singularity, where r is a distance from the crack tip. However, a major deviation occurs when the crack tip actually impinges on the interface. At this point, the stress singularity is characterized by a $r^{-\lambda}$ singularity, where λ is no longer equal to one-half if E_2 is not equal to E_1. This peculiar behavior of the stress field was originally pointed out by Zak and Williams [28] by solving the eigenvalue problem, which predates the work of Cook and Gordon [25]. The stress fields were further analyzed by Swenson and Rau [29] and Cook and Erdogan [30]. Thus, λ is less than 0.5 when E_2/E_1 is greater than 1, and vice versa; λ is approximately 0.45 for $E_2/E_1 = 4$. More significantly, from a stress intensity perspective, *the stress intensity factor goes to zero for $E_2/E_1 > 1$, when the crack tip hits the interface.* Note that the stresses are still singular, but there is the anomalous problem of the stress intensity factor going to either zero (for $E_2/E_1 > 1$) or infinity (for $E_2/E_1 < 1$), because λ is no longer one-half. It is useful to note that if the crack lay parallel to the interface, then the dominant singularity would be 0.5, independent of the first Dundurs parameter but negligibly dependent on the second Dundurs parameter [15,16,27], i.e., $\lambda \neq 0.5$ only is relevant for perpendicular or oblique incidence of the crack on the interface. Erdogan [27] attempted to rationalize the problem of $\lambda \neq 0.5$ in terms of the strain energy release rates, where they were defined in terms of the nonsquare root singularity parameter. However, the problem was relating it to the energy release rates, G_{Ic}, of the constituent materials, where G_{Ic} is based on the square root singularity of the stress and strain fields of a crack in a homogeneous material.

It is here that He and Hutchinson's work [31] becomes extremely relevant. By considering equal and small perturbations of the crack for debonding versus penetration, it was shown that, to satisfy debonding, the ratio of energy release rate along the interface, G_d, to that for propagation, G_p, had to be below a critical value. This critical value was independent of the length of the perturbation and dependent only on the elastic properties of the two medium [31]. The critical ratio of G_d/G_p is 0.25 when the two materials are identical. Correspondingly, for the SiC-Ti system ($\alpha \sim 0.53$), this ratio is approximately 0.5 (see Figure 8). Thus, the model predicts that when

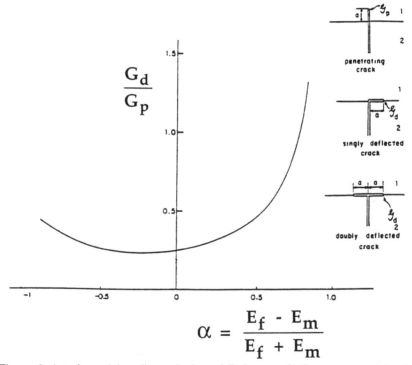

$$\alpha = \frac{E_f - E_m}{E_f + E_m}$$

Figure 8. Interface debonding criterion of Reference [31], represented by the plot of G_d/G_p versus the first Dundurs parameter, α. α is approximately 0.53 for the SiC-Ti-alloy system. The curve represents both singly and doubly deflected cracks (Reprinted from Reference [31] with permission.)

$G_c(\sim26°)$ for the interface crack is greater than half the G_{Ic} of SiC, then crack bridging will not occur. The angle of 26° is obtained from Reference [31] and represents the value of the phase angle when a Ti-matrix crack, perpendicular to the interface, just impinges on the interface [Figure 4(b)] and creates an infinitesimal debond crack. Notable here is the fact that this phase angle is significantly less than 75 to 90°, which is typically associated with pull out and push out type of loading. Partial validation of the toughness ratio can be found in the work of Kendall [32] on bonded ethylene propylene rubber, although Kendall derived and partially validated a G_d/G_p ratio of $1/\{4\pi(1 - v^2)\}$, based on energy concepts; interestingly, his experiments well predate the analyses of He and Hutchinson [31]. Additional validations have been attempted on CMC systems, although it has not been established rigorously whether interface debonding occurs prior to arrival of a matrix crack at the interface, or whether it occurs after the crack arrives at the interface. Residual stresses can alter the critical ratios, as was shown in Reference [33], with compressive axial residual stress in medium 2 allowing for higher ratio of G_d/G_p for assuring debonding. However, the analyses [33] incorporate length parameters that are difficult to estimate and thus reduces applicability and experimental validation somewhat. It is also important to remember that not only is debonding necessary for protecting the fiber but that debonding should occur over a certain length scale. Otherwise, as may be interpreted from Figure 7, the stress in the fiber may not be sufficiently attenuated to prevent its failure.

The toughness of SiC typically ranges between 3.5 and 5.5 MPa$\sqrt{m}$ [34]. However, simultaneous measurements of flaw size (using fracture-mirror radii) and strength of 140-μm diameter SiC monofilaments (specifically SCS-6 fibers) suggested a K_{Ic} of approximately 2.2 MPa$\sqrt{m}$ [35]; the fiber modulus is approximately 400 GPa. Thus, G_p is approximately 11 J/m^2. For Nicalon SiC ($\sim$10 μm diameter, $E\sim200$ GPa), the number is reported to be in the range of 4 to 8 J/m^2 [36], with K_{Ic} reported to be approximately 2.3 MPa$\sqrt{m}$ [37]. The He and Hutchinson criterion [31] would therefore predict no debonding for the 140 μm-SiC/Ti-alloy MMC if G_d (G_c at Ψ of $\sim$26°) of the interface is above approximately 5.5 J/m^2. Although numbers lower than this have been reported in the CMC literature for SiC-glass systems, recent push out tests on the SCS-6/Ti-15-3 system [14] show $G_c(\sim75°)$ values of 20 to 40 J/m^2; pull out tests suggest a G_c of about 20 J/m^2 [38]. These numbers appear quite high for the interface and may be a result of high radial compression ($\sim$250 MPa) at the interface and local inelastic processes. At a phase angle of 26°, G_c is anticipated to be lower, and using the trend in Figure 5, it would lie in the range of 10 to 14 J/m^2 for a G_c ($\sim 75°$) of 40 J/m^2. Additional evidence of high G_c is obtained from the work of Gupta et al. [39], who used double-cantilever beam tests to obtain a G_c ($\approx 15°$) of approximately 60 J/m^2 for SiC coating on Pitch-55 carbon; this carbon has some similarity with the SCS-6 coating, in terms of the layered carbon structure next to an SiC surface. All these tests involved macroscopic test specimens. In contrast, Argon et al. [40] analyzed spontaneous delamination of SiC coatings on a carbon substrate to obtain an intrinsic G_c ($\sim 15°$) of 5.5 J/m^2. This value is low compared with the other data and suggests that macroscopic mechanical loading can induce large inelastic processes, and such processes may occur during loading of bulk MMC samples. In any case, the available data suggest that G_c (26°) for the SCS-6/Ti system probably lies in the range of 5 to 15 J/m^2, although further experimentation is clearly necessary. Then, according to the He and Hutchinson criterion [31], crack bridging would be marginal or may not occur for the SCS-6/Ti-alloy system. On the other hand, crack bridging is well established for this system. Consequently, it is possible that the criterion may be too restrictive in terms of the G_d requirement for the interface crack. Additional complications arise because there is a number of carbon and interphase layers with their own elastic properties, so that there also are problems of defining the interface of significance; experiments tend to suggest that interface failure often occurs between the two carbon layers of the SCS-6 coating [13,14].

Before discussing the toughness aspect any further, it would be useful to consider the stress criterion for debonding. In this approach, the criterion for debonding is posed in the form of the inequality $\sigma_b/\sigma_f > \sigma_{xx}/\sigma_{yy}$, where σ_{xx} is the transverse stress across the interface (see Figure 6), σ_{yy} is the stress perpendicular to the crack plane in material 2 immediately ahead of the interface, σ_b is the transverse bond strength of the interface, σ_f is the strength of the fiber, and the ratio on the right side of the inequality is a function of the Dundurs parameters. Interface debonding can occur either before the crack tip impinges on the interface, such that the ratio on the right side is unity for $E_1 = E_2$ (which is different from one-half according to the elliptic crack model of Cook and Gordon [25]), or when the crack tip actually impinges on the interface, in which case the ratio is 0.3536 {= $1/(2\sqrt{2})$} for $E_1 = E_2$. These numbers derive directly from standard fracture mechanics formulae for homogeneous solids and illustrate that there is greater propensity for debonding when the interface is approached. Swenson and Rau [29] provide values of the stress ratios when the crack tip hits the interface with $E_1 \neq E_2$. The results were further generalized by Gupta et al. [41] for anisotropic media with the crack tip just impinging on the interface, and the requirement for debonding was posed in terms of the ratio $\Phi = \sigma_{xx}(\pi/2)/\sigma_{yy}(0)$, which is a function of the two Dundurs parameters. The primary effect is that of α, which is illustrated in Figure 9.

For a given material combination (i.e., a specific α), as long as the ratio of σ_b to σ_f lies below the curve, debonding is predicted to occur in preference to fiber fracture. For the SiC-Ti-alloy system, the ratio is approximately 0.45. The Weibull strength of SCS-6 fibers, etched out of a Ti-matrix, is around 4500 MPa [42-44] for a gauge length of 25.4 mm; the Weibull modulus is approximately 14. Also, transverse tests performed on single-fiber SCS-6/Ti-6Al-4V MMCs using a cruciform geometry [12] suggest a bond strength of approximately 100 MPa. Adding a radial residual stress of approximately 250 MPa to the bond strength and an axial residual stress of 700 MPa to the Weibull strength, the σ_b/σ_f ratio is 0.067, which is well below 0.45; if the axial residual stress is neglected, the ratio is still low, being 0.078. Thus, this approach does provide a satisfactory explanation for the widely observed crack-bridging behavior with SCS-6 fibers. In the case of SCS-0 fibers, where the outer C-rich coatings are absent, transverse

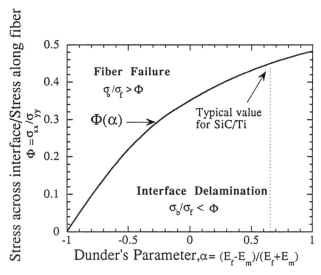

Figure 9. The stress-based debonding criterion. (Figure based on the results of Gupta, Argon, and Suo [41].)

tests indicate a bond strength of about 380 MPa [12]. The strength of these fibers is low, with Weibull strength and modulus of 1500 MPa and 6, respectively, for 25.4-mm-long fibers extracted out of a consolidated SCS-0/Ti-6A1-4V composite [42]. Accordingly, σ_b/σ_f ratio is 0.29 for SCS-0 fibers in a Ti-alloy matrix, which is not much less than 0.45 necessary for debonding. Thus, crack bridging in SCS-0 fibers would be marginal. Recent FCG tests [43] indicate that crack bridging, as interpreted from crack arrest conditions and microscopy, does occur for the SCS-0/Ti-6A1-4V system, provided that the applied stress is kept low. However, additional experiments are needed to firmly establish the existence of bridging at low stresses. The lower stress requirement stems mainly from the low fiber strength, with indications that fibers break in the crack wake [43]. At higher applied loads, although there were periods when the crack appeared to go toward arrest, the crack growth rates suddenly increased, indicative of loss of bridging conditions. The marginal bridging conditions for SCS-0 fibers are in general agreement with the strength criterion for debonding. Although the interface toughness for the SCS-0/Ti-6A1-4V system has not been determined, it is estimated to be much larger than that for SCS-6 fibers, based simply on the bond strength of the interfaces. Consequently, it may be difficult to explain crack bridging in SCS-0/Ti-6A1-4V using the He and Hutchinson criterion [31]. The ability of the strength-based model to predict debonding also was verified for a sapphire-γ-TiA1 system with Nb-Y coatings [45], where Nb provided reasonable bond strength between the sapphire and the rest of the matrix, and Y served as a diffusion barrier for Ti; Ti was extremely damaging to the fiber. However, although debonding was observed in this system, the length of debonding was small and very little pull out occurred in tensile tests of single-ply specimens.

Referring to Figure 7, it may be noted that a reasonable length of debonding is necessary, such that the maximum stress in the fiber in the plane of the crack is kept below the fiber strength. Weibull statistics would modify the situation and is neglected for the moment. The stress-based analysis can be used to predict the length of debonding, with the specification that debonding would occur over a length where the shear stress at the interface is greater than the frictional sliding stress of the interface. However, a fracture mechanics-based approach provides more useful insight here. In Reference [46], He and Hutchinson considered a semiinfinite crack along the interface and derived the local G for a small kink crack emerging at different angles out of the interface into the fiber. The G of the interface crack was kept constant, whereas the ratio K_2/K_1 was changed to determine its effect on the local G of the kinked crack. This parameter K_2/K_1 is important for the crack-bridging problem because it was already mentioned that Ψ is approximately $26°$ when a mode I crack in the Ti-alloy matrix impinges on the interface, and thereafter Ψ increases for longer length of debond. The results in Reference [46] show that when $\alpha = 0.53$ (the SiC-Ti-alloy system), the original debond crack will be trapped at the interface and cannot branch into the fiber, so long as the interface G_d is less than approximately 0.8 times the fiber toughness, G_p. This requirement is less stringent on the interface toughness than the original He and Hutchinson criterion [31] (a G_d/G_p value of 0.5 required for debonding, as discussed earlier). More significantly, the value of 0.8 is in better accord with observed crack bridging in SCS-6/Ti-alloy composites. In addition, this instability behavior may explain why SCS-0 fibers with presumably higher interface toughness have marginal crack-bridging behavior.

Adequate validation of the different criteria for crack bridging in the SiC-Ti-alloy system does not appear to exist. Until that is done, a scenario that is in agreement with experimental observations appears to be as follows. Debonding occurs either ahead of the matrix crack tip or as soon as the tip hits the interface, based on the strength criterion discussed earlier [25,41]; computational results on the stress distributions are provided in Reference [47]. In this regard, past fatigue studies on an SCS-6/Ti-15-3 MMC [48] and an SCS-6/Ti-24Al-11Nb MMC [49] provide microstructural evidence of normal interface separation even before the main matrix crack tip arrives at the fiber-matrix interface. More recently, interface separation ahead of the crack tip was observed in a model PMMA-PMMA system [50], where interface defects were considered important for the debonding behavior. The He and Hutchinson criterion [31] for predicting the occurrence of debonding is also a good one, but it may be more conservative and is based on the assumption that debonding is delayed until the crack impinges on the interface. In any case, when debonding does occur, then its phase angle is known, which increases as the debond length increases. Further growth of the interface crack is then controlled by the ratio of interface toughness to the fiber toughness [46], the ratio being approximately 0.8 for the SiC-Ti-alloy system.

A number of factors, however, can alter the scenario. First, if the applied stress is reasonably large, then the fiber could break after some length of debonding, because of greater volume of fiber being sampled at a high stress. Second, and more importantly, interface fracture in MMCs is likely to be associated with inelastic processes. Recall G_c values of 20 to 40 J/m^2 from pull out and push out tests for

SCS-6/Ti-alloys and about 60 J/m^2 for pyrolytic graphite-SiC. For an alumina-alumina joint with a Cu interlayer, $G_c(\sim75°)$ was approximately 20 J/m^2 [20]. A G_c of about 50 J/m^2 was reported for an alumina-Au joint [51]. All these values are high, certainly much larger than the intrinsic surface energy based on thermodynamic considerations of the work of adhesion. The high toughness numbers indicate the difficulty of extracting true interface toughness at different levels of mode mixity, without introducing inelastic effects in the experiments. Also, inelastic effects can alter the fine balance between debonding and fiber fracture, because the inelastic component of G for the interface can become higher than the toughness of the ceramic. Thus, fiber bridging may ultimately have to be assessed based on actual FCG testing of composites. This approach is currently being pursued with meaningful results [43], using single-fiber and single-ply microcomposites. At the same time, it is important to extend the debonding predictions of linear elastic fracture mechanics to include inelastic effects. In this regard, the elastic-plastic analysis of Reference [52] provides a concise description of the issues involved. Whether the elastic-plastic approach will yield improved predictions is not clear at this stage. Based on past experience with inelastic fracture mechanics, new fracture criteria may have to be established for predicting the influence of effective interface toughness on crack deflection at ceramic-metal interfaces.

INTERFACE MEASUREMENTS

Shear Properties

The mechanical properties of the fiber-matrix interface under shear loading include the shear strength, interface toughness, fictional sliding stress, and the coefficient of friction. These properties are evaluated from fiber pull out and push out tests, using various models to interpret the experimental data. In this section, those tests and models are briefly described, and important issues and problem-areas are pointed out.

The three important tests for shear characterization of the interface are:

1. The fiber pull out test
2. The fiber push out test
3. The fiber fragmentation test

In all of these tests, Ψ values are high (70 to 90°). For lower Ψ values, the delaminating bend-bar test of Reference [53] is useful. However, although it has been applied successfully for planar interfaces (in laminated structures), there are difficulties in adapting it for the fiber-matrix interface.

THE PULL OUT TEST

This test, illustrated in Figure 10, most closely approaches conditions experienced in the wake of a bridged fatigue crack. The test has been used extensively in polymer matrix composites but poses experimental difficulties for MMC and CMC systems. A number of models have been developed for the pull out test [54–60]. At one end of the spectrum, a strength-based shear-lag approach [54] is used to obtain a shear strength (τ^*), given approximately by:

$$\tau^* = (n/2) \cdot \sigma^* \tag{6}$$

where σ^* is the fiber stress corresponding to the onset of debonding, and:

$$n^2 = E_m/\{E_f \cdot (1 + v_m) \cdot ln(R/r)\} \tag{7}$$

where E_m and E_f are matrix and fiber modulus, respectively, v_m is the Poisson's ratio of the matrix, r is the fiber radius, and R is an outer radius, which can be approximated from the fiber volume fraction, $v = (r/R)^2$. For the single-fiber case, R may be taken as 10 times r. The approach has worked well for polymeric matrix composite (PMC) systems, where debond stresses are low compared with fiber strengths and residual stresses are generally small or negligible compared with

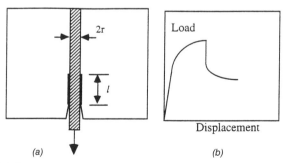

Figure 10. (a) Schematic of the pull out test. (b) A typical load-displacement plot.

the interface strength. In these systems, the flow stress of the matrix is low, so that inelastic processes are anticipated to dominate around the crack tip, making those contributions significantly larger than the inherent bond-breaking energy. Thus, although one can determine an effective G_c for the interface, it may not provide any additional insight compared with characterization in terms of a shear strength; in fact, as shown in References [55,56], the defined strength and toughness are related.

In the case of Ti-alloy based MMCs at RT, the axial residual stress in the fiber is high, which generates a large fiber-matrix shear stress at the interface close to the free surface; for SCS-6/Ti-alloy systems, these shear stresses are typically around 250 to 300 MPa [57]. These values are high compared with friction stresses of around 80 to 150 MPa (see Table 1), which can be directly measured following complete debonding in pull out or push out tests. The consequence of a high residual shear stress at the interface close to the free surface is that the interface is either debonded or is very close to debonding even before application of a pull out load. Thus, any strength determination is dominated by the thermal and specimen preparation history, rather than by loads imposed during the test. That is why a strength-based approach does not appear to be attractive for MMCs when large residual stresses are present. On the other hand, when residual stresses

Table 1. Interface shear properties in some SiC/Ti-alloy systems from push out tests at room temperature.

Fiber	Matrix	Specimen Thickness, t μm	τ_{debond}, MPa	$\tau_{friction}$, MPa	Reference
Sigma 1200+	Ti-6Al-4V	20-140	—	80-100	[70,72,75]
SCS-6	Ti-6Al-4V	150-450	156	88	[76][a]
SCS-6	Ti-6Al-4V	700	160-190	140-160	[42][b]
SCS-6	Ti-15-3[c]	220-550	124	82	[76][a]
SCS-6	Ti-15-3	740-800	120-160	100-130	[14]
SCS-6	Ti-15-3	620	105-120	70-80	[14]
SCS-6	Ti-15-3	200-300	115	50	[13][d]
SCS-6	Ti-15-3	410	164	126	[65]
SCS-6	Ti-24Al-11Nb	150	70	40	[63]
SCS-6	Ti-24Al-11Nb	280-470	100	50-60	[63]
SCS-6	Ti-24Al-11Nb	470-580	110-140	60	[63]

[a]Refer to chapter to see how τ_d and τ_f were defined, since load-displacement was not recorded.
[b]Single-ply, very low volume fraction (0.04) composite.
[c]Ti-15-3 represents the Ti-15V-3Cr-3Al-3Sn metastable beta titanium alloy
[d]Bottom groove width was large (0.36 mm)

are low, such as in the case of a sapphire-Nb system, a strength-based approach may prove useful.

A number of models have been proposed for interpreting the fracture mechanics of the pull out test [58–61]. They differ in terms of emphasizing Coulomb friction versus a constant friction stress and in terms of completeness of the analysis. With regard to the former, it is important to point out that in SiC-Ti-alloy and SiC-A1 systems, except for a distance of about one to two fiber radius from the loading end, where contact stresses are important, the *residual* radial compressive stress dominates the radial stress at the interface along the entire length of the debonded crack [14,59]. Thus, in these systems, a constant friction stress assumption provides a simpler analysis of the problem, from which a friction coefficient can be determined using straightforward calculation of the residual radial stress at the interface.

Hutchinson and Jensen's fracture mechanics analysis [59], specialized for the case of $v_f = v_m = 0.25$, $E_f = 4E_m$ (this ratio is typical of many TMCs), and v (volume fraction of fibers) approaching zero (i.e., single-fiber pull out in a number of SiC-Ti-alloy systems), provides the following useful equations:

$$\delta = 0.99 \left(\frac{4G_c}{E_f r} \right)^{1/2} \ell + 0.98 \left(\frac{\tau \ell^2}{E_f r} \right) \tag{8}$$

$$\left(\frac{\ell}{r} \right) = \frac{\sigma_f - \sigma_f^i}{2\tau} \tag{9}$$

$$\left(\frac{\sigma_f^i}{E_f \varepsilon_T} \right) = 1.01 \left(\frac{4G_c}{E_f r \varepsilon_T^2} \right)^{1/2} - 1.11 \tag{10}$$

Here, G_c is the interface toughness, ℓ is the length of the debond crack, τ is the friction stress of the interface, σ_f is the applied tensile stress on the fiber, σ_f^i is the applied tensile stress in the fiber when a debond starts propagating, δ is the relative displacement between the matrix and the fiber, $\varepsilon_T = (\alpha_f - \alpha_m) \cdot (T_{processing} - T_{testing})$ is the misfit thermal strain, and the axial residual stress in the fiber far from the free surface is $1.11 E_f \varepsilon_T$.

The load-displacement data in principle can be used to determine the two constants, i.e., G_c, and τ. However, one problem is that σ_f^i is either negative or zero, because an interface crack may already exist because of thermal residual stresses before any application of a load; attempts to measure that metallographically has generally been unsuccessful [14]. If a crack already exists, then σ_f^i also becomes another parameter that has to be determined, and the measured displacements have to be corrected to take into account the existing debond crack. These problems make a unique evaluation of the three parameters slightly difficult, particularly with regard to the evaluation of G_c. Pull out data are limited, because of experimental difficulties. In Reference [38], a G_c of 20 J/m^2 was estimated for an SCS-6/Ti-24A1-11Nb MMC, although the model used [61] was slightly different from the analysis of Hutchinson and Jensen [59]. Overall, based on the literature survey, it appears that additional experimental work is needed to establish the suitability of the pull out test for extracting interface shear parameters.

THE PUSH OUT TEST

The push out test is, by far, the most common test used for characterizing interface shear behavior in MMCs. The reason is the inherent simplicity and robustness of the test technique, although this comes at the expense of some complications in the analysis.

The test was initially developed by Marshall for measuring interface properties in CMC specimens [62]. The setup is sketched in Figure 11 in which a tungsten carbide or a diamond-tipped indenter is used to push out the fiber from a thin slice of a CMC or an MMC specimen. For fibers of 100 to 150 μm diameter, the fiber is located above a slit of 220 to 300 μm width. Typical load-displacement plots are sketched in Figure 12, based on which two average parameters are commonly defined. One parameter is the debond strength, $\tau_d = P_d/(2\pi rt)$, where P_d is the load at which the load-drop (instability) occurs. Additional discussion of this parameter is provided later. The

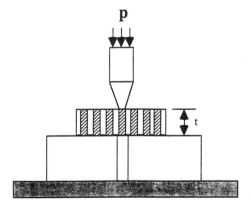

Figure 11. Schematic of the push out test.

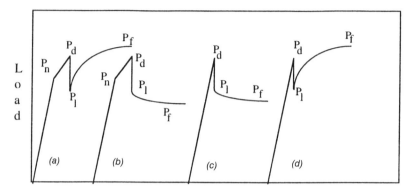

Displacement of Indenter with Respect to the Top-Face of the Specimen

Figure 12. Typical load-displacement curves for the push out test. Curves (a) and (b) show a nonlinearity at load P_n, before instability at load P_d. This is an evidence of push in, i.e., interface crack initiation and propagation from the top face. Curves (c) and (d) do not show any nonlinearity and are associated with crack propagation from the bottom face. Curve (c) is common with SiC-Ti-24Al-11Nb MMCs, and curve (d) is common with SiC-Ti-alloy systems, such as SCS-6/Ti-6Al-4V, SCS-6/Ti-15-3, etc.

other parameter is the friction stress, τ_f, which is an average shear stress following debonding. SiC/Ti-alloy MMCs exhibit a wide variation in the post-debond response, depending upon the alloy and specimen thickness. Thus, in SiC/Ti-15V-3Cr-3Al-3Sn specimens at room temperature with $t \geq 0.4$ mm, the load is observed to increase [see Figure 12(d)] immediately following debonding [13,14,63,65], possibly because of local damage processes in the graded carbonaceous coatings [13,63]. Correspondingly, in SCS6/Ti-24Al-11Nb, the load shows a steady, but slow, decrease [Figure 12(c)] following debonding [63]. Because of such dependence on the associated fiber displacement, the post-debond frictional shear stress is often based on the minimum load, P_l, immediately after the load drop [i.e., $\tau_f = P_l/(2\pi rt)$], although it has also been based on the final minimum or maximum load, P_f [63]. The definition based on P_l has the advantages that it can be used to understand the fracture mechanics of the push-out test, and that it likely involves minimum coating damage. This definition $\{\tau_f = P_l/(2\pi rt)\}$ is used in the rest of the discussion in this section.

A wide range of issues are involved in push out testing, as has been pointed out by a number of investigators [14,57–58,64–74]. From an experimental standpoint, one important variation has been the wide range of specimen thicknesses that were utilized. The effect of this parameter on interface strength is summarized in Table 1, although a direct comparison between the results of different investigators is difficult because of different fibers, matrix alloys, and fiber volume fractions involved. Although the debond strength is anticipated to depend on the thickness, the data in Table 1 indicate that the friction stress also may be influenced by thickness. Possible reasons for the latter include a modest reduction in the radial clamping stress on the average (through the thickness), coupled with its significant reduction (possibly even tensile) near the bottom face, for very thin specimens.

The effect of the radial clamping stress on frictional sliding behavior was nicely established by Clyne and co-workers [70,71,75], who used transverse loads during push out testing to show that the friction stress decreased in direct proportion to the reduction in the average radial clamping stress (in the ratio of 1:2). A similar effect of clamping stress on interface shear characteristics was illustrated in Reference [57], in which the variation in the local shear stress at the onset of debonding (usually at the bottom face) as a function of temperature was used to emphasize the strong dependence of interface shear properties on the clamping stress and suggest the greater mechanical versus chemical nature of the interface bond. However, the debond (maximum) load is a more difficult one to interpret than the friction stress, because it involves both frictional sliding and interface fracture.

The fracture mechanics of the push out test is similar to the pull out test, as long as crack initiation occurs from the indenter end (also referred to as the top face). Under this scenario, the same equations as the pull out case can be used. They are provided below, for the same material constants as used earlier for the pull out case, with v approaching zero and with the notation that the stress, p, is positive for compressive stresses:

$$\delta = 0.99 \left(\frac{4G_c}{E_f r} \right)^{1/2} \ell + 0.98 \left(\frac{\tau \ell^2}{E_f r} \right) \tag{11a}$$

$$= 0.98 \, (p - p_r) \frac{\ell}{E_f} - 0.98 \left(\frac{\tau \ell^2}{E_f r} \right) \tag{11b}$$

$$\left(\frac{\ell}{r} \right) = \frac{p - p_i}{2\tau} \tag{12}$$

$$p = p_r + 1.01 \left(\frac{4 G_c E_f}{r} \right)^{1/2} \tag{13}$$

$$p_r = -1.11 \, E_f \varepsilon_T \tag{14}$$

$$p_i - p_r = 1.01 \left(\frac{4 G_c E_f}{r} \right)^{1/2} \tag{15}$$

As before, p_i corresponds to the applied compressive stress on the fiber when a debond crack starts propagating from the top face; p is the applied compressive stress on the fiber. One important difference from the pull out problem is the existence of an instability [66] before the crack from the top face reaches the other end of the specimen, called the bottom face. As shown by Liang and Hutchinson [66], this instability occurs at approximately $1.5r$ from the bottom face. Using the same material constants as used previously, the stress drop in the fiber is given by:

$$\Delta p = 1.01 \left(\frac{4 E_f G_c}{r} \right)^{1/2} + p_r - 3\tau \tag{16}$$

and $\Delta p \ \{=(P_d - P_l)/\pi r^2\}$ is the stress drop on debonding, and P_l is illustrated in Figure 12.

The residual axial fiber stress, p_r, can be calculated from the thermal history, and τ can be determined from P_l (see Figure 12). Then, using the load drop, it is possible to calculate G_c. Although this approach overcomes the problem of estimating G_c from the load-displacement traces, such traces are nonetheless important for checking that the material parameters have been estimated correctly. In Reference [65], Equation (16) was used to obtain a G_c of 4 to 7 J/m^2 for the SCS-6/Ti-15-3 system, using 0.2- to 0.4-mm-thick specimens. However, as discussed below, push in does not occur in such samples, so that a more appropriate analysis is required for determining the toughness value.

The push out test has worked very well for CMCs in which residual stresses are generally low and fiber push in is evidenced by a nonlinearity in the load-displacement plot prior to attainment of the debond load[see (a) and (b) in Figure 12]. In MMCs, however, such a nonlinearity is generally nonexistent [see (c) and (d) in Figure 12]. A likely reason for the difference is the large axial compressive stress in the fiber for the MMC, which opposes push in imposed by the indenter. More specifically, the shear stress from thermal mismatch opposes that from the applied load. Koss et al. [67,68] showed for the first time in a sapphire-NiA1 MMC that crack initiation occurred from the bottom face rather than from the top face. More recently, it was shown by Majumdar and Miracle [14], using 2.5-mm-thick SCS-6/Ti-15-3 specimens, that crack initiation and propagation from the top face were absent or negligible, even for applied stresses on the fiber in excess of 4.5 GPa. Whatever nonlinearity that did occur could only be explained by assuming exceedingly high values of friction stress (>1000 MPa). When the axial residual stress was reduced to almost zero level, by predeforming a 0° MMC prior to slicing, the nonlinearity in the load-displacement record (associated with push in) was observed at anticipated reasonable values of applied load [14]. When specimen thickness was reduced to 0.8 mm and below, push out occurred for the as-received samples. However, the load-displacement traces from both the top and bottom faces provided clear evidence of crack initiation from the bottom face [14]. Evidence of bottom face initiation has also been obtained from 0.4-mm-thick SCS-6/Ti-24Al-11Nb composite, where interrupted tests followed by scanning electron microscopy showed interface cracks at the bottom face and none at the top face [57,67]. Considering that most push out tests in the past have been performed on 0.1- to 0.4-mm-thick specimens, the recent evidences of crack propagation from the bottom face suggest that the push in analysis may not be valid for those cases. It also may be noted here that when 0.4- to 0.8-mm-thick predeformed samples were used for push out testing [14], then a load-displacement behavior similar to CMC specimens was observed, reinforcing the idea that high axial residual stress in the fiber is to a large part responsible for bottom face crack initiation in most as-received MMC samples.

In Reference [57], bottom face initiation was analyzed based on the local shear stress at the bottom and top faces. Using this approach, a local shear strength of 350 MPa was estimated near the bottom face for the SCS-6/Ti-24Al-11Nb system at RT; at the onset of debonding, the local shear stress at the top face was significantly lower. Similar numbers were obtained for the SCS-6/Ti-15-3 system. However, these shear stresses were only marginally higher than the residually induced shear stresses near the free surfaces. Such small differences between the residual shear stress and the shear strength appear coincidental.

In References [73,74], a more elaborate shear strength-based approach was used to model the push out test, utilizing FEM procedures. Interface failure was allowed to occur when the local shear stress (τ_{local}) at the interface near the free surface (before crack initiation) or at the crack tip (after initiation) was such that:

$$\tau_{local} \geq \tau_o + \mu \cdot s_r \qquad (17)$$

where τ_o is the interface shear strength, μ is the friction coefficient, and s_r is the radial interface stress at that location. Using this approach, bottom face initiation was predicted for a number of SiC-Ti-alloy systems, for thicknesses less than 0.75 mm. In the case of SCS-6/Ti-15-3 MMC, τ_o was shown to be approximately 4 times the average shear strength (τ_d) based on maximum load (i.e., $\tau_o \approx 4 \times p_d$), with μ assumed to be 0.2. Load drops also were predicted by this computational model, which provided a stress description of events associated with interface crack initiation and propagation. In particular, the results suggest that mode-mixity may be changing at the crack tip, with the opening component (mode I) increasing slightly with increasing length of the crack from the bottom face. This mixity change, which needs verification using a fracture mechanics formulation, is likely a manifestation of the friction stress and is opposite to what occurs during pull out tests.

A problem with the local shear strength-based approaches is that they neglect the singular stress fields that are generated for the bimetallic problem, as indicated in the work of Bogy [9]. Thus, the computed maximum stresses may be only as good as the mesh that is used, although they do provide a good overall picture of the relative values of the different stress components. Recently, Majumdar and Miracle [14] analyzed the bottom face initiation problem using a fracture mechanics approach. Starting with the fundamental fracture mechanics equation:

$$GdA = dW - dU = Pd\delta - dU \qquad (18)$$

where dA is the incremental crack area, G is the crack-driving force, $d\delta$ is the load-point displacement due only to crack growth, and dU is the strain energy of the system, the following crack growth equation was derived:

$$G_c + \frac{\tau^2 l^2}{r_f E_f} = \frac{r_f}{4E_f} p^2 g(\eta) + \frac{r_f}{4E_f} p_r^2 \qquad (19a)$$

where G_c is the local crack-driving force, assumed constant throughout the crack propagation stage, $\eta = (t - l)/r$ is the normalized remaining ligament, t is the specimen thickness, r is the fiber radius, l is the crack length measured from the bottom, p is the magnitude of the compressive stress on the fiber, p_r is the compressive residual stress in the fiber, and $g(\eta)$ is a function of the normalized ligament. The function $g(\eta)$ is close to zero for $\eta > 15$, and it rises quite sharply for $\eta < 6$ [14]. More recent finite element analysis [77] shows that $g(\eta)$ can be represented as $g(\eta) = 6.408 - 6.975\eta + 3.375\eta^2 - 0.90\eta^3 + 0.1406\eta^4 - 0.01278\eta^5 + 0.00625\eta^6 - 0.0000127\eta^7$, for $\eta < 10$. Equation

(19a) is arranged in such a form that the left side represents material properties, and the right side represents effects of the applied load and the residual stress. The friction stress manifests itself as a crack-driving force that opposes crack growth. The crack growth from the bottom face can also be represented by the following stress-intensity formulation:

$$K_{local} = f_1 \cdot p\sqrt{r} + f_2 \cdot p_r\sqrt{r} - f_3 \cdot \tau\sqrt{r} \qquad (19b)$$

where $f_1 = \sqrt{\{E' \cdot g(\eta)/4E_f\}}$, $f_2 = 0.5$, $f_3 = \{0.62 + (l/r)\}$, $2/E' = \{(1 - v_f^2)/E_f\} + \{(1 - v_m^2)/E_m\}$, $K_{local} = \sqrt{\{E'G_c\}}$, and v_f and v_m are, respectively, the Poisson's ratio of the fiber and the matrix. The analysis shows the existence of a peak load associated with crack growth; i.e., an instability also exists for the bottom-face initiation problem. In addition, the drop in load at complete debonding is also a strong function of the interface toughness, similar to the push-in analysis of Reference [66]. When the bottom-face analysis is applied to data for the SCS-6/Ti-15-3 MMC, it provides toughness (G_c) ranging between 20 J/m^2 and 40 J/m^2 [14,77], values that are high compared with intrinsic bond energies. They are also high compared to push-in data for SiC/glass CMCs, which have low compressive or even tensile radial clamping stresses. The higher toughness values for the case of MMCs suggest inelastic effects in the C-rich coating region and the surrounding metal, possibly associated with high radial clamping stress at the interface.

The above discussion shows that great care needs to be taken when characterizing interface shear properties using the push out test. The face from which crack propagates needs to be monitored, so that the appropriate analysis can be used. One method is to simultaneously monitor both top face and bottom face displacements with a high accuracy, possibly also relying on acoustic emission signatures to interrogate any nonlinearity in the load displacement trace. An example is shown in Figure 13, where there is a nonlinearity associated with the bottom displacement, but none for the top face displacement, prior to attainment of the maximum load. Another approach is to perform interrupted tests, followed by microscopy and profilometric measurements, to determine the failure initiation location. This approach has been used in References [14,57,67–69].

THE FIBER FRAGMENTATION TEST

The classical form of this technique is to strain a single-fiber composite to high strain levels, such that fiber fragmentation proceeds to a saturation limit, controlled by the frictional characteristics of the

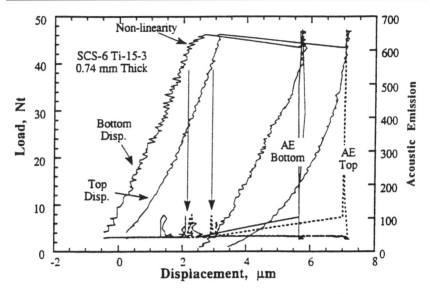

Figure 13. Load versus displacement plots for a fiber in an SCS-6/Ti-15-3 push out specimen. The top face displacement shows no nonlinearity, whereas the bottom face displacement shows a distinct nonlinearity before the onset of debonding. The small acoustic emission signature also coincides (see long arrows) with the nonlinearity in the bottom face displacement plot. (Reprinted from Reference [14] with permission.)

interface and the fiber strength. A shear lag analysis is then used to deduce a friction stress, according to:

$$\tau_i = \sigma_F \cdot r/L_c \qquad (20)$$

where τ_i is the friction stress, σ_F is the strength of fibers, and L_c is the average fragment length at saturation.

The fragmentation has been used successfully in PMCs. However, in MMCs, fragmentation tests show significant damage of fiber coatings, and it is not clear whether a saturation limit is actually reached. Average fragment lengths as low as 2 to 3 fiber diameters are reported for SCS-6/Ti-6A1-4V MMCs [78,79], and significant amounts of secondary cracking are observed next to the primary fiber breaks. Friction stresses values from fragmentation tests of SiC-Ti-alloy MMCs range from 180 to 400 MPa [42,78–80], which are much larger than values determined using the push out or pull out tests.

The simple equation provided above is clearly an oversimplification, and a number of refinements have been suggested to account for the statistics of fiber fracture. Among these, Curtin's fragmentation model [81] is an attractive one. The model allows the fiber to follow weak-link Weibull statistics and provides analytical equations to determine the extent of fragmentation at any point during the test. Thus, the experiment can be interrupted or stopped at any level of strain (such that damage of coatings is kept low), and the stress corresponding to each fiber break (monitored by acoustic emission signals) in conjunction with fiber fragment distribution (measured by ultrasonic imaging) can be used not only to derive the friction stress but also to determine the in situ fiber strength statistics (i.e., Weibull modulus and strength). This approach, which relies on two relatively independent sets of data (i.e., stresses at fiber breaks and fragment length distribution) was used successfully in Reference [42] to derive the material parameters in Ti-6A1-4V-based MMCs reinforced with SCS-6 and SCS-0 (no outer carbon coatings) fibers. Figure 14 provides an example of the two sets of data, with the solid lines being predictions based on best-fit values of Weibull parameters and friction stress. In addition, simulation was used to model fiber breakage, and good agreement was obtained between the simulation and Curtin's fragmentation model. However, even with these modifications, the friction stress was larger than 300 MPa for SCS-6 fibers [42], which have weak carbon coatings. The reason for the high numbers is suspected to be high

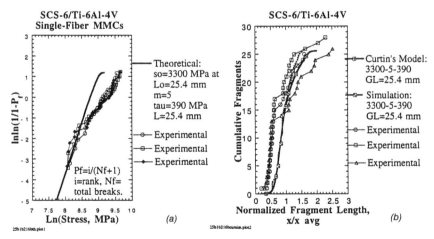

Figure 14. Results from single-fiber fragmentation tests for the SCS-6/Ti-6A1-4V system. (a) Weibull plot of the stress at fiber break and (b) fragment length distribution plot. Experimental data [42] are indicated by symbols, and the solid lines represent predictions from the Curtin model [81] based on best-fit values of the Weibull parameters and the friction stress.

radial clamping (pinching) of fibers at fiber breaks. The occurrence of such stresses was pointed out recently in Reference [82] using a computational approach, in agreement with similar findings in the early 1970s [83], where the local radial clamping stresses at the break appeared to be controlled by the flow stress of the matrix.

Although additional experimental and analytical work on fiber fragmentation are continuing, it appears that the high-friction stress is a real effect. The implication is that when a fiber does break in a high volume fraction ductile-matrix composite, the behavior can be quite different from that predicted, based on friction stress data from push out tests. Thus, with lower extent of debonding, the elastic stress concentration on neighboring fibers may be larger. Additional complications arise because of generation of intense plastic slip lines from a fiber break, which can actually dominate the local stress concentration behavior, leading to fracture of neighboring fibers. A brief discussion on such local load sharing behavior is provided later. In summary, the single-fiber fragmentation test is a useful one for understanding composite behavior and does provide in situ values of interface friction stress, which can then be used to model the ultimate tensile strength of MMCs.

OTHER INTERFACE SHEAR TESTS

Two additional interface shear tests are worthy of mention here. One is the *slice compression test*, in which the polished cross section of a composite is pressed onto a soft material, such as brass. The different moduli of the fiber and matrix result in greater penetration of the fibers into the brass. The extent of fiber protrusion is then related to the interface friction stress [84,85]. Work performed with an SiC-Ti-6Al-4V MMC [86] indicate problems associated with uniform loading of the MMC on the soft substrate and difficulties in deriving interface parameters from a complex elastic-plastic punch problem. On the other hand, the technique has the obvious advantage of not requiring individual loading of fibers and easier adaptability to high-temperature measurements.

The other test worth mentioning is the *torsion test* [87], in which a single-fiber MMC is subjected to torsional loading along the fiber axis. Essentially, a strain-gauged flat specimen with a fiber running axially down the middle is held between aligned grips at both ends, and one grip is rotated with respect to the other. In this configuration, interface failure is initiated near one or both ends of a torsion specimen, by the circumferential shear stress at the fiber-matrix interface. The crack then propagates axially along the fiber axis, driven by the energy stored in the torsion specimen. Crack initiation is usually associated with an acoustic emission signature, and propagation is manifested in the form of a nonlinear plot of the applied torque versus the angle of twist. This geometry appears to be the only one whereby the circumferential shear strength of the interface can be determined, because push out and pull out specimens have interface failure caused by shear stresses parallel to the fiber axis. Results [87] show that the circumferential shear strength is lower than the friction stress derived from push out tests. For example, the circumferential shear strength was found to be approximately 50 MPa for a single-fiber SCS-6/Ti-6A1-4V composite, whereas push out tests showed a friction stress of approximately 150 MPa.

The circumferential shear strength is important when considering fiber-matrix separation under transverse loading. Here, the circumferential shear stress at the interface at $\theta = 45°$ [see Figure 1(a)] is comparable with the normal stress at $\theta = 0°$. If shear failure does occur before interface separation, then it increases the normal stress at $\theta = 0°$, leading to fiber-matrix normal separation at lower loads. Recent calculations indicate [88] that even if the interface normal strength were low, a higher shear strength or a higher friction coefficient can aid in postponing the onset of fiber-matrix debonding under normal loads.

Normal Strength

The normal (transverse) strength of the interface has significant influence on the transverse composite response. A straight forward approach for measuring the normal strength would involve the following: (1) loading an uniaxial MMC in the transverse (90°) direction, (2) accurately determining the onset of interface separation, and (3) using a suitable analysis to determine the local normal stress at the interface. A number of methods have been used to determine the onset of debonding in transverse tests of multiple-ply specimens. They include a sharp decrease in the slope of the stress-strain curve [89,90], a drop in the Poisson's ratio [90], and replication of the edge of the sample for evidence of fiber-matrix debonding [89,90]. The same approaches can be used for single-ply or single-fiber specimens.

Past research on multi-ply specimens suggested that the interface normal strength for SCS-6/Ti-alloy MMCs was negligible or zero [91]. A similar conclusion was obtained on testing of uniform-width, single-fiber SiC/Ti-6A1-4V composites [92]. In this latter case, edge replication was used to detect the onset of debonding, and fiber coatings, such as SCS-6, SCS-0, soft carbon-coated Trimarc (made by Amercom) fibers, did not have any significant influence on the far-field applied stress at the onset of debonding. However, ultrasonic evaluation using the shear back reflection (SBR) technique showed that interface separation initiated at the interface next to the two free edges and subsequently progressed rapidly to the center of the specimen [93]. These experimental results were consistent with Kurtz and Pagano's theoretical analysis [94], which shows that the residual radial stress at the interface switches from a compressive value in the specimen interior to a tensile value at the free surface [in fact, a tensile singularity, see Figure 15(a)].

Thus, a normal interface debond can preexist near the free surface; the local K_1 and K_2 for specific values of crack length and thermal misfit strain can be estimated using equations provided in Reference [23]. During subsequent transverse loading of the MMC, the crack could propagate, so that there is ambiguity as to what parameter would be actually measured by such a test. Even without an existing crack, the singular nature of the tensile radial interface stress near the free surface makes a simple determination of interface normal strength difficult.

The above problems were nicely addressed [95] using the cruciform specimen geometry, shown in Figure 15(b), where the wings prevent any preexisting crack from propagating into the center section of the

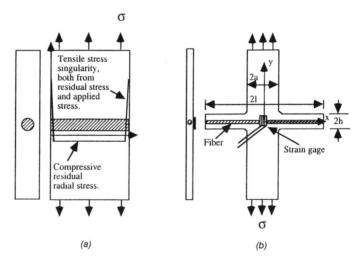

(a)   (b)

Figure 15. (a) A transverse specimen of uniform width illustrating stress singularity at the interface near the free edges. (b) The cruciform specimen geometry [12,95] that avoids the stress singularity and allows determination of a normal interface strength.

specimen. A strain gauge is used to monitor strains in the center of the cross. The onset of debonding is determined from the knee in the stress-strain curve, because interface failure is accompanied with a drop in specimen stiffness. The test is suitable for both single-fiber [12,94–97] and multiple-fiber multiple-ply specimens [14]. For a monolithic sample, the stress field along the centerline of the cross was derived in Reference [95] using an Airy stress function approach:

$$\sigma_{yy} = \left[\frac{\sigma a}{l}\right] + \left[\frac{2\sigma}{\pi}\right] \sum_1^\infty \sin(\beta_n a) \cdot \frac{\cos(\beta_n x)}{n \cdot \{\beta_n h + (0.5)\sinh(2\beta_n h)\}}$$

$$[(\beta_n h)\cosh(\beta_n h) \cdot \cosh(\beta_n y) - (\beta_n y) \cdot \sinh(\beta_n y) \cdot$$
$$\sinh(\beta_n h) + \cosh(\beta_n y) \cdot \sinh(\beta_n h)] \tag{21}$$

where $\beta_n = n\pi/l$.

The tensile stress along the centerline of the cross is illustrated in Figure 16 for one particular set of geometrical parameters and at different positions, y/h.

There are two important points to note here. First, the tensile stress goes to zero very rapidly in the wings of the cross, so that even if interface debonding initiated at the free edges due to the thermal residual stresses, it cannot propagate inward. Second, for reasonable

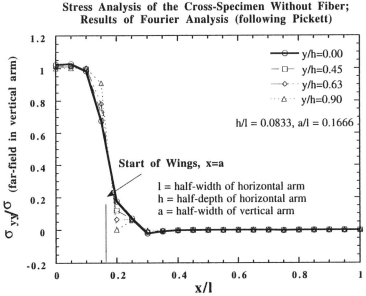

Stress Analysis of the Cross-Specimen Without Fiber; Results of Fourier Analysis (following Pickett)

Figure 16. Stress distribution in the cruciform specimen for a monolithic material [Equation (21)].

dimensions of the cruciform specimen, the stress along the centerline is constant over a substantial portion, so that a debond strength can be directly interpreted from the load at debonding. Moreover, the stress in the central section is almost identical to the stress in the uniform (narrow) part of the specimen, so that no special analysis is necessary for analyzing the test data.

The additional pieces of information required for bond strength (σ_b) determination are the residual radial stress at the interface and the relationship between the radial stress at the fiber-matrix interface and the far-field applied stress. Mathematically, this is represented as:

$$\sigma_b = k \; \sigma_a + \sigma_r \qquad (22)$$

where k is a stress concentration factor (SCF), defined as the ratio of the local radial stress at $\theta = 0°$ [see Figure 1(a)], to the far-field applied stress (σ_a), and σ_r is the residual radial stress at that location. The reason that this particular location is considered is because almost all analyses suggest that normal interface separation occurs here. k ranges between 1.2 and 1.3 for single- and multiple-fiber SiC-Ti-alloy MMCs [96,98]. For other systems, Mushkellishvilli's elastic analysis [99] can be used to determine k. The value of k is weakly dependent on the volume fraction of fibers up to approximately 40 volume percent but thereafter increases fairly rapidly [100,101].

The stress concentration factor can be larger (1.35 to 1.6) if tangential interface shear failure at $\theta \approx 45$ to $60°$ occurs prior to normal interface separation, because of high values of the tangential shear stress at those locations [96–98]. These issues of shear failure and the effects of tangential sliding on normal interface separation have been addressed in a recent study [88], in which it is shown that the stress concentration is dependent both on the coefficient of friction and on the sliding length, the latter being a function of the applied stress less the contribution of the residual stresses. It also is useful to note that k is not significantly affected by plasticity, as demonstrated in References [3,100,101] for an elastic-plastic solid; the fully plastic case with rigid fibers has been addressed in Reference [4], where plastic strain concentration factors and hydrostatic stresses also are provided.

The cruciform geometry has been used with success to determine interface normal strengths in a number of MMC systems. Typical stress-strain curves for SCS-6/Ti-6Al-4V and SCS-0/Ti-6Al-4V are reproduced in Figure 17. Approximate bond strengths reported [12,95,97] are 115 MPa for SCS-6/Ti-6Al-4V, 380 MPa for SCS-0/Ti-

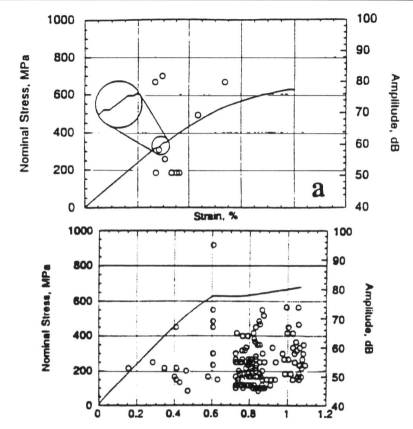

Figure 17. Transverse stress-strain of Ti-6Al-4V specimens reinforced with single SCS-6 and SCS-0 fibers. Both show strain jumps at debonding that are consistent with finite bond strengths, determined from the applied stress and Equation (22) (Reprinted from Reference [12] with permission from Woodhead Publishing Ltd.)

6Al-4V, and approximately zero for an Amercom Trimarc SiC-Ti-6Al-4V MMC, where it may be noted that the Trimarc fiber has graded carbon coatings without silicon additions (unlike the SCS-6 coating) and has a push-out frictional stress of only 30 MPa [42]. The high strength of the SCS-0/Ti-6Al-4V MMC is noteworthy. The strength also is comparable with similar strengths obtained for SiC-Ti-SiC joints [102,103] formed by a diffusion bonding technique. However, the good fiber-matrix bond strength of SCS-0/Ti-alloy must be considered in the context that fiber strengths are reduced by local reactions; the dominant reaction product is Ti_5Si_3 at the SiC interface, along with formation of some TiC [102,103].

Another specimen that also appears promising for evaluating the normal strength of the interface is the *embedded fiber* geometry [104]. In this geometry, the fiber ends are completely embedded in the matrix, so that stress reversal at fiber ends is avoided.

EFFECTS OF INTERFACE PROPERTIES ON BULK MMC BEHAVIOR

Effects of the Normal Interface Strength

The normal interface strength has the most significant effect on the transverse strength and creep resistance of the MMC. Once normal separation occurs at the poles [$\theta = 0°$, see Figure 1(a)] of the fiber, high strains are generated at θ between 60 and 90°, leading to damage and crack nucleation in the matrix at those sites. Models of damage following debonding have been developed [105,106]. For intermetallic based MMCs with low strain-to-failure, fiber-matrix debonding can quickly lead to failure at applied strains less than 0.3%. This can limit application in systems that need some damage tolerance in the transverse direction. That is why matrices with a reasonable degree of elongation are desirable, particularly for systems with weak fiber-matrix interfaces.

At high temperatures, the loss of load-carrying capability of the fiber, along with the strain concentrations, results in creep rates that are much larger than the matrix. This type of behavior was illustrated in Figure 2. When thermal cycling is also present, a ratchetting is observed even at low stresses, such that the effective creep rate greatly exceeds the isothermal creep rate at the highest temperature [107]. This is because residual stresses are regenerated every cycle. Increased interface strength through high-temperature fiber-matrix reactions have been shown to reduce the thermally induced creep rate [72].

In summary, significant benefits in the transverse MMC response can be obtained by increasing the normal interface strength. However, any strength improvement will have to be considered in the context of possible loss of crack bridging in the presence of a matrix crack.

A direct approach for increasing the transverse MMC response is to increase the bond strength. This entails chemistry considerations, adequate emphasis on lattice mismatch between the reaction products and the constituents, and possible reductions in the fiber strength.

From a mechanism perspective, it is also important to consider tangential shear failure of the interface. As indicated earlier, a shear failure prior to interface separation can increase the stress concentration factor [Equation (22)] at the poles of the fiber, leading to premature fiber-matrix separation. The friction coefficient also is important, because analyses [88,98] show that a high-friction coefficient is almost equivalent to no tangential shear failure prior to interface separation. Thus, if prior shear failure is suspected, then a higher friction coefficient can be used to increase the applied stress at debonding.

Another approach that may be considered is to introduce an interlayer of very low modulus between the fiber and the matrix. This layer can reduce the stress concentration factor, even to values less than unity [108]. However, too low a modulus can enhance stress concentration at the equatorial positions ($\theta = 90°$), leading to strain concentrations in the matrix. Thus, there are issues of optimization.

Longitudinal Strength of Unnotched Material

In MMCs with a brittle matrix (strain-to-failures less than 3%), cracks are easily nucleated in the matrix due to stress concentration arising from a fiber break or from defects, such as machining flaws, at the specimen surface. A strong interface is undesirable for such systems because crack bridging may be inhibited, and breakage of a defective fiber can precipitate premature failure of the MMC. However, the situation is slightly different from CMCs, in that the ductility and toughness of the matrix provides some protection against stress concentrations, such as a premature fiber break, arising at low applied stresses. Thus, the requirement for a weak interface may not be as demanding as in the case of CMCs. Even with a weak interface, such as the reasonably weak carbon-Si coatings for an SCS-6 fiber, through-thickness multiple cracks, which are characteristic of CMCs, are not observed with relatively brittle intermetallic matrices [109]. Rather, failure is generally precipitated by one dominant crack, which occurs at an applied stress when the bulk of the matrix is mostly past yielding. At this stress level, all the fibers are quite highly stressed, and any fiber break can generate enough stress concentration to fracture the adjacent fiber. Thus, for these systems, MMC failure may be controlled by the occurrence of the first random fiber break, and Weibull statistics can be conveniently used to estimate the tensile strength. This approach provided good agreement with experimental strength data in the case of an SCS-6/Ti-25Al-17Nb MMC, where the matrix had a strain-to-failure of only about 1.5% [109].

In ductile matrix composites subjected to monotonic or creep loads, matrix cracks are not present, and the interface becomes important only in the presence of fiber breaks. Here, the primary concern is one of global versus local load sharing, with the former being more effective in utilizing the full strength of the fibers. *Global load sharing* (gls) implies that a fiber break at a given plane is shared equally among the remaining intact fibers on that plane. On the other hand, when fracture of a fiber leads to significant stress concentration on the neighboring fiber, then that adjacent fiber could fail and set up a chain of failures leading to ultimate failure of the MMC. This type of failure mode is termed *local load sharing* (lls), and obviously, it is not as efficient as global load sharing in utilizing the full strength of fibers.

The effects of interface toughness and friction stress on global versus local load sharing in ductile matrix composites has not been explored in sufficient detail. It is often implicitly assumed that a high-friction stress is detrimental to tensile strength, although there is very little data to back up this assumption. Tensile tests performed on carbon-P-55/Al and carbon-P-55/SiC coating/Al MMCs showed that the latter was actually slightly stronger than the former, although the latter had a higher interface strength and toughness [110]. This result is contrary to the assumption that higher interface strength is detrimental to longitudinal strength. In Reference [111], Sigma (SiC)-Ti-6A1-4V composites were heat treated at 865°C for different times to alter the friction stress and to evaluate its effects on longitudinal strength. The results showed a decrease in strength with heat treatment. However, it is likely that fiber strengths may have been reduced by the heat treatments, so that a definitive correlation between the interface strength and tensile strength cannot be established.

Reference [112] provides a methodology to predict the effect of friction stress on global versus local load sharing behavior. For E_f/E_m values that are characteristic of SiC-Ti-alloy MMCs and for volume fractions that are normally considered, global load sharing was predicted when the friction stress was less than about 10% of the fiber strength. For SCS-6 fibers with a Weibull strength of about 4500 MPa (Weibull modulus is 14) for a gauge length of 25.4 mm, this would imply that global load sharing would occur in SiC-Ti-alloy MMCs as long as the friction stress remained less than about 450 MPa. Experimental verification of this prediction requires use of a model that can predict both the tensile strength and fragment distribution and includes the contribution of the friction stress. Curtin's gls model [6,113] appears to be the most comprehensive, to date, for evaluating global load sharing behavior, and it predicts that at tensile fracture of the MMC, the average stress (σ_u) carried by the fibers (includes broken and unbroken ones) is:

$$\sigma_u = \sigma_c \left(\frac{2}{m+2}\right)^{\frac{1}{m+1}} \left(\frac{m+1}{m+2}\right) \tag{23}$$

$$\sigma_c = \left(\frac{\sigma_o^m \tau \, Lo}{r}\right)^{\frac{1}{m+1}} \tag{24}$$

$$\delta_c = r\sigma_c/\tau \tag{25}$$

where m is the Weibull modulus, σ_o is the Weibull reference strength, τ is the friction stress, r is the radius of the fiber, and L_o is the gauge length on which the Weibull parameters are based.

In addition, Curtin's gls model predicts an average fiber fragment length of $(m+2)*l_f$ at the point of MMC fracture, where $l_f = [\{2/(m+1)\}^{1/(m+1)}]*\delta_c/2$. Alternately, if a strip of width $2*l_f$ is cut out from the gauge length, it will contain $2*N/(m+2)$ fiber breaks, where N is the number of fibers at any cross section. To provide quantitative insight into the length scales and the number of fragments involved, let us assume $\sigma_o = 3500$ MPa, $m = 10$, $L_o = 25.4$ mm, $\tau = 100$ MPa, and $r = 70$ μm. Then the average fragment length is 15.6 mm; also, if there are 100 fibers in a cross section, then a 2.6-mm-wide strip will contain approximately 17 fiber breaks, or a 25.4 length of the sample will contain approximately 166 random breaks. Clearly, a fairly large number of fiber breaks are anticipated if global load sharing is present. Attempts at experimental verification of global load sharing prediction based on Reference [112] and using Curtin's model [6,113] have been extremely limited. In Reference [114], gls was suggested to have occurred in a Sigma (SiC)-Ti-15-3 MMC, although the measured fiber fragments were much larger than predicted by the theory; there appears to be an error in the study regarding the predicted length of fragments. Recently, fiber extraction following tensile failure of a heat-treated SCS-6/Ti-25Al-17Nb MMC showed less than 10 fiber breaks in the entire gauge length (=25.4 mm) of a sample that had a total of about 80 fibers in the cross section [109]. This was significantly less than that predicted using the gls model, although the ratio of the interface friction stress (measured from push out tests) to the fiber strength was well below 0.1, which satisfied the gls requirement according to Reference [112]. In addition, the tensile strength of the MMC was also less than that predicted by Curtin's model. Rather, the strength data appeared to correlate well with a local load sharing model based on the criterion of the second fiber, according to the work of Zweben and Rosen [115]. Longitudinal tensile tests in Reference [116] on a SiC-Ti-1100 MMC also showed no evidence of the many fiber breaks needed to validate a gls behavior. Here, too, the strength predictions based on Curtin's model well exceeded the experimental strengths. In Reference [113], the fragment distribution in Reference [117] was used to validate the occurrence of gls in an SCS-6/Ti-6Al-4V MMC. However, the fiber breaks in Reference [117] were revealed by polishing, and cracks may have been introduced during polishing if a residual tensile stress was present in the fibers at the time of polishing.

Thus, a number of investigations appear to suggest that gls may not be occurring in SiC-Ti-alloy MMCs, in apparent contradiction to the predictions of Reference [112]. However, it must be realized that a friction stress to fiber strength ratio of 0.1 was essentially based on an E_f/E_m of approximately 3. On the other hand, during almost all of the tensile tests, the matrix is fully yielded. This effectively increases E_f/E_m to high values, producing significant stress concentration in neighboring fibers [112]. The occurrence of high-stress concentrations in a yielded composite material were also predicted in Reference [83] a number of years ago. In addition to the role of plasticity in elevating the stress concentration factors, it is also useful to remember that single-fiber fragmentation tests typically show friction stresses that are at least 2 to 3 times the value from push out tests. It is likely that radial clamping following fiber fracture can increase the shear stress required for interfacial sliding. Thus, these conditions may make gls difficult.

Recent tensile tests of single-ply MMCs provide additional insight into local load sharing behavior [14,42]. Figure 18(a) illustrates slip bands observed on the face of a single-ply Trimarc (SiC)-Ti-6Al-4V MMC. Figure 18(b) is an NDE shear back reflection micrograph of the same specimen and provides details on the location of the fiber breaks. Comparison of the two figures clearly shows a direct correlation between the intersection of slip bands and the fiber breaks. The feature that is extremely interesting is that even with a weak soft-carbon interface for this system (a friction stress of 30 MPa measured by push out tests [42]), there is a significant stress concentration effect, arising mainly from localized plasticity associated with fiber breaks.

It is important to point out that the macroscopic slip bands are not observed for this Ti-6Al-4V material if fibers are not present, so that

Trimarc SiC/Ti-6Al-4V

(a) Slip Bands Observed on Face of 15-Fiber Specimen

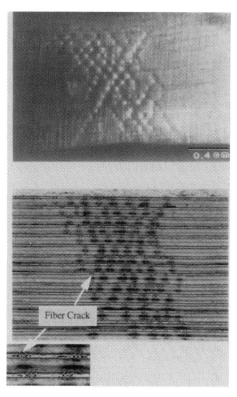

(b) Fiber Cracks at Slip Band Intersections, as Revealed by Ultrasonic SBR

Figure 18. (a) Intense slip bands associated with fiber breaks in a Trimarc (SiC)-Ti-6A1-4V single-ply MMC containing 15 fibers approximately 10r apart. The specimen was loaded longitudinally to approximately 1.5% strain [14,42]. (b) NDE shear back reflection (SBR) image of the same specimen, the arrow pointing to a fiber break location. (Reprinted from Reference [14] with permission.)

the slip bands result from high-strain concentration arising from fiber breaks. At the same time, based on interrupted tests, it appears that when the slip band from a broken fiber impinges on the adjacent fiber, it provides enough stress concentration to break that fiber. The net result is the crisscross type of arrangement of slip bands and fiber breaks.

Statistical analyses of the stresses at which fiber breaks occurred and analysis based on fiber fragment distribution also pointed to a local load-sharing phenomenon. Although not shown here, SCS-0 fibers in a Ti-6A1-4V matrix (this system has a strong interface) showed extremely limited length of localized slip bands, and breaks in neighboring fibers were less correlated, i.e., they suggested less stress concentration effects. These differences, arising from the effect of friction stress on extent of matrix plasticity, are schematically illustrated in Figure 19.

Ongoing research by the present author shows that when the fiber spacing is reduced to 0.2 mm for the SCS-0 fibers to be more representative of high-volume fraction composites, fractures became more correlated. However, here too, neighboring fiber fractures lie along sharp slip bands of the matrix, rather than cracks being coplanar and lying along planes perpendicular to the loading direction. These results suggest that plasticity-induced stress concentration is dominant across the entire range of interface strengths, from low to high interface shear strengths. More importantly, these single-ply experiments [118] provide an explanation of why MMCs with SCS-6 fibers show strengths that are better predicted by local load sharing models, such as that of Zweben and Rosen [115], than by global load-sharing models, i.e., a plasticity-dominated mechanism appears to exist that may explain local load-sharing behavior in high-volume fraction composites with weak interface coatings.

In summary, the effects of interface friction stress on the longitudinal strength of MMCs are not very clear, and additional experiments and analyses are certainly necessary. Until that is done, the survey of available literature suggests that the effect of interface strength on

Mechanisms of Fiber Fracture, Illustrating the Dominant Role of Plasticity in Modifying Local Versus Global Load Sharing Behavior in MMCs with Weak and Strong Interfaces.

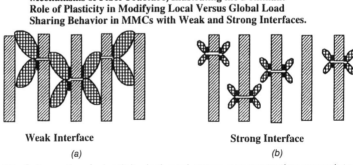

Weak Interface	**Strong Interface**
(a)	(b)

Figure 19. Schematic of plasticity-induced stress concentration associated with fiber breaks. The interface is weaker for case (a), which produces greater length of debonding and zone of plasticity. The associated stress concentration can break the adjacent fiber. With a stronger interface (case b), the plastic zone associated with the original fiber break is well contained.

longitudinal strength may not be very significant. The property that may be more affected by interface strength would be the fracture toughness, which is discussed below.

Creep Deformation

The effects of the interface strength on longitudinal and transverse creep behavior of the MMC were discussed at the beginning of this chapter. Under longitudinal creep conditions and under isothermal and thermomechanical fatigue, the effect of the interface friction stress is primarily felt when fiber fractures start occurring, i.e., at high enough applied stresses. An example of the effect of friction stress on creep life was illustrated in Figure 3. Under transverse loading, the interface can significantly reduce creep rates if fiber-matrix interface separation is prevented through an inherent bond strength of the interface. This was illustrated in Figure 2.

Fracture Toughness

Notched tensile tests performed on MMCs by Kelly and co-workers [119] a number of years ago demonstrated that the bulk of the energy absorbed before failure is associated with plasticity of the matrix between cracked fibers. A weaker interface allows a larger volume of matrix on either side of the crack plane to participate in the fracture process and produce an increase in toughness. Interestingly, the toughness was shown to also increase with the fiber radius, because this allowed a greater volume of matrix plasticity near the crack plane. In recent years, work on a carbon/Al MMCs [110] showed that the total work of fracture increased with a decrease in interface strength, although the maximum load in notched specimens was negligibly affected; in fact, the stronger interface exhibited a slightly higher notched strength. The fracture energy did not, however, increase monotonically with increase in the length of debond. This is because beyond a certain debond length, the matrix participating in the fracture process is also limited by the distance between the fibers. Thus, too weak an interface may not contribute any extra amount to fracture toughness, but at the same time it can significantly reduce the transverse strength and slightly decrease the tensile strength. In Reference [120], a novel approach employing engineered precipitate morphology was used to enhance debonding at fiber-matrix interfaces. Although this approach has potential for toughness improvement, it necessarily involves some reduction in transverse strength. The approach is perhaps best suited for cast MMCs, because deposited fiber coatings often cannot survive the extreme reactions imposed by molten alloys.

Fatigue Crack Growth

The fatigue crack growth rate in an MMC is different from the base matrix alloy in two primary ways, both being associated with crack tip shielding. First, a high-modulus fiber ahead of the crack tip acts to take up a larger fraction of the load and thus reduces the local stress intensity factor at the matrix crack tip. This shielding reduces the crack growth rate, with the effect increasing with higher values of E_f/E_m. The modulus effect on FCG rate can be quite significant and is independent of whether or not crack bridging occurs. The interface plays only a minor role here. The second crack-shielding mechanism is through crack bridging, which can dramatically reduce the crack growth rate and even arrest crack growth. The fiber-matrix interface plays an important role here. A sizable literature exists on crack-bridging behavior during crack growth in CMCs and MMCs [7,121–127]. The conditions required for crack bridging have already been discussed earlier and will not be considered here.

The crack-bridging model of McMeeking and Evans [121] has been widely used for characterizing the fatigue crack growth (FCG) behavior of MMCs. In this and other models, the interface affects the crack growth rate through a friction stress term. Typically, this friction stress is essentially a parameter that best fits the crack deceleration data, and is small (1 to 30 MPa), being many times smaller than the friction stress from push out tests of untested specimens. A higher friction stress, provided crack bridging is not compromised, lowers the crack-opening displacement and hence is more effective in reducing the crack growth rate. Thus, from an interface design standpoint, if the failure criterion is based on the attainment of a critical crack length (being associated with dimensional stability of the component), then a higher friction stress would be desirable. However, according to the model in Reference [121], a higher friction stress also increases the fiber stress in the crack wake. Thus, if failure is governed by the attainment of a critical fiber stress in the crack wake, then a weak interface friction stress would be desirable. Clearly, there are issues of compromise based on design considerations.

A concern in some of the models [7,121,122,127] is that the smeared traction on the fracture surface in the crack wake (due to bridged fibers) asymptotically approaches zero at vanishingly small distance behind the crack tip. This implies that the stress in the fiber goes from an extremely large value ahead of the crack tip to zero immediately behind the crack tip. Similarly, the stress in the fiber just behind the crack tip goes from a reasonable value far from the crack plane to zero at the crack plane, i.e., a phenomenon of reverse load transfer to the matrix. It is not clear if these stress changes can occur within the small span that a fiber goes from immediately ahead of the crack tip to behind the crack tip. Majumdar et al. [125] proposed an alternate methodology, whereby the stress in the fiber in the bridged region was the far-field stress in the fiber *plus* an additional amount that resulted from load pickup (transfer) from the cracked matrix; the magnitude of the load pickup was a function of relative fiber-matrix sliding, i.e., the crack face displacement at that location. The model was subsequently refined by Danchaivijit and Shetty [126], and their model appears to be an attractive one. The feature of both these models [125,126] is that the crack face displacements have a flatter profile than in the models of References [7,121,122,127], and, in addition, the fiber stresses change less rapidly with distance behind the crack tip, starting with a nonzero value immediately behind the crack tip. The fiber pressure model of Reference [124] also predicts a reasonably flat fiber stress profile, although the friction stress does not appear in the model. Recent measurements [128,129] of crack face displacements appear to validate the low rate of stress increase in the bridged fibers, with a nonzero value immediately behind the crack tip; they show that the models of References [121,127] greatly overestimate the fiber stress, particularly far behind the crack tip. However, additional experiments are necessary to fully understand the mechanism of load transfer, as fibers pass from ahead of the crack tip to behind the crack tip.

SUMMARY

This review was an attempt to summarize current understanding in the area of fiber-matrix interfaces in MMCs. Significant progress has been made over the last few years, particularly with regard to the mechanics and mechanisms of interface failure, the characterization of interfaces, and how interfaces influence bulk MMC response. However, there are areas where additional understanding is needed. These have been pointed out. It is important to realize that the requirements of interfaces in MMCs can be significantly different from the CMC and PMC systems. Thus, although basic knowledge of interface mechanics can be derived from those systems, development efforts will have to concentrate on satisfying the key requirements of MMCs in aerospace structures. Chemical aspects associated with interfaces have not been addressed here for the interest of space. For additional reading on interfaces, one may refer to the books by Metcalfe [130], Kelly and Macmillan [131] and Clyne and Withers [75].

ACKNOWLEDGMENTS

This work was performed at the USAF Materials Directorate, Wright Patterson Air Force Base, Ohio, under by Air Force contract number F33615-91-C-5663 to UES, Inc. The discussions with Drs. D. B. Miracle, S. G. Warrier, D. B. Gundel, M. L. Gambone, B. Muryama, T. Nicholas, R. Dutton, and N. Pagano at the WPAFB are sincerely acknowledged.

REFERENCES

1. F. W. Crossman, R. F. Karlak, and D. M. Barnett, "Creep of B/Al composites as influenced by residual stresses, bond strength, and fiber packing geometry", *Failure Modes in Composites—II, AIME Symp. Proceedings,* TMS, Eds. J. N. Fleck and R. L. Mehan, pp. 8–31 (1974).

2. P. K. Wright, *Titanium Matrix Composites,* Ed. P. R. Smith and W. L. Revelos, Air Force Rep. WL-TR-92-4035, pp. 251–276 (1992).

3. R. F. Karlak, F. W. Crossman, and J. J. Grant, "Interface failures in composites", *Failure Modes in Composites—II, AIME Symposium Proceedings,* TMS, Eds. J. N. Fleck and R. L. Mehan, pp. 119–130 (1974).

4. D. B. Zahl, S. Schmauder, and R. M. McMeeking, "Transverse strength of metal matrix composites reinforced with strongly bonded continuous fibers in regular arrangements", *Acta Metall.,* Vol. 42, No. 9, pp. 2983–2997 (1994).

5. Z. Z. Du and R. M. McMeeking, "Creep models for metal matrix composites with long brittle fibers", *J. Mech. Phys. Sol.,* Vol. 43, No. 5, pp. 701–726 (1995).

6. W. A. Curtin, "Theory of mechanical properties of ceramic matrix composites", *J. Am. Cer. Soc.,* Vol. 74, p. 2837 (1991).

7. D. B. Marshall, B. N. Cox, and A. G. Evans, "The mechanics of matrix cracking in brittle matrix composites", *Acta Metall.,* Vol. 33, No. 11, pp. 2013–2121 (1985).

8. J. F. Despres and M. Monthioux, "Mechanical properties of C/SiC composites as explained from their interfacial features", *J. European Cer. Soc.,* Vol. 15, pp. 209–224 (1995).

9. D. B. Bogy, "Two edge bonded elastic wedges of different materials and wedge angles under surface tractions", *ASME J. Applied Mechanics,* Vol. 38, pp. 377–386 (1971).

10. B. S. Majumdar and J. A. Ahmad, "Fracture of ceramic-metal joints: zirconia/nodular cast iron system", *Metal-Ceramic Joining, Proc. of TMS Symposium,* Eds. P. Kumar and V. A. Greenhut, TMS, pp. 67–96 (1991).

11. J. Ahmad and B. S. Majumdar, "An engineering fracture mechanics analysis of metal-ceramic and ceramic-ceramic joints", *ASTM STP 1131: Fracture Mechanics: 22nd Symp.* (Volume I), Eds. H. A. Ernst et al., ASTM (1991).

12. D. B. Gundel, B. S. Majumdar, and D. B. Miracle, "The intrinsic transverse response of several SiC/Ti-6A1-4V composites to transverse tension", *Proc. of ICCM-10, Vol. 2,* Ed. A. Poursartip and K. Street, pp. 703–710 (1995).

13. P. Kantzos, J. Eldridge, D. A. Koss, and L. J. Ghosn, "The effect of fatigue loading on the interfacial shear properties of SCS-6/Ti-based MMCs", *MRS Symposium Proceedings,* Vol. 273, pp. 325–330 (1992).

14. B. S. Majumdar and D. B. Miracle, "Interface measurements and applications in fiber-reinforced MMCs", *J. Key Engg. Materials,* Vol. 116–117, pp. 153–172 (1996).

15. J. W. Hutchinson, "Mixed mode fracture mechanics of interfaces", Harvard Technical Report MECH-139, Division of Applied Sciences, Harvard University (1989).

16. J. R. Rice, "Elastic fracture mechanics concepts for interface cracks", *ASME J. of Applied Mechanics,* Vol. 55, pp. 98–103 (1988).

17. D. Singh and D. K. Shetty, "Fracture toughness of polycrystalline ceramics in combined Mode-I and Mode-II loading", *J. Am. Cer. Soc.,* Vol. 72, pp. 78–84 (1989).

18. J. A. Ahmad, "Combined Mode I and Mode II fracture toughness of brittle materials and interfaces", *ASME Journal of Materials and Technology,* Vol. 115, pp. 101–105 (1993).

19. A. R. Rosenfield and B. S. Majumdar," Fracture toughness evaluation of ceramic bonds using a chevron-notch disk specimen", *Chevron Notch Fracture Test Experience: Metals and Non-Metals, ASTM STP 1172,* pp. 63–73 (1992).

20. J. S. Wang, "Interfacial fracture toughness of a copper/alumina system and the effect of the loading phase angle", *Mech. of Materials,* Vol. 20, pp. 251–259 (1995).

21. A. G. Evans and J. W. Hutchinson, "Effects of non-planarity on the mixed mode fracture resistance of bimaterial interfaces", *Acta Metall.,* Vol. 37, No. 3, pp. 909–916 (1989).

22. Z. Suo and J. W. Hutchinson, "Interface crack between two elastic layers", *Int. J. of Fracture,* Vol. 43, pp. 1–18 (1990).

23. P. G. Charalambides and A. G. Evans, "Debonding properties of residually stressed brittle-matrix composites," *J. Am. Ceram. Soc.,* Vol. 72, No. 5, pp. 746–753 (1989).

24. K. S. Chan, "Failure diagrams for unidirectional fiber metal-matrix composite", *Metall. Trans.,* Vol. 24A, pp. 1531–1542 (1992).

25. J. Cook and J. E. Gordon, "A mechanism for the control of crack propagation in all-brittle systems", *Proc. Roy. Soc. London, A 282,* pp. 508–520 (1964).

26. E. Orowan, *Zertschrift fur Physik,* verlag von Julius Springer, Berlin (1933).

27. F. Erdogan, "Fracture problems in composite materials", *Engineering Fracture Mechanics,* Vol. 4, pp. 811–840 (1972).

28. A. R. Zak and M. L. Williams, "Crack point stress singularities at a bimaterial interface", *J. Applied Mechanics, Trans. ASME,* Vol. 30, pp. 142–143 (1963).

29. D. O. Swenson and C. A. Rau, "The stress distribution around a crack perpendicular to an interface between materials", *Int. J. Fract. Mech.,* Vol. 6, pp. 357–360 (1970).

30. T. S. Cook and F. Erdogan, "Stresses in bonded materials with a crack perpendicular to the interface", *Int. J. of Engg. Science,* Vol. 10, pp. 677–697 (1972).

31. M. Y. He and J. W. Hutchinson, "Crack deflection at an interface between dissimilar elastic materials", *Int. J. Sol. and Structures,* Vol. 25, pp. 1053–1067 (1989).

32. K. Kendall, "Transition between cohesive and interfacial failure in a laminate", *Proc. Roy. Soc. London,* Vol. A 344, pp. 287–302 (1975).

33. M. Y. He, A. G. Evans, and J. W. Hutchinson, *Int. J. Sol. and Structures,* Vol. 31, p. 3443 (1994).

34. *Ceramic Source, Vol. 6,* American Ceramic Society, Westerville, OH, p. 351 (1991).

35. M. L. Gambone and F. E. Wawner, "The effect of elevated temperature exposure of composites on the strength distribution of the reinforcing fibers", *MRS Symposium Proc.,* Vol. 350, pp. 111–118 (1994).

36. A. G. Evans, "The mechanical performance of fiber-reinforced ceramic matrix composites", *J. Materials Sc. and Engg.,* Vol. A107, pp. 227–239 (1989).

37. I. S. Raju and J. C. Newman, *Fracture Mechanics: Seventeenth Symposium, ASTM STP 905,* Eds. J. H. Underwood et al., pp. 789–805 (1986).

38. D. B. Marshall, M. C. Shaw, and W. L. Morris, "Measurement of interfacial properties in intermetallic composites", *J. Key Engg. Materials,* Vol. 116–117, 209–228 (1996); Acta Metall., Vol. 40, No. 3, p. 443 (1992).

39. V. Gupta, A. S. Argon, and J. A. Cornie, "Interfaces with controlled toughness as mechanical fuses to isolate fibers from damage", *J. Mat. Sc.,* Vol. 24, pp. 2031–2040 (1989).

40. A. S. Argon, V. Gupta, H. S. Landis, and J. A. Cornie, "Intrinsic toughness of interfaces between SiC coatings and substrates of Si or C fibre", *J. Mat. Sc.,* Vol. 24, pp. 1207–1218 (1989).

41. V. Gupta, A. S. Argon, and Z. Suo, "Crack deflection at an interface between two orthotropic media", *J. Appl. Mech.,* Vol. 59, pp. S79–S87 (1992).

42. B. S. Majumdar, T. E. Matikas, and D. B. Miracle "Experiments and analysis of fiber fragmentation in single and multiple-fiber SiC/Ti-6A1-4V MMCs", *J. of Composites: B,* in press (1997).

43. S. G. Warrier, B. S. Majumdar, and D. B. Miracle, "Interface effects on crack deflection and fiber bridging behavior during fatigue crack growth in titanium matrix composites", *Acta Metall.* in press (1997).

44. M. L. Gambone, "SiC fiber strength after consolidation and heat treatment in Ti-22Al-23Nb matrix composite", *Scripta Metall.,* Vol. 34, No. 3, pp. 507–512 (1996).

45. B. S. Majumdar and D. B. Miracle, "Fiber coatings for a sapphire/gamma-TiAl composite utilizing ductile and barrier metallic layers", *Proc. ICCM-10, 2,* Eds. A. Poursartip and K. Street, Woodhead Publishing Limited, pp. 747–754 (1995).

46. M. Y. He and J. W. Hutchinson, "Kinking of a crack out of an interface", *ASME J. Applied Mechanics,* Vol. 56, pp. 270–278 (1989).

47. J. A. Ahmad, "A micromechanics analysis of cracks in unidirectional composites", *ASME J. Appl. Mech.,* Vol. 58, pp. 964–972 (1992).

48. B. S. Majumdar and G. M. Newaz, "Constituent damage mechanisms in metal matrix composites under fatigue loading, and their effects on fatigue loading", *Mat. Sc. and Engg.,* Vol. A 200, pp. 114–129 (1995).

49. P. K. Brindley and P. A. Bartolotta, "Failure mechanisms during isothermal fatigue of SiC/Ti-24Al-11Nb composites", *Mat. Sc. and Engg.,* Vol. A 200, pp. 55–67 (1995).

50. W. Lee and W. J. Clegg, "The deflection of cracks at interfaces", *J. Key Engg. Materials,* Vol. 116–117, pp. 193–208 (1996).

51. I. E. Reimanis, B. J. Dalgleish, M. Brahy, M. Ruhle, and A. G. Evans, "Effects of plasticity on the crack propagation resistance of a metal/ceramic interface", *Acta Metall.,* Vol. 38, No. 12, pp. 2645–2652 (1990).

52. R. J. Asaro. N. P. Dowd, and C. F. Shih, "Elastic-plastic analysis of cracks on bimaterial interfaces: interfaces with structure", *Mat. Sc. and Engg.,* Vol. A162, pp. 175–192 (1993).

53. P. G. Charalambides, J. Lund, A. G. Evans, and R. M. McMeeking, "A test specimen for determining the fracture resistance of bimaterial interfaces", *ASME J. Applied Mechanics,* Vol. 56, pp. 77–82 (1989).

54. P. S. Chua and M. R. Piggott, "The glass fiber-polymer interface. I. Theoretical considerations for single fiber pullout tests", *Compos. Sc. and Tech.,* Vol. 22, pp. 33–42 (1985).

55. J. K. Wells and P. W. R. Beaumont, "Debonding and pull-out processes in fibrous composites", *J. Materials Sc.,* Vol. 20, pp. 1275–1284 (1985).

56. J. D. Outwater and M. C. Murphy, Paper 11c, *24th Annual Conf. on Composites,* Society of Plastics Ind., NY (1969).

57. L. J. Ghosn, J. I. Eldridge, and P. Kantzos, "Analytical modeling of the interfacial stress state during pushout testing of SCS-6/Ti-based composites", *Acta Metall.,* Vol. 42, No. 11, pp. 3895–3908 (1994).

58. R. J. Kerans and T. A. Parthasarathy, "Theoretical analysis of the fiber pull-out and push-out tests", *J. Amer. Cer. Soc.,* Vol. 74, No. 7, pp. 1585–1596 (1991).

59. J. W. Hutchinson and H. M. Jensen, "Models of fiber debonding and pull-out in brittle matrix composites with friction", *Mechanics of Materials,* Vol. 9, pp. 139–163 (1990).

60. C. Y. Leung, "Fracture based two-way debonding model for discontinuous fibers in elastic matrix", *ASCE J. of Engg. Mech.,* Vol. 118, No. 11, pp. 2298–2318 (1992).

61. D. B. Marshall, M. C. Shaw, and W. L. Morris, "Measurement of debonding and sliding resistance in fiber reinforced intermetallics", *Acta Metall.*, Vol. 40, No. 3, pp. 443–454 (1992).

62. D. B. Marshall, "An indentation method for measuring matrix/fiber frictional stresses in ceramic composites", *J. Amer. Cer. Soc.*, Vol. 67, No. 12, pp. C259–260 (1984).

63. J. I. Eldridge, "Fiber push-out testing of intermetallic matrix composites at elevated temperatures", *MRS Symposium Proceedings*, Vol. 273, pp. 325–330 (1992).

64. D. B. Marshall and W. C. Oliver, "Measurement of interfacial mechanical properties in fiber-reinforced ceramic composites", *J. Amer. Cer. Soc.*, Vol. 70, No. 8, pp. 542–548 (1987).

65. P. D. Warren, T. J. Mackin, and A. G. Evans, "Design, analysis and application of an improved push-through test for the measurement of interface properties in composites", *Acta Metall.*, Vol. 40, No. 6, pp. 1243–1249 (1992).

66. C. Liang and J. W. Hutchinson, "Mechanics of the fiber pushout test", *Mechanics of Materials*, Vol. 14, pp. 207–221 (1993).

67. D. A. Koss, M. N. Kallas, and J. R. Hellman, "Mechanics of interfacial failure during thin-slice pushout tests", *MRS Symposium Proceedings*, Vol. 273, pp. 303–313 (1992).

68. M. N. Kallas, D. A. Koss, H. T. Hahn, and J. R. Hellman, "On the interfacial stress state present in a 'thin-slice' fiber push-out test", *J. Materials Sc.*, Vol. 27, p. 3821 (1992).

69. D. A. Koss, J. R. Hellman and N. M. Kallas, *J. of Metals, No. 3*, pp. 34–37 (1993).

70. M. C. Watson and T. W. Clyne, "The tensioned pushout test for measurement of fiber-matrix interfacial toughness under mixed mode loading", *Mat. Sc. and Engg.*, Vol. 160, pp. 1–5 (1993).

71. A. F. Kalton, C. M. Ward-Close, and T. W. Clyne, "Development of the tensioned push-out test for study of fibre/matrix interfaces", *Composites*, Vol. 2, No. 6 (1994).

72. T. W. Clyne, P. Feillard, and A. F. Kalton, "Interfacial mechanics and macroscopic failure in titanium-based composites", *ASTM Symposium on Life Prediction Methodology for Titanium Matrix Composites*, Hilton Head (March 1994).

73. N. Chandra and C. R. Ananth, "Analysis of interfacial behavior in MMCs and IMCs by the use of thin-slice push-out tests", *Compos. Sc. and Tech.*, Vol. 54, pp. 87–100 (1995).

74. C. R. Ananth and N. Chandra, "Numerical modeling of fiber push-out test in metallic and intermetallic matrix composites—mechanics of the failure process", *J. Compos. Materials*, Vol. 29, No. 2, pp. 1488–1514 (1995).

75. T. W. Clyne and P. J. Withers, *An Introduction to Metal Matrix Composites*, Cambridge University Press, Great Britain (1993).

76. C. J. Yang, S. M. Jeng, and J. M. Yang, "Interfacial properties measurement for SiC fiber-reinforced titanium alloy composites", *Scripta Metall.*, Vol. 24, pp. 469–474 (1990).

77. B. S. Majumdar, unpublished research.

78. M. C. Waterbury, P. Karpur, T. Matikas, S. Krishnamurthy, and D. B. Miracle, "In situ observation of the single fiber fragmentation process in metal matrix composites by ultrasonic imaging", *Compos. Sc. and Tech.*, Vol. 52, No. 2, pp. 261–266 (1994).

79. M. C. Waterbury and S. Krishnamurthy, "Evaluation of interfacial and fiber properties in titanium matrix composites by single fiber fragmentation testing", *Proc. ICCM-10*, Eds. A. Poursartip and K. Street, Woodhead Publishing Limited, Cambridge, England, pp. v.

80. Y. L. Petticorps, R. Pailler, and R. Naslain, "The fiber-matrix interfacial shear strength in titanium alloy matrix composites reinforced by SiC or B CVD filaments", *Compos. Sc. and Tech.*, Vol. 35, pp. 207–214 (1989).

81. W. A. Curtin, "Exact theory of fiber fragmentation in a single-filament composite", *J. Mat. Sc.*, Vol. 26, pp. 5239–5253 (1991).

82. T. Nicholas and J. A. Ahmad, "Modeling fiber breakage in a metal-matrix composite", *Compos. Sc. and Tech.*, Vol. 52, pp. 29–38 (1994).

83. M. J. Iremonger and W. G. Wood, *J. Strain Anal.*, Vol. 5, pp. 212–222 (1970).

84. N. Shafry, D. G. Brandon, and M. Terasaki, "Interfacial friction and debond strength of aligned ceramic matrix composites", *Euro Ceramics*, Vol. 3, p. 3.453 (1989).

85. C. H. Hsueh, "Analysis of slice compression tests for aligned ceramic matrix composites", *Acta Metall.*, Vol. 31, No. 12, pp. 3585–3593 (1993).

86. M. C. Waterbury, D. Tilly, W. Kralik, and D. B. Miracle, "Evaluation of TMC interface properties by the slice compression test", *Proc. ICCM-10*, Ed. A. Poursartip and K. Street, Woodhead Publishing Limited, Cambridge, England, pp. VI-719-726 (1995).

87. S. G. Warrier, B. S. Majumdar and D. B. Miracle, "Determination of interface failure mechanism during transverse loading of single fiber composites from torsion tests", *Acta Metall.*, 45, pp. 1275–1284 (1997).

88. S. G. Warrier, B. S. Majumdar, D. B. Gundel, and D. B. Miracle, "Implications of tangential failure during transverse loading", *Acta Metall.* in press (1997)

89. W. S. Johnson, S. J. Lubowinski, and A. L. Highsmith, "Mechanical characterization of unnotched SCS-6/Ti-15-3 MMC at room temperature", *ASTM STP 1080*, p. 193 (1990).

90. B. S. Majumdar and G. M. Newaz, "Inelastic deformation of metal matrix composites: plasticity and damage mechanisms", *Phil. Mag., Vol. 66*, No. 2, pp. 187–212 (1992).

91. R. P. Nimmer, R. J. Bankert, E. S. Russell, G. A. Smith, and P. K. Wright, "Micromechanical modeling of fiber/matrix interface effects in transversely loaded SiC/Ti-6-4 metal matrix composites". *ASTM J. Compos. Tech. Res., JCTRER*, Vol. 13, pp. 3–13 (1991).

92. L. L. Shaw and D. B. Miracle, *Lightweight Alloys for Aerospace Applications*, Eds. E. W. Lee and K. Jata, TMS (1995).

93. T. E. Matikas, P. Karpur, N. J. Pagano, S. Hu, and D. B. Miracle, *Review*

of Progress in Quantitative NDE, Ed. D. O. Thompson, Plenum Press, p. 14 (1995).

94. R. D. Kurtz and N. J. Pagano, *Composites Engineering,* Vol. 1, pp. 13–27 (1991).

95. D. B. Gundel, B. S. Majumdar, and D. B. Miracle, "Evaluation of the intrinsic transverse response of fiber-reinforced composites using a cross-shaped sample geometry", *Scripta Metall.,* Vol. 33, p. 2057 (1995).

96. S. G. Warrier, D. B. Gundel, B. S. Majumdar, and D. B. Miracle, "Stress distribution in a transversely loaded cross-shaped single fiber SCS-6/Ti-6A1-4V composite", *Scripta Metall.,* Vol. 34 No. 2, p. 293 (1995).

97. S. G. Warrier, D. B. Gundel, B. S. Majumdar, and D. B. Miracle, "Interface Effects on Transversely Loaded Single-Fiber SCS-6/Ti-6A1-4V", *Metall. Trans.,* 27A, pp. 2035–2044 (1996).

98. R. P. Nimmer, "Fiber-matrix interface effects in the presence of thermally induced residual stresses", *J. Compos. Tech. and Research, JC-TRER,* Vol. 12, No. 2, pp. 65–75 (1990).

99. N. I. Muskhelishvilli, *Some Basic Problems of the Mathematical Theory of Elasticity,* Noordhoff, The Netherlands (1963).

100. L. J. Ebert and P. K. Wright, "Mechanical aspects of the interface", *Interfaces in Metal Matrix Composites,* Ed. A. G. Metcalfe, Academic Press, New York (1974).

101. D. F. Adams and D. R. Doner, *J. Compos. Materials,* Vol. 1, pp. 152–164 (1967).

102. S. Morozumi, M. Ondo, M. Kikuchi, and K. Hamajima, "Bonding mechanism between SiC and thin foils of reactive metals", *J. Mat. Sc.,* Vol. 20, pp. 3976–3982 (1995).

103. T. Iseki and T. Yano, *Mater. Sc. Forum,* Vol. 34–36, p. 421 (1988).

104. N. Pagano et al., Unpublished research (1995–1996).

105. S. R. Gunawardena, S. Jansson, and F. A. Leckie, "Modeling of anisotropic behavior of weakly bonded fiber reinforced MMCs", *Acta Metall.,* Vol. 41, pp. 3147–3156 (1993).

106. S. Jansson and F. A. Leckie, "Global-micro mechanical relationships in metal-matrix composites", *Hitemp Review 1992,* NASA Conf. Pub. 10104, pp. 45.1–45.14 (1992).

107. F. H. Gordon and T. W. Clyne, "Thermal cycling creep of Ti-6A1-4V/SiC monofilament composites under transverse loading", *Residual Stresses in Composites, Modeling, Measurements, and Effects on Thermomechanical Properties,* Eds. E. V. Barrera and I. Dutta, TMS, Warrendale, pp. 293–304 (1993).

108. G. P. Carman, R. C. Averill, K. L. Reifsnider, and J. N. Reddy, "Optimization of fiber coatings to minimize stress concentrations in composite materials", *J. Compos. Materials,* Vol. 27, No. 6, pp. 589–611 (1993).

109. C. Boehlert, B. S. Majumdar, S. Krishnamurthy, and D. B. Miracle, "Role of matrix microstructure on RT tensile properties and fiber-strength utilization of an orthorhombic Ti-alloy based composite", *Metall. Trans.,* 28A, pp. 309–323 (1997).

110. J. B. Friler, A. S. Argon, and J. A. Cornie, "Strength and toughness of carbon fiber reinforced aluminum matrix composites", *Mat. Sc. and Engg.*, A162, pp. 143–152 (1993).

111. M. C. Watson and T. W. Clyne, "Reaction induced changes in interfacial and macroscopic mechanical properties of SiC monofilament-reinforced titanium", *Composites*, Vol. 24, No. 3, pp. 222–228 (1993).

112. M. Y. He, A. G. Evans, and W. A. Curtin, "The ultimate tensile strength of metal and ceramic-matrix composites", *Acta Metall.*, Vol. 41, No. 3, pp. 871–878 (1993).

113. W. A. Curtin, "Ultimate strengths of fiber-reinforced ceramics and metals", *Composites*, No. 2, pp. 98–102 (1993).

114. C. H. Weber, X. Chen, S. J. Connell, and F. W. Zok, "On the tensile properties of a fiber reinforced titanium matrix composite—I, unnotched behavior", *Acta Metall.*, Vol. 42, No. 10, pp. 3443–3450 (1994).

115. C. Zweben and B. W. Rosen, "A statistical theory of material strength with application to composite materials", *J. Mech. Phys. of Solids*, pp. 189–206 (1970).

116. D. B. Gundel and F. W. Wawner, "Experimental and theoretical investigation of the tensile strength of SiC/Ti-alloy composites", *Compos. Sc. and Tech.*, in press (1996).

117. S. L. Draper, P. K. Brindley, and M. V. Nathal, "Effect of fiber strength on the room temperature tensile properties of SiC/Ti-24A1-11Nb", *Metall. Trans.* A, pp. 2541–2548 (1992).

118. B. S. Majumdar, T. E. Matikas, and D. B. Miracle, "Effects of the interface on local versus global load sharing behavior in metal matrix composites under longitudinal tension", published in *Proc. of Int. Conf. of Comp. Mat., ICCM-11*, Australia (1997).

119. G. A. Cooper and A. H. Kelly, "Tensile properties of fiber-reinforced metals: fracture mechanics", *J. Mech. Phys. of Solids*, Vol. 15, pp. 279–297 (1967).

120. M. A. Seleznev, A. S. Argon, and J. A. Cornie, "Control of toughness in MMCs through controlled precipitation of a weak intermetallic phase on the fiber-matrix interface", *Proc. First Int. Conf. on Composites Engineering*, ICCE-1, University of New Orleans (1994).

121. R. M. McMeeking and A. G. Evans, *Mechanics of Materials*, Vol. 9, p. 217 (1990).

122. L. N. McCartney, "Mechanics of matrix cracking in brittle matrix fiber reinforced composites", *Proc. Roy. Soc. London*, Vol. A409, pp. 329–350 (1987).

123. J. Aveston, G. A. Cooper, and A. Kelly, "Single and multiple fracture", in *The Properties of Fiber Composites, Conf. Proc.*, pp. 15–26, National Physical Laboratory, IPC Science and Technology Press, UK (1971).

124. L. Ghosn, P. Kantzos, and J. Telesman, "A fiber-pressure model", *Int. J. Fracture*, Vol. 54, p. 345 (1992).

125. B. S. Majumdar, G. M. Newaz, and A. R. Rosenfield, "Yielding behavior of ceramic matrix composites", *Advances in Fracture Research, Proc.*

Seventh Int. Conf. on Fracture, ICF-7, Eds. Salama et al., Pergamon Press, pp. 2805–2814 (1989).

126. S. Danchaivijit and D. K. Shetty, "Matrix cracking in ceramic matrix composites", *J. Amer. Cer. Soc.,* Vol. 76, No. 10, pp. 2497–2504 (1993).

127. B. N. Cox and C. S. Lo, "Simple approximations for bridged cracks in fibrous composites", *Acta Metall.,* Vol. 40, No. 7, pp. 1487–1496 (1992).

128. T. Zhang, D. Zhang, and H. Ghonem, "COD measurements and fiber bridging stress distribution in FCG in TMCs at elevated temperatures", *Proc. First Int. Conf. on Composites Engineering,* ICCE-1, University of New Orleans, Ed. D. Hui, pp. 159–160 (1994).

129. D. J. Buchanan, R. John, and D. A. Johnson, "Crack face displacements and calculation of bridging stresses in the crack wake of TMCs", *Int. J. Fract.,* in press (1997).

130. *Interfaces in Metal Matrix Composites,* Ed. A. G. Metcalfe, Academic Press, New York (1974).

131. A. H. Kelly and N. H. Macmillan, *Strong Solids,* Third Edition, Clarendon Press, Oxford, England (1986).

5

Fatigue Failure Mechanisms

TIMOTHY P. GABB AND JOHN GAYDA

NASA Lewis Research Center
21000 Brookpark Road
MS49-3
Cleveland, OH 44135

INTRODUCTION

THE FATIGUE FAILURE mechanisms of titanium alloy matrix composites and titanium aluminide matrix composites have many commonalities. Isothermal fatigue failure mechanisms associated with fiber-dominated failures, crack initiation at damaged fibers on the specimen edges, and preferential matrix cracking with fiber bridging are activated in 0° fiber-reinforced laminates of both classes of titanium matrix composites (TMCs). Also, fiber-matrix debonding and crack initiations at fiber-matrix interfaces are common in 90° fiber-reinforced laminates of both classes. In both classes, similar environment-assisted surface cracking and fiber overload failure processes can be activated by thermomechanical out-of-phase and in-phase cycling, respectively. But in some cases, the higher strength and environmental resistance of titanium aluminide matrices can allow modest increases in use temperature.

Many load-controlled fatigue studies of TMCs have been conducted in recent years [1]. In these studies, stress is cycled to a constant stress amplitude with a cyclic stress ratio of minimum to maximum stress (R_σ) typically equal to 0 or 0.1. However, only a few strain-controlled fatigue studies have been undertaken [2–4], because of the limitations of thin TMC specimens in supporting compressive loads. Therefore, this review of failure mechanisms draws mostly on load-controlled test data, unless otherwise noted.

169

PHYSICAL METALLURGY

Silicon carbide and boron fibers have been applied as continuous reinforcement in TMCs with some success. The boron fibers are made by chemical vapor deposition onto tungsten cores. Outer carbide coatings of B_4C or SiC have been applied to boron fibers to prevent excessive fiber-matrix reaction [5]. Silicon carbide fiber reinforcements have seen much wider use and development in TMCs. These fibers usually are produced by physical vapor deposition [6] of β-SiC onto pyrolitic carbon cores for SCS (Textron Specialty Materials) fibers or tungsten cores for Sigma fibers [7]. They usually are coated with protective carbon-rich outer layers. The coatings protect the fibers from

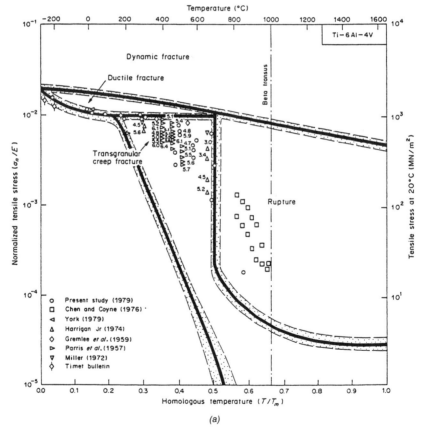

(a)

Figure 1. Fracture mechanism maps for: (a) α + β titanium alloy, and (b) $α_2$ titanium aluminide alloy. (Reprinted from Reference [3], with kind permission from Elsevier Science Ltd, The Boulevard, Langford Lane, Kidlington OX5 1GB, UK.)

abrasion and damage during fiber and composite manufacturing, and they minimize fiber-matrix reactions within the composite. SiC fibers have strengths near 3400 MPa and an elastic modulus near 400 GPa, but they can only withstand strains of about 1.0%. The high-strength, high-stiffness, and low-failure strain of these fibers have major influences on composite mechanical and thermal response.

A variety of titanium and titanium aluminide alloys have been employed as matrices in TMCs. These include the metastable β-titanium alloys Ti-15V-3Cr-3Al-3Sn weight percent (Ti-15-3) [8] and Ti-15Mo-2.7Nb-3.0Al-0.2Si weight percent (Timetal 21S) [9], the α + β titanium alloy Ti-6Al-4V weight percent (Ti-6-4), the α_2 + β titanium aluminide alloys Ti-14Al-21Nb weight percent (Ti-24Al-11 Nb atomic percent, Ti-

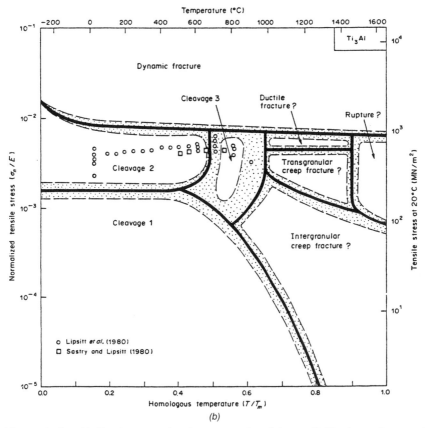

Figure 1. (cont.) Fracture mechanism maps for: *(a)* α + β *titanium alloy, and (b)* α_2 *titanium aluminide alloy. (Reprinted from Reference [3], with kind permission from Elsevier Science Ltd, The Boulevard, Langford Lane, Kidlington OX5 1GB, UK.)*

24-11) and Ti-25Al-10Nb-3V-1Mo atomic percent (Ti-25-10) [10], and the alloys containing orthorhombic α_2 + β phases, such as Ti-22Al-23Nb atomic percent [11]. Matrix microstructures are strongly affected by composite consolidation, which is usually performed at temperatures above the β-transus temperatures of these alloys. The microstructures of these titanium and titanium aluminide alloys can be varied significantly by heat treatment, even after composite consolidation. However, many TMC fatigue properties are not strongly affected by significant differences in matrix microstructure and mechanical properties, provided composite integrity is maintained and the matrix can withstand elongations of at least 1.5 to 2.0% [12]. Monolithic titanium alloys and titanium aluminides have significantly different failure modes (Figure 1) [13], with titanium aluminides having significantly lower ductility at low temperatures and higher strength at higher temperatures than titanium alloys. But the fatigue behaviors of TMCs with these matrices are generally quite similar [14]. This similarity appears to be related to the early fatigue crack initiation and the extended crack propagation with fiber bridging that are observed in many of these TMCs.

The fiber-matrix interface reaction products in TMCs have been extensively characterized. Uncoated SiC fibers consolidated with Ti-6Al-4V tend to form a relatively thin reaction zone, composed mainly of titanium silicides, such as $(Ti,Nb,Al)_5Si_3$. SCS-6 carbon-coated SiC fibers consolidated with Ti-6Al-4V have been shown to form a more complex reaction zone with carbides, such as $(Ti,Nb)C$ and $(Ti,Nb)_3AlC$, along with the silicides [15]. These reaction products also have been identified in titanium aluminide TMCs, such as Ti-24-11 (Figure 2) [16]. Researchers have suggested that the coated boron fibers also produce these carbide and silicide reaction products [5]. During consolidation and subsequent exposures, these reaction zones grow by diffusion-controlled parabolic growth [15]. The reaction zone grows more rapidly in high β-titanium alloys and slowest in orthorhombic and α_2-rich alloys with low β-phase content [17,18]. (B_4C/B)/titanium alloys have been reported to have the highest interfacial shear strength (240 to 275 MPa) among TMCs [15]. SiC-titanium alloys have shear strengths of 120 to 150 MPa, whereas SiC-titanium aluminide TMCs have shear strengths of 100 to 120 MPa [18,19]. The lower interfacial strengths of SiC-reinforced TMCs allows fiber-matrix debonding at low loads in laminates containing fibers oriented at 30° to 90° to the load, but also facilitates the bridging of fatigue cracks by 0° oriented fibers which can impede crack growth.

Titanium and titanium aluminide matrix composites have been most commonly fabricated by foil-fiber mat layup [6], powder cloth-fiber mat

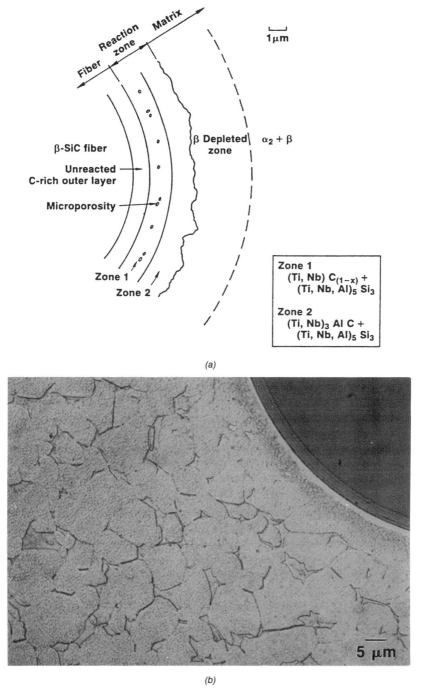

Figure 2. Fiber-matrix reaction zone of SCS-6/Ti-24Al-11Nb (atomic percent): (a) schematic [16], (b) optical micrograph.

173

layup [3,16], and tape cast monotape layup [7]. Hot isostatic pressing or diffusion bonding has then been employed for final consolidation. Comparisons of TMCs with the same fiber and alloy chemistries but produced by these different layup techniques with comparable consolidation conditions generally have shown comparable mechanical properties, when the fiber content is held constant [19–22]. Fiber content in these composites is usually 25 to 40 volume percent.

MONOTONIC RESPONSE AND THERMAL EFFECTS

Monotonic failure mechanisms are considered in depth in the chapter on monotonic behavior. However, several basic aspects of monotonic failure mechanisms in TMCs also apply in fatigue loading and are briefly reviewed here.

The monotonic behavior of 0° fiber-reinforced TMCs is strongly dictated by the fibers. The fibers make the composite very strong, but they limit the composite strains to near 1.0%. In fact, 0° composite strength has been directly correlated with fiber strength, where fiber strength was determined before and after composite processing [23]. However, fiber strength is often reduced by composite consolidation [1,23,24]. The matrix also strengthens the composite somewhat, but this contribution decreases after matrix yielding. Matrix yielding caused by slip band formation at the fiber-matrix interfaces has been demonstrated at tensile strains above 0.4% in 0° fiber-reinforced, 8-ply ($[0]_8$) SCS-6/Ti-15-3 [25,26]. The matrix's contribution to composite strength further decreases with increasing temperature, as matrix yield strength decreases and stress relaxation increases.

Tensile and creep strengths of TMCs reinforced by 90° fibers are limited by the very weak fiber-matrix interfaces of these TMCs [1,27]. The interfaces fail at very low transverse loads and allow very little fiber strengthening; thus, interface and matrix properties ultimately control 90° composite strength. Therefore, 90° composites have very low strengths: 200 to 400 MPa. Matrix yielding caused by slip band formation at fiber-matrix interfaces has been observed here also in 90° fiber-reinforced, 8-ply ($[90]_8$) SCS-6/Ti-15-3 (Figure 3) [26]. Laminates of 0, 90, and 30 to 45° fiber plies have intermediate response, with the 90° fiber plies providing very little strength, the plies with intermediate angle fibers providing some strength, and the 0° fiber plies providing most of the laminate's strength.

The effects of thermal exposures on TMC strength have been briefly assessed in several studies. Exposures at 650°C for 100 hours and at

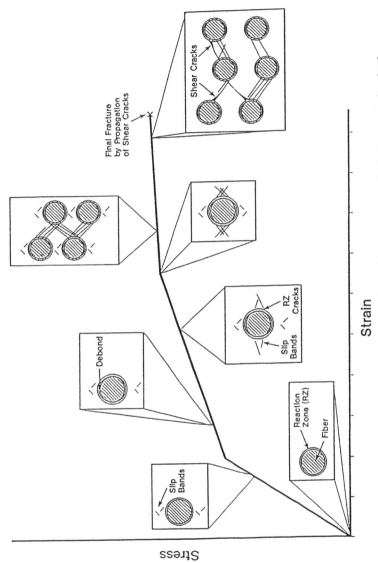

Figure 3. Schematic of deformation and cracking sequence in 90° composite [26].

175

815°C for 10 hours in air degraded the tensile strengths of 0 and 90° fiber SCS-6/Ti-15-3 specimens [28]. Strength reductions of $[0]_8$ specimens were attributed to oxidation and oxygen embrittlement of the matrix, whereas strength reductions of $[90]_8$ specimens appeared related to the degradation of the fiber-matrix reaction zone. Exposure at 800°C for 50 to 100 hours in air also lowered the bending strength of $[0]_8$ SCS-6/Ti-6-4 and SCS-6/Ti-15-3 [29]. Exposures at 900°C for 1 to 100 hours in vacuum reduced the tensile strength of $[0]_8$ SCS-6/Ti-6-4 [30]. Increasing the thickness of the fiber's initial outer carbon layer improved the retention of fiber and composite strength, but it reduced the initial transverse strength. Exposures at 760°C for 500 hours in air did not reduce $[0]_4$ SCS-6/Ti-24-11 tensile strength, but they reduced $[90]_8$ transverse strength [24].

The effects of unconstrained thermal cycling on TMCs have also been scrutinized. Thermal cycling of $[0]_4$ SCS-6/Ti-24-11 in air to maximum temperatures greater than 600°C reduced tensile strength to near zero after 1000 cycles [31]. This degradation was attributed to the formation of surface cracks in the matrix transverse to the fiber direction. The air environment embrittled the matrix; later, strength retention was demonstrated in this TMC when it was cycled over the same temperatures in vacuum [32,33]. $[0]_8$ laminates have had smaller reductions in strength than $[0]_4$ laminates, because a smaller proportion of the cross-sectional area was subjected to environmental embrittlement [34] and less specimen warping induced by thermal cycling occurred [35]. Orthorhombic TMC has shown better resistance to such thermal cycling environmental damage than other TMCs have [11,36]. The fiber-matrix interfaces near the specimen surfaces have shown signs of enhanced oxidation after such cycling in air, often because the fibers intersect with surface cracks. Thermal cycling at 65 to 815°C or exposure at 800°C greatly reduces the fatigue crack growth resistance of $[0]_4$ SCS-6/Ti-25-10 [40], possibly because interface properties are degraded. Away from the environmentally affected regions, reaction zone growth and the initiation of small radial cracks from the reaction zone were observed after both air and vacuum thermal cycling [37–39]. These small radial cracks had little effect on tensile strengths.

The effects of unconstrained thermal cycling have been assessed on other laminates also. Subjecting TMC made up of symmetrically opposed inner plies containing 90° oriented fibers and outer plies containing 0° oriented fibers ($[0/90s]$) SCS-6/TIMETAL®21S to thermal cycles of 150 to 815°C reduced strength by 50 to 80% [41]. Thermal cycles of 150 to 815°C in air and vacuum on [90] and [0/90] laminates of

SCS-6/Ti-24-11 produced smaller strength reductions than for the 0° fiber composite [34]. Surface cracks grew normal to the outer row fibers in each laminate. This indicates that surface cracking induced by unconstrained thermal cycling was apparently related to the combined effects of matrix surface embrittlement and thermal stresses induced in the matrix because of the mismatch between the coefficients of thermal expansion for the matrix and fibers. Cracks also emanated from oxidized 90° fiber-matrix interfaces. Strength was retained when these laminates were thermally cycled in vacuum.

In summary, [0°] TMC strength is strongly dictated by the fibers, whereas [90°] composite strength is controlled by properties of the fiber-matrix interfaces and matrix. Thermal exposures can reduce composite strength through environmental embrittlement of the matrix and growth of the fiber-matrix reaction zone. Thermal cycling reduces composite strength through environmental embrittlement of the matrix and generation of surface cracks normal to the fiber direction. Fiber-matrix interfaces also can be degraded by thermal cycling.

ISOTHERMAL FATIGUE

Longitudinal ([0°]) TMCs

The fatigue life response of longitudinal ([0°]) TMCs as a function of maximum strain or stress often has the general features illustrated in Figure 4 [42]. In contrast to monotonic materials, the lifelines are often not continuous. At maximum strains above 0.65% (Region I), fatigue life is highly variable, lying between 1 and 1000 cycles. At maximum cyclic strains between 0.30 to 0.65% (Region II), fatigue life varies less and consistently increases with decreasing strain. Life at maximum strains below 0.30% (Region III) can exceed 10^6 cycles. This general life response has been observed for SCS-6/Ti-6-4 [1,2,43], SCS-6/Ti-15-3 [44,45], and SCS-6/Ti-24-11 [24,42,46,47].

Similar lifelines have been reported for many polymer matrix composites [48]. In polymer composites, Region I—catastrophic failure— was chiefly due to fiber failures, whereas Region II failure was due to matrix and fiber cracking. Region III was associated with matrix cracking only. In several of the cited TMC studies, fatigue failure mechanisms in Regions I, II, and III have been related to the polymer matrix composite failure modes. Fiber-dominated failures have been observed in Region I for several SiC-titanium composites. Fiber failure was exclusively observed in Region I in $[0]_8$ SCS-6/Ti-15-3 tested

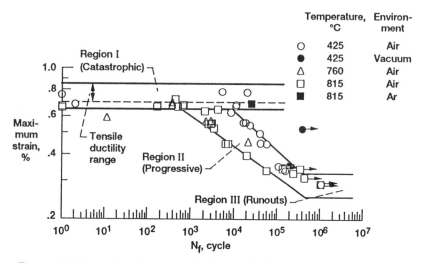

Figure 4. Fatigue life diagram for isothermal fatigue of SCS-6/Ti-24-11 [42].

at 300 to 550°C [44,49], and in [0]$_8$ SCS-6/Ti-6-4 tested at 25 to 370°C
[43] and at 316°C [2]. After sufficient fiber failures occurred in such
tests, the matrix subsequently failed because of ductile overload (Figure 5). Preferential matrix cracking also has been observed in Region
II in these SiC-titanium composites. Preferential matrix cracking occurred with uncracked fibers bridging matrix cracks near the crack
tips and localized fiber-matrix interface debonding in SCS-6/Ti-6-4
[2,43] and SCS-6/Ti-15-3 [50,51]. Fibers in the crack wake were sometimes cracked [52], although not coplanar with the matrix crack (Figure 6). However, matrix and fiber cracking were observed in Regions
I, II, and III for titanium aluminide TMCs, such as SCS-6/Ti-24-11
[20,47,53], often as shown in Figure 7. The titanium aluminide matrix
cracking observed in Region I may be due, in part, to the lower ductility of titanium aluminide alloys in comparison with titanium alloys.
In summary, Region I failure is generally more closely associated with
fiber failure, but Region II failure is related to matrix, fiber, and interface failures at different stages of fatigue crack initiation and propagation. Region III response appears to be often related to preferential matrix cracking.

Fatigue studies have shown that composites have a longer life as a
function of maximum stress than the unreinforced titanium matrix
does. However, composites usually have significantly shorter lives as
a function of maximum strain than the unreinforced matrix does

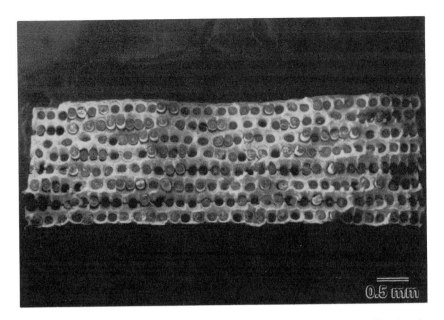

Figure 5. Fatigue fracture surface of $[0]_8$ SCS-6/Ti-15-3 tested in Region I at 550°C [44].

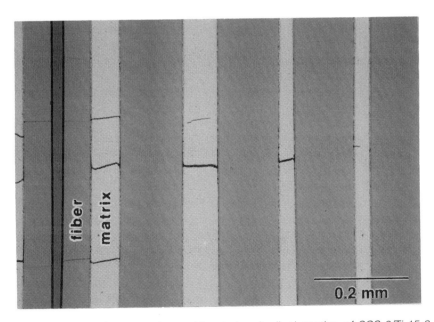

Figure 6. Predominant matrix cracking in longitudinal section of SCS-6/Ti-15-3 tested in Region II at 300°C [44].

179

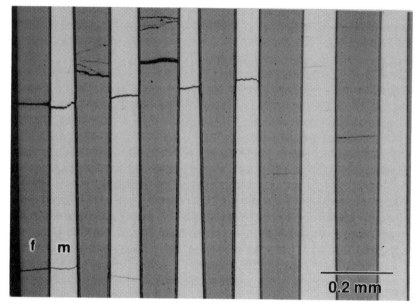

Figure 7. Matrix and fiber cracking in longitudinal section of [0]₄ SCS-6/Ti-22-23 fatigue tested in Region I at 427°C [92].

[1,54]. In many studies, this shorter life has been related to the specific crack initiation mechanisms in the TMCs. Fatigue cracks often initiate at damaged fibers along the machined specimen edges in SiC-Ti-6-4 [1,2,43] and SCS-6/Ti-15-3 [55,44,45]. In isothermal tests performed from room temperature to 650°C, these cracks initiated at the interfaces of damaged fibers on the specimen edges. Then, they preferentially grew in the matrix. From room temperature to 650°C, initiations at damaged edge fibers also predominate in isothermal fatigue of titanium aluminide matrix composites, such as SCS-6/Ti-24-11 [20,47,53,56], SCS-6/Ti-25-10-3-1 [43,56], and SCS-6/Ti-22-23 [22] (Figure 8). The edge fibers can be considered to be large defects promoting early crack initiations that are not present in the monolithic titanium alloys.

However, fatigue cracks do not initiate only at the edges of TMC specimens. They can initiate on specimen sides at surface imperfections in SCS-6/Ti-15-3 from 25 to 300°C [52,57]. Crack initiations also have been observed along specimen sides in SCS-6/Ti-24-11 at higher temperatures of 760 and 815°C [20,58]. These cracks are largely due to environmental effects. Crack initiations at internal fiber-matrix in-

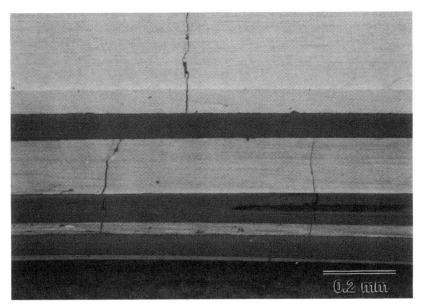

Figure 8. Fatigue cracks initiated at specimens edges in [0]₄ SCS-6/Ti-22-23 tested at 650°C [22].

terfaces have been observed for SCS-6/Ti-15-3 at 300 to 550°C [44,45,49] (Figure 9).

Detailed metallographic evaluations of fatigue-tested SCS-6/Ti-15-3 [26,49] indicate that matrix plastic deformation is often concentrated as slip bands near fiber-matrix interfaces (Figure 10). Composite processing induces radial cracks in the fiber-matrix reaction zones [38], which can initiate such slip bands. These cracks were postulated to propagate from the fiber matrix interface along these slip bands. A similar, enhanced matrix deformation and cracking process can be expected to occur preferentially at damaged fibers on the specimen edge, where machining produces many fiber-matrix reaction zone cracks.

Evaluations of interrupted tests indicate that fatigue cracks can initiate in the first 10% of life in $[0]_8$ SCS-6/Ti-15-3 at 150°C [52]. Similar evidence for early crack initiations has been observed by counting striations [56] and by evaluating interrupted fatigue tests of $[0]_4$ SCS-6/Ti-24-11 tested at maximum stresses of 700 to 1000 MPa and temperatures of 650 to 815°C [20]. Therefore, it appears that cracks often initiate very early in TMCs and that a large portion of TMC isothermal fatigue life is associated with crack propaga-

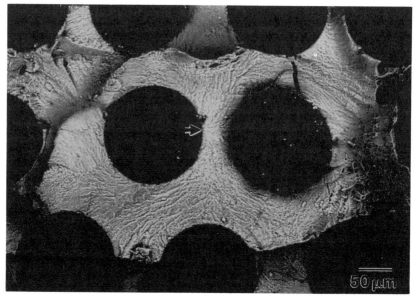

Figure 9. *Fracture surface of [0]₈ SCS-6/Ti-15-3 fatigue tested at 550°C, showing internal fatigue cracks initiated at fiber-matrix interfaces [52].*

tion, even for low applied stresses that produce lives greater than 10^5 cycles.

Specific mechanisms of isothermal fatigue crack propagation and associated models are covered in the chapter on fatigue crack growth. However, because cracks tend to initiate early in TMCs, a brief review is necessary to understand composite fatigue life and failure mechanisms. Early work showed that cracks can propagate in a continuous manner ("self similar") through both the matrix and fibers in B_4C-B-titanium alloys [55]. Conversely, cracks often grow preferentially through the matrix in SiC-titanium composites. This has been observed by metallographic sectioning of fatigue cracks in interrupted or failed test specimens of SCS-6/Ti-6-4 [2,43], SCS-6/Ti-15-3 [44,45, 49,57], and SCS-6/Ti-24-11 [20,47,53,56]. In addition, this crack growth behavior has been inferred from numerous other studies, where observations of fiber fracture surfaces show that they are noncoplanar with matrix fracture surfaces. The difference between the crack propagation behaviors of (B_4C-B)-titanium and SiC-titanium composites can be attributed to the 50% lower shear strength of the fiber-matrix interfaces in SiC-Ti composites [18]. The interfaces in

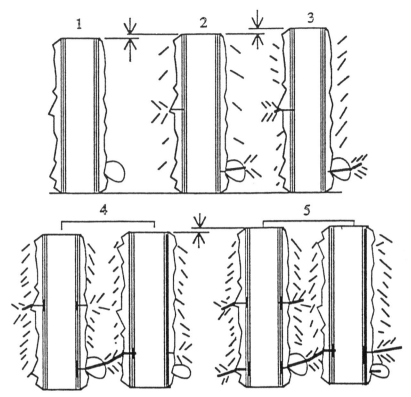

Figure 10. Schematic illustration of fatigue cracks initiating at fiber matrix interfaces in 0° TMC [49].

SiC-Ti composites are not strong enough to support crack growth into the stronger fibers. Instead, a localized area of the weak interface between the fiber and matrix debonds. A matrix crack can then grow around and past the fiber, whereas the uncracked fiber remains behind the crack tip, bridging the crack and absorbing a large portion of the crack growth-driving force. Crack growth rates in [0°] SiC-titanium composites are much slower than in the unreinforced matrix, and total crack arrest can occur [59,60]. Therefore, although TMC fatigue cracks often initiate early, crack propagation life can be very long [61]. It should be noted that such fatigue cracking can significantly reduce TMC stiffness [57], and this could limit useful life in some applications.

TMC fatigue life decreases with increasing temperature, typically as shown in Figure 4. Region I—catastrophic life response—may not be inherently temperature dependent in strain-controlled tests [3]. However, in load-controlled tests, Region I response can occur at decreasing maximum stresses with increasing temperature [2,44]. Cyclic stress relaxation in the matrix increases with increasing temperature. This relaxation shifts a larger portion of the applied load to the fibers, and at high applied stresses, can cause fiber-dominated failures with no matrix cracking. Temperature dependence of Region II life has been demonstrated for SiC-Ti-6-4 [1], SCS-6/Ti-15-3 [44], and SCS-6/Ti-24-11 [42]. This dependence could be due in large part to environmental effects activated by increasing temperature. Local areas of the matrix can be embrittled by oxygen in solid solution when additional α-phase is formed in titanium alloys [62]. More generally, oxygen reduces the ductility and increases the hardness of titanium and titanium aluminide alloys [32,63,64]. The reduced ductility promotes surface fatigue crack initiation in the matrix. In addition, the fiber-matrix interfaces are susceptible to oxidation at high temperatures. This can promote crack initiations at the exposed interfaces on specimen edges, shortening initiation life.

Propagation life also could decrease with increasing temperature. Interface shear strength, friction strength, and fiber strength have all been shown to decrease for fibers near macroscopic TMC cracks at high temperatures [65] because of the reaction of the carbon layer, abrasion during fatigue, and oxidation. These factors could reduce the effectiveness of fiber bridging in slowing crack growth. It is interesting that there have been successful correlations of TMC life with maximum fiber stress [43,55,57], even though there is no reported evidence of fatigue damage or cracking within fibers. Rather, the fiber stress-life relationship appears to be due to oxidation, abrasion, and matrix attack on fiber surfaces, which all can occur during the fatigue of TMCs. These factors can influence fiber strength and TMC life at high applied stresses.

Several studies have shown that cycle frequencies below 0.1 Hz and dwells at constant stress can reduce isothermal fatigue life in SCS-6/Ti-24-11 [47,53]. Dwells and lower frequencies at minimum stress allow more time per cycle for oxidation and environmental embrittlement of the matrix that enhance surface crack initiation. Dwells and lower frequencies at maximum stress can also encourage more cyclic stress relaxation in the matrix, shifting more load to the fibers. In load-controlled tests, this causes the maximum strain to ratchet upward with little decrease in composite stiffness. Each of these time-dependent mechanisms can reduce TMC life. However, the resulting TMC lives are not purely time dependent but are both cycle and time dependent in SCS-6/Ti-15-3 [66] and SCS-6/Ti-24-11 [47].

Transverse ([90°]) and Cross-Ply Laminates

Transverse ([90°]) TMC life is several orders of magnitude shorter than [0°] life in SCS-6/Ti-6-4 [1,2], SCS-6/Ti-15-3 [67,68], and SCS-6/Ti-24-11 [24]. [90°] TMC life is also lower than monolithic matrix life. This low life is due to the weak transverse fiber-matrix interfaces in [90°] TMCs. As previously observed in monotonic loading, the fiber-matrix interface has very low strength, and transverse fibers debond at very low applied loads [27,67]. Fatigue cracks often initiate in the matrix perpendicular to the loading direction at these weak interfaces (Figure 11). This crack initiation mechanism is enhanced at higher temperatures in air by preferential oxidation of the fiber-matrix interfaces. Rapid oxygen transport can occur along fiber-matrix interfaces, allowing oxidation of the fiber-matrix interfaces across the entire specimen width in some cases [68]. Crack propagation life can be much shorter in [90°] TMCs as well because the easily debonded fibers allow cracks to propagate more rapidly [59]. Oxygen accelerates fatigue crack growth rates in such conditions [70]. Therefore, although relatively weak fiber-matrix interfaces can help extend the [0°] TMC crack propagation life through fiber bridging, the weak in-

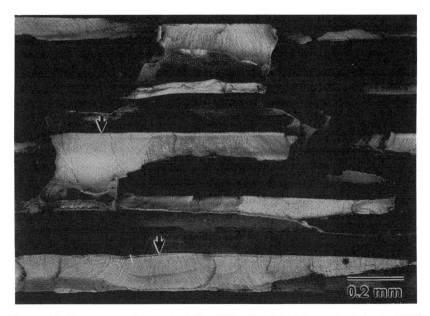

Figure 11. Fracture surface of [90]₈ SCS-6/Ti-15-3 fatigue tested at 427°C, showing internal fatigue cracks initiated at fiber-matrix interfaces [67].

terfaces in [90°] TMCs promote crack initiation and growth, thereby limiting life. Stronger or more compliant outer fiber coatings of silver and tantalum have not improved [90°] fatigue crack growth resistance [70], because the carbon coatings in SCS-6 fibers also are very weak and still allow easy debonding.

The isothermal fatigue failure mechanisms of laminates, such as $[0/90]_s$ and $[0/45/90]_s$, are related to the fatigue failure processes of individual plies. The failure response of plies with 90° fibers is dictated by rapid debonding of the 90° fibers, oxidation of the fiber-matrix interfaces, and early initiation of cracks in the matrix at transverse fiber-matrix interfaces. Plies with fibers at intermediate angles, such as 45°, are subject to some fiber debonding at intermediate stress levels. This damage and the associated early reductions in stiffness have been reported for $[0/90]_s$ and $[0/45/90]_s$ laminates of SCS-6/Ti-15-3 [57] and for $[0/90]_s$ SCS-6/TIMETAL®21S [71]. The remaining plies with 0° fibers ultimately provide the residual composite strength and fatigue crack growth resistance. Laminate lives have been correlated with the maximum effective stresses in 0° fibers in SCS-6/Ti-15-3 [55,57]. However, laminate life may be ultimately limited by the

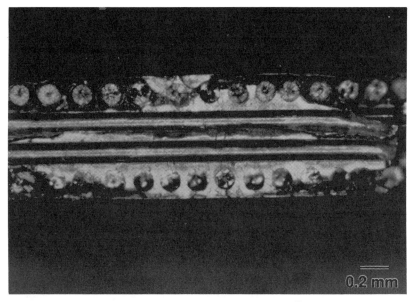

Figure 12. Fracture surface of $[0/90]_s$ SCS-6/TIMETAL®21S fatigue tested at 650°C [71].

growth of cracks initiated in the matrix at the transverse fiber-matrix interfaces [71] (Figure 12). Eventually, the 0° fiber plies can be overcome by tensile overload, environmental attack at adjacent cracks from 90° fibers, and matrix relaxation.

In summary, isothermal fatigue life response as a function of decreasing maximum applied stress and strain can be divided into three regions. Region I—catastrophic failure—is associated with fiber failure. Region II—progressive failure—is associated with matrix, fiber, and interface cracking. Finally, Region III—runout—is associated with matrix cracking. The fatigue life of [0°] TMCs as a function of maximum applied stress generally exceeds that of the unreinforced matrix, because the TMCs are much stronger. However, the fatigue life of [0°] TMCs as a function of maximum applied strain is usually lower than that of the matrix, because of early crack initiations at weak locations, such as damaged edge fibers. Yet, subsequent crack propagation is slower than in the unreinforced matrix, because the crack is bridged by fibers. TMC life decreases with increasing temperature because of increasing environmental embrittlement and matrix relaxation. The lives of [90°] and cross-ply laminate TMCs are limited by enhanced debonding, crack initiation, and crack propagation along 90° fiber-matrix interfaces.

THERMOMECHANICAL FATIGUE

The failure mechanisms of thermomechanical fatigue (TMF) will be described here and compared with isothermal fatigue failure mechanisms reviewed in the previous section. The lives and mechanisms associated with in-phase (IP) and out-of-phase (OP) TMF will be discussed. Stress and temperature are increased simultaneously for IP-TMF, with maximum stress occurring at maximum temperature. Stress is increased as temperature is decreased for OP-TMF, with maximum stress occurring at minimum temperature. In the similar OP nonisothermal cycle, a stress cycle is first applied at minimum temperature. The stress is then maintained at zero as a temperature cycle is applied. In this way, the OP nonisothermal cycle also applies maximum stress at minimum temperature.

Both the OP and IP failure mechanisms are temperature activated. At sufficiently low temperatures, the life responses and failure mechanisms of TMF tests are comparable with those of isothermal tests performed at the same maximum temperature. However, above a transition temperature, TMF lives are often lower than isothermal fatigue lives as more damaging TMF failure mechanisms emerge. This transition temperature depends on the specific TMC system but often lies between 450 and 550°C. Above the transition temperature, both isothermal and TMF lives also become more time dependent.

Longitudinal ([0°]) TMCs

At temperatures above approximately 500°C and low applied stresses, OP-TMF cycles produce shorter lives than isothermal fatigue and IP-TMF. This poor OP-TMF life response has been attributed to the extensive environment-assisted surface cracking produced in the matrix, as shown in Figure 13. Such behavior has been observed in SCS-6/Ti-15-3 [51], SCS-6/TIMETAL®21S [72,73], SCS-6/Ti-6-4 [74], and SCS-6/Ti-24-11 [75]. The same behavior has been observed in OP nonisothermal tests of SCS-6/Ti-15-3 [52] and SCS-6/Ti-24-11 [53]. The OP-TMF and OP nonisothermal lives were shown to be quite comparable in these two systems [14]. Both TMF and non-isothermal tests are typically performed at a low frequency near 0.005 Hz, which allows time for matrix embrittlement. Moreover, when OP nonisothermal tests of SCS-6/Ti-15-3 were performed in vacuum [76] and OP-TMF tests of SCS-6/Ti-24-11 were performed in gettered argon [77], the surface cracking was suppressed and lives were greatly increased. These inert-environment results indicate an environment-dependent process strongly influences OP-TMF and OP nonisothermal fatigue lives.

In titanium and titanium aluminide alloys, the matrix embrittlement mechanism appears related to oxygen-related, solid solution embrittlement along with phase precipitation in titanium and titanium aluminide alloys [63,79,80]. Several [0°] SCS-6-reinforced titanium and titanium aluminide composites had enhanced surface cracking after out-of-phase nonisothermal fatigue loading at a maximum temperature of 815°C [63]. On all these composites, the microhardness was about 3 times higher, and the oxygen levels were elevated near the specimen surface and surface crack faces. This variation could be partly attributed to precipitation of additional α-phase near the specimen surface of $\alpha + \beta$ alloys and of additional α_2-phase near the specimen surface of titanium aluminide alloys [81] due to oxygen stabilization of these precipitate phases. However, titanium alloys with initial α-content of 17 and 78 volume percent had comparable surface hardening and oxygen enrichment. Titanium aluminide alloys with initial α_2-contents of 25 and 87 volume percent also had similar surface hardening response and oxygen enrichment. Therefore, the em-

Figure 13. Fracture surface of [0]₈ SCS-6/Ti-15-3 tested in OP-TMF at 93 to 538°C [14].

brittlement appears to be associated with increased oxygen in solid solution of titanium and titanium aluminide alloy phases.

This OP failure mechanism is more dependent on maximum temperature than temperature range in studies of SCS-6/Ti-15-3 [52], SCS-6/TIMETAL®21S [73], and SCS-6/Ti-24-11 [58,75]. Lower cycle frequencies also decreased OP life in these. Figure 14 shows a typical example of thermomechanical fatigue life response versus maximum applied stress [82]. The slope of the OP lifeline is quite steep, with only modest improvements in life with decreasing stress. This signifies the strong environment-related time dependence of this mechanism. However, OP-TMF life was both cycle and time dependent in these studies.

During OP-TMF cycling, this surface cracking occurs early, consistent with observations of rapid matrix embrittlement at high temperatures. Oxide spikes and cracks formed along the matrix surface in the first 10% of OP life for SCS-6/Ti-15-3 [52], SCS-6/TIMETAL®21S [72], and Ti-24-11 [77]. These surface cracks produced early reductions in composite stiffness and thermal expansion. The surface cracks propagated preferentially through the matrix and debonded

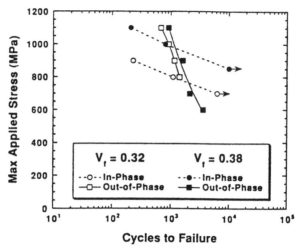

Cycles to Failure

Figure 14. Thermomechanical fatigue life for $[0]_4$ SCS-6/TIMETAL®21S tested at 150 to 650°C. (Reprinted from Reference [82], copyright ASTM, with permission.)

the fiber-matrix interfaces that were in the path of the matrix cracks. The cracks then grew around and past the fibers, and the fibers bridged the cracks, typically as shown in Figure 15.

These results have emphasized the strong impact of environmental embrittlement of the matrix on OP-TMF life. However, OP-TMF life has been shown to be shorter than isothermal life at the same maximum temperature and frequency in SCS-6/Ti-15-3 [51] and SCS-6/Ti-24-11 [53]. Therefore, the load-temperature phasing of OP-TMF cycles also appears to contribute to the shorter life. This phasing has been related to thermal mismatch stresses in many discussions of TMF in TMCs. Tensile axial stresses are generated in the matrix at low temperature because of the mismatch between the coefficients of thermal expansion of the matrix and fibers. An axial tensile stress of about 260 MPa would be generated in the matrix upon cooling SCS-6/Ti-15-3 from consolidation temperature to room temperature [54], and similar stresses have been determined for SCS-6/Ti-24-11 [78]. For OP cycling, the applied tensile stress and thermal stress in the axial direction are additive in the matrix.

However, the collective experimental results suggest that above 500°C the OP failure mechanism of TMCs requires at a minimum the following: (1) oxygen uptake at high temperatures, to reduce low-temperature matrix ductility and (2) the application of a high stress at

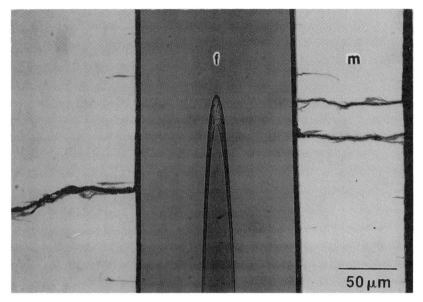

Figure 15. Surface fatigue cracks in longitudinal section of [0]₈ SCS-6/Ti-24-11 tested under OP-nonisothermal fatigue at 150–815°C [58].

low temperature, to fracture the embrittled matrix [14]. These two requirements, repetitive time-dependent high temperature exposure followed by cycle-dependent high stress application at low temperature, are unique to OP-TMF and OP-nonisothermal loading. This process is both time dependent and cycle dependent, as observed for OP-TMF and OP-nonisothermal lives. Although stresses caused by thermal mismatch between the fiber and matrix may add to the problem, testing in vacuum and argon has shown that this factor is of lesser importance.

At temperatures above 500°C, IP-TMF lifelines in TMCs have shallower slopes than OP-TMF lines. These TMF lifelines intersect, such that IP-TMF cycling limits TMC life at higher applied stresses (Figure 14). This was first observed in SCS-6/Ti-15-3 [51,83] and later in SCS-6/Ti-6-4 [74], SCS-6/TIMETAL®21S [82,72], and SCS-6/Ti-24-11 [75]. IP-TMF damage at high applied stresses has been associated mainly with fiber failures and subsequent ductile overload of the matrix, often as shown in Figures 16 and 17. The mechanism encouraging this failure mode has been fairly well characterized. With in-phase cycling, cyclic stress relaxation occurs in the matrix at high

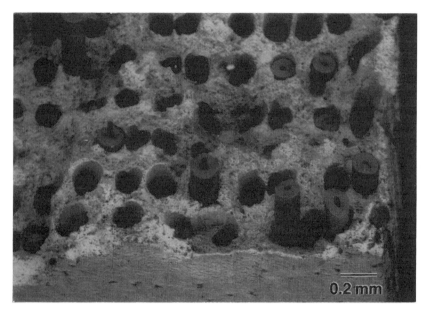

Figure 16. Fracture surface of $[0]_8$ SCS-6/Ti-24-11 IP-TMF tested at 150 to 650°C [14].

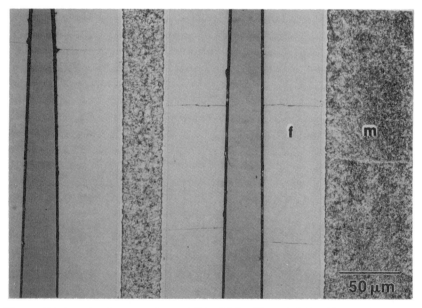

Figure 17. Longitudinal section of $[0]_8$ SCS-6/Ti-15-3 IP-TMF tested at 93 to 538°C [14].

temperatures, in a manner similar to that of isothermal fatigue tests at low frequencies or with dwells at maximum load. The high loads occurring at high temperatures produce considerable matrix relaxation during in-phase loading near peak stress and temperature. In load-controlled tests, this process increases the maximum stress and strain in the fibers. This is indicated by an increase in maximum strain during IP-TMF cycling due to the uniformity of axial strain in the composite. Matrix relaxation occurs fairly rapidly at the high temperatures that have been typically examined, so that a rapid initial increase in maximum strain is nearly stabilized by 10% of life in SiC-Ti-15-3 [51]. Lower cycle frequencies further enhance this initial response [83]. Interrupted tests and acoustic emission measurements during IP-TMF tests [84] indicate fiber fractures occur throughout life. The acoustic emission tests indicate a high initial fiber fracture rate and then a lower steady-state rate until near failure (Figure 18). Strain range and modulus were relatively unaffected by cycling until late in IP-TMF life, in spite of the fiber cracking [72,75]. Debonding has only been observed locally at fiber cracks. Modeling [85] has indicated that full-load transfer can still occur between the matrix and the cracked fiber segments, for the segment lengths observed during

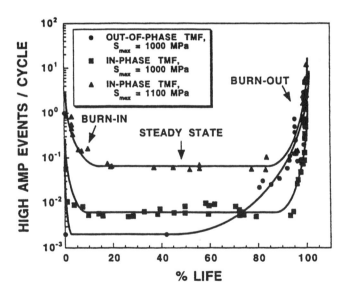

Figure 18. High-amplitude acoustic emission events associated with fiber failures during TMF of [0]₄ SCS-6/TIMETAL®21S at 150 to 650°C. (Reprinted from Reference [84], with kind permission from Elsevier Science Ltd, The Boulevard, Langford Lane, Kidlington OX5 1GB, UK.)

IP-TMF cycling. In this fiber-dominated failure regimen, IP-TMF life varies directly with fiber content in SCS-6/Ti-15-3 [83] and SCS-6/TIMETAL®21S [82]. Accordingly, IP-TMF life probably also varies with fiber strength, although this has not been verified.

To summarize the TMF behavior of [0°] TMCs, OP-TMF cycling produces shorter TMC life than IP-TMF and isothermal cycling at low applied stresses. The critical mechanism in OP-TMF is enhanced surface cracking in the matrix, which is driven by the combination of environmental embrittlement of the matrix at high temperatures and the application of tensile stresses to crack the embrittled matrix at low temperatures. However, IP-TMF cycling produces a shorter life at high applied stresses, due to stress relaxation in the matrix that leads to fiber overload.

Transverse ([90°]) and Cross-Ply Laminates

Few TMF studies on [90°] laminates have been published. The available data suggest TMF life of [90°] is not strongly degraded from isothermal [90°] life. In fact, isothermal life was slightly lower than both IP and OP-TMF lives at the same maximum temperature for SCS-6/Ti-15-3 [68]. IP and OP-TMF lives were comparable in this study, and the fiber-matrix interface debonded at low applied loads in initial loading for both IP and OP cycles. Cracks initiated at fiber-matrix interfaces in both TMF cycles at the 9 and 3 o'clock positions and grew normal to the loading axis. Isothermal fatigue tests of [90]$_8$ SCS-6/Ti-15-3 previously indicated these same damage mechanisms and no frequency dependence of cyclic life [67,68], indicating that this damage may be more cycle dependent than time dependent.

In cross-ply laminates, such as [0/90]$_s$ the TMF lives lie between those of the [0°] and [90°] laminates. The TMF damage mechanisms are consistent with the failure mechanisms of individual plies. Initial loading again produces early debonding of many transverse fibers, and higher loads debond some fibers oriented at 45° to the loading axis [86,87]. OP-TMF cycling promotes crack initiations at the interfaces of the transverse fibers and at the surface of the matrix in SCS-6/TIMETAL®21S and SCS-6/Ti-15-3 [88], typically as shown in Figure 19. These cracks begin initiating within only 7% of OP life in SCS-6/TIMETAL®21S [86]. All transverse fibers appeared to be debonded at this early point in life. This produced early reductions in stiffness and thermal expansion. The cracks growing from the transverse interfaces and the surface often converged at the 0° fibers. Hence, the

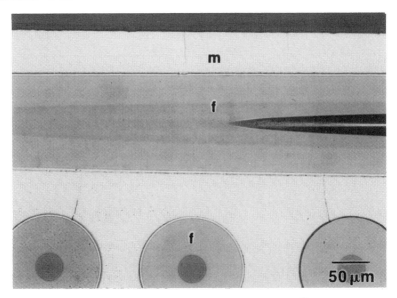

Figure 19. Longitudinal section of [0/90]$_s$ SCS-6/TIMETAL®21S OP-TMF tested at 150 to 650°C, showing matrix cracks initiated at both surface and 90° fiber-matrix interface locations [86].

plies containing 0° fibers ultimately influence OP life, and OP stress-life response has a similar slope to that of [0°] unidirectional TMC.

IP-TMF cycling of the laminates has not produced significant matrix cracking at high applied stresses in [0/90] SCS-6/Ti-15-3 [88] and [0/90]$_s$ and [0/145/90]$_s$ SCS-6/TIMETAL®21S [82,86,87]. Debonding of 90° fibers still occurs, and this can affect modulus and thermal expansion. But matrix cracking can be suppressed by cyclic stress relaxation of the matrix during IP-TMF cycling. This cyclic relaxation again increases 0° fiber stresses, and 0° fiber overload ultimately produces failure. Due to the predominance of fiber overload, the IP-TMF life of laminates is shorter than OP life at high applied stresses, with a slope similar to that observed for 0° laminates. Cracking of 0° fibers begins early and has been proven to occur in the first 20% of cyclic life in SCS-6/TIMETAL®21S [0/90]$_s$ [86]. In these studies, this produced an early increase in maximum mechanical strain during cycling in these studies.

To summarize the TMF behavior of [90°] laminates, OP-TMF and IP-TMF cycling causes fiber-matrix debonding and matrix cracking at fiber-matrix interfaces as in isothermal cycling. Thus, the TMF and isothermal lives were roughly equivalent.

Summarizing [0/90] laminate behavior, OP-TMF causes the 90° fibers to debond, and matrix cracks initiate at both the 90° fiber interfaces and the exterior surface. IP-TMF also causes 90° fiber debonding, but matrix cracking is suppressed by matrix stress relaxation. Because of these factors, both OP and IP-TMF lives are eventually limited by the plies containing 0° fibers. This could explain why OP and IP stress-life response lines have similar slopes to [0°] unidirectional laminates.

MISSION CYCLES

Eventual airframe and engine service applications of TMCs will impose complex combinations of cyclic load and temperature changes. The failure mechanisms that limit life in such applications may be difficult to infer from simple isothermal or conventional TMF tests. Simulated mission cycle tests were performed on [0/90]$_s$ SCS-6/Ti-15-3 [89] to simulate a hypersonic airframe application. Such an application would require a life of only several hundred cycles. This mission cycle had IP-TMF loading characteristics of maximum stress applied at maximum temperature, combined with a long dwell near maximum stress at a maximum temperature of 593°C. Tests of this mission and different components within the mission indicated that the isothermal dwell was the most damaging component of the mission. The isothermal dwell induced excessive debonding and matrix crack initiations at 90° fibers, limiting life as previously observed in isothermal tests. [0/90]$_s$ SCS-6/TIMETAL®21S was later subjected to this same simulated hypersonic cruise mission preceded by a simulated hypersonic ascent mission [90]. The combined experimental missions (Figure 20) illustrate how complex service missions can be. The orbital mission had OP-TMF characteristics, with a maximum stress cycle applied at near minimum temperature, and a slow thermal cycle through 815°C at low stress. Separate mission tests indicated that the ascent mission was more damaging than the cruise mission. This study and independent work on [0]$_8$ SCS-6/Ti-24-11 [58] indicated that the ascent mission was primarily limited by the load cycle near the minimum temperature in combination with the slow thermal cycle to 815°C. The associated damage mechanism was environment-assisted surface cracking in the matrix of [0]$_8$ SCS-6/Ti-24-11, which could be reproduced by OP-nonisothermal cycling at the same fre-

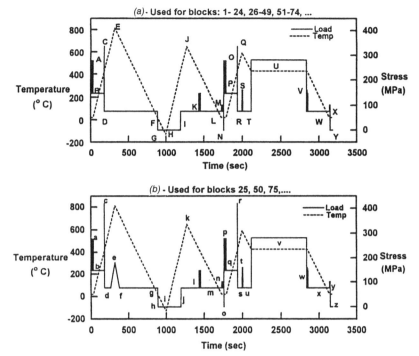

Figure 20. Schematic plot of combined hypersonic ascent-cruise mission cycle tests performed on [0/90]ₛ SCS-6/TIMETAL®21S [90].

quency. The mission produced considerable oxidation along the 90° fibers in $[0/90]_s$ SCS-6/TIMETAL®21S, and it promoted matrix crack initiations at these interfaces and the specimen surface.

Simulated mission cycle tests have also been performed to assess applications of [0°] TMCs as reinforcing rings in compressor rotors of military gas turbine engines. Such applications require lives of at least several thousand cycles at maximum stresses of 900 to 1100 MPa. $[0]_8$ SCS-6/Ti-6-4 was subjected to mission cycles that simulated a TMC ring in a compressor impeller of an advanced helicopter engine [91]. Maximum stress was applied at minimum temperature, and the maximum temperature was 428°C. Both stress and temperature dwells were present. Cracks initiated at damaged fibers on the specimen edges and at surface defects such as thermocouple spot welds. Environmental embrittlement was not dominant at this temperature, and strain ratcheting due to matrix relaxation was also minimal. $[0]_4$ SCS-6/Ti-22-23 was tested in simulated mission cycles to assess a

TMC ring application in a compressor rotor of an advanced turbine engine [92,93]. These mission cycles had maximum stress applied at low to intermediate temperatures and maximum temperatures of 482 to 593°C. Repetitive stress and temperature dwells were present in this mission. Cracks initiated at the damaged edge fibers in these tests as in the previous simulated mission. Additionally, matrix stress relaxation promoted ratcheting of the maximum strain and reduced fatigue cracking in comparison with isothermal tests at the same maximum stress and temperature. Fiber stresses apparently increased to encourage intervening fiber-dominated failure.

In summary, cyclic tests that simulate service missions can produce shorter lives than isothermal or TMF cycles. The operative fatigue damage modes are similar to those of isothermal and TMF cycles but can be accentuated because of the longer dwells and complex combined temperature-stress cycling of service missions.

CONCLUDING REMARKS: IMPROVING THE FATIGUE RESISTANCE OF TMCS

In their present state of development, TMCs are probably limited to maximum temperatures of near 550°C for man-rated aerospace applications that require thousands of fatigue cycles in air. TMF and mission cycle damage mechanisms greatly curtail durability at higher temperatures, producing unrealistically short lives at practical applied stresses.

Identification of the dominant fatigue failure mechanisms can help in developing strategies to improve the fatigue resistance and temperature capability of TMCs. The overall crack initiation life of the matrix would be expected to improve with increasing matrix tensile strength and ductility. However, heat treatments that significantly varied the mechanical properties of the matrix in $[0]_8$ SCS-6/Ti-15-3 and the resulting fatigue resistance of the matrix did not appear to strongly affect the composite fatigue resistance [12]. This apparent contradiction appears to be related to the early initiation of cracks in TMCs and the longer propagation life brought on when cracks are bridged by fibers. Thus, specific TMC fatigue initiation mechanisms and sites need to be altered. Fully encasing the TMC within an outer layer of matrix to eliminate damaged fibers along specimen edges might improve life. Yet, other competing surface and internal crack initiation mechanisms may not allow much improvement in life [94,95]. In such encased specimens, surface cracks might propagate more rapidly in the matrix encasement than in the TMC core. Crack initiations at fiber-matrix interfaces, both at the specimen edges and elsewhere, suggest that alterations in the mechanical properties of fiber coatings and reaction zones would be beneficial. Conversely, higher strength fiber-matrix interfaces could reduce the effectiveness of the fiber process in impeding crack growth and could allow self-similar crack growth. Therefore, any potential improvement in fiber-matrix interface properties should be screened with both fatigue crack initiation and crack growth tests. Fiber-dominated failures could be hindered by higher strength fibers and a stronger, more relaxation-resistant matrix. Stronger SiC fibers [96] and titanium aluminide matrices [97] are currently being pursued. Development of a titanium matrix more resistant to environmental embrittlement could reduce environment-assisted cracking in the matrix. Fatigue-resistant protective coatings might also be considered to reduce environmental interactions.

The titanium and titanium aluminide matrix composites reviewed in this chapter possess matrices with a considerable range of strengths, ductilities, and environmental resistances. The many observed commonalities in fatigue failure mechanisms suggest that these mechanisms might not be easily suppressed by incremental improvements in matrix, fiber, or interface properties. Major changes in matrix and fiber properties may be necessary to significantly improve the fatigue resistance of TMCs, especially at high temperatures.

REFERENCES

1. Bhatt, R. T. and Grimes, H. H. 1981. "Fatigue Behavior of Silicon-Carbide Reinforced Titanium Composites," *Fatigue of Fibrous Composite Materials, ASTM STP 723,* Philadelphia, PA: American Society for Testing and Materials, pp. 274–290.

2. El-Soudani, S. M., and Gambone, M. L. 1990. "Strain Controlled Fatigue Testing of SCS6/Ti-6Al-4V Metal Matrix Composite," *Proc. on Fundamental Relationships Between Microstructure and Mechanical Properties of Metal Matrix Composites,* Liaw, P. K. and Gungor, M. N. eds. Warrendale, PA: The Mining, Metallurgy, and Materials Society, pp. 669–704.

3. Bartolotta, P. A. and Brindley, P. K. 1990. "High Temperature Fatigue Behavior of a SiC/Ti-24Al-11Nb Composite," *Composite Materials: Testing and Design, ASTM STP 1120,* Philadelphia, PA: American Society for Testing and Materials, pp. 192–203.

4. Lerch, B. and Halford, G. R. 1994. "Control Mode and R-Ratio Effects on the Fatigue Behavior of a [0] MMC," HITEMP Review 1994, NASA CP-10146, Vol. II, Washington, DC: National Aeronautics and Space Administration, pp. 40-1–40.11.

5. Jeng, S. M., Kai, W., Shih, C. J. and Yang, J.-M. 1989. "Interface Reaction Studies of B_4C/B and SiC/B Fiber-reinforced Ti_3Al Matrix Composites," *Materials Science and Engineering,* 114:189–196.

6. Mittnick, M. A. and McElman, J. 1987. "Continuous Silicon Carbide Fiber Reinforced Metal Matrix Composites," *Damage Mechanics and Composites,* Wang, A. S. D. and Haritos, G. K., eds. New York, NY: American Society for Mechanical Engineers, pp. 395–404.

7. Spear, S. 1991. "Evaluation of Silicon Carbide Fiber for Use in Titanium Matrix Composites for NASP Applications," *Titanium Aluminide Composites,* WL-TR-91-4020, Smith, P. R., Balsone, S. J. and Nicholas, T., eds., Wright-Patterson Air Force Base, OH: Wright Laboratory, pp. 73–95.

8. Bania, P. J., Lenning, G. A., and Hall, J. A. 1983. "Development and Properties of Ti-15V-3Cr-3Sn-3Al (Ti-15-3)," *Beta Titanium Alloys in the 80's,* Boyer, R. R. and Rosenberg, H. W., eds., Warrendale, PA: The Mining, Materials, and Metallurgical Society, pp. 209–230.

9. Bania, P. J. and Parris, W. M. 1990. "Beta-21S—A High Temperature Metastable Beta Titanium Alloy," *Titanium 1990: Products and Applications,* Proc. of the Int. Conf., Buena Vista, FL, Sept. 30–Oct. 3, 1990, Vol. 2, pp. 784–793.

10. Larsen, J. M., Revelos, W. C., and Gambone, M. L. 1992. "An Overview of Potential Titanium Aluminide Composites in Aerospace Applications," *Materials Research Society Symposium Proceedings,* Vol. 273, Pittsburgh, PA: Materials Research Society, pp. 3–16.

11. Smith, P. R., Graves, J. A. and Rhodes, C. G. 1992. "Evaluation of an SCS-6/Ti-22Al-23Nb 'Orthorhombic' Composite," *Materials Research Society Symposium Proceedings,* Vol. 273, Pittsburgh, PA: Materials Research Society, pp. 43–52.

12. Gabb, T. P., Gayda, J., Lerch, B. and Halford, G. R. 1991. "The Effect of Matrix Mechanical Properties on $[0]_8$ Unidirectional SiC/Ti Composite Fatigue Resistance," *Scripta Metallurgica et Materialia,* 25:2879–2884.

13. Krishnamohanrao, Y., Kutumbarao, V. V., and Rama Rao, P. 1986. "Fracture Mechanism Maps for Titanium and Its Alloys," *Acta Metallurgica,* 34 (9):1783–1806.

14. Gabb, T. P., Bartolotta, P. A., and Castelli, M. G. 1993. "A Review of Thermomechanical Fatigue Damage Mechanisms in Two Titanium and Titanium Aluminide Matrix Composites," *International J. Fatigue,* 15 (5):413–422.

15. Yang, J.-M. and Jeng, S. M. 1989. "Interfacial Reaction Kinetics of SiC Fiber-Reinforced Ti_3Al Matrix Composites," *Scripta Metallurgica et Materialia,* 23:1559–1564.

16. Baumann, S. F., Brindley, P. K. and Smith, S. D. 1990. "Reaction Zone Microstructures in a Ti_3Al + Nb/SiC Composite," *Metallurgical Transactions A,* 21:1559–1569.

17. Gundel, D. B. and Wawner, F. E. 1991. "Interfacial Reaction Kinetics of Coated SiC Fibers with Various Titanium Alloys," *Scripta Metallurgica et Materialia,* 25:437–441.

18. Yang, J.-M., Jeng, S. M., and Yang, C. J. 1991. "Fracture Mechanisms of Fiber-Reinforced Titanium Alloy Matrix Composites. Part I. Interfacial Behavior," *Materials Science and Engineering,* A138:155–167.

19. Eldridge, J. I. and Brindley, P. K. 1989. "Investigation of Interfacial Shear Strength in a SiC Fiber/Ti-24Al-11Nb Composite by a Fiber Push-Out Technique," *J. Materials Science Letters*, 8:1451–1454.

20. Brindley, P. K. and Bartolotta, P. A. 1995. "Failure Mechanisms During Isothermal Fatigue of SiC/Ti-24Al-11Nb Composites," *Materials Science and Engineering*, A200:55–67.

21. MacKay, R. A., Draper, S. L., Ritter, A. M. and Siemers, P. A. 1994. "A Comparison of the Mechanical Properties and Microstructures of Intermetallic Matrix Composites Fabricated by Two Different Methods," *Metallurgical and Materials Transactions A*, 25A:1443–1455.

22. Gabb, T. P. and Gayda, J. 1993. "Fatigue Behavior of Unidirectionally Reinforced Orthorhombic Matrix Composite," HITEMP Review 1993, NASA CP-19117, Vol. II, Washington, DC: National Aeronautics and Space Administration, pp. 33-1–33-10.

23. Draper, S. L., Brindley, P. K. and Nathal, M. V. 1992. "Effect of Fiber Strength on the Room Temperature Tensile Properties of SiC/Ti-24Al-11Nb," *Metallurgical Transactions A*, 23A:2541–2548.

24. Gambone, M. L. 1989. "Fatigue and Fracture of Titanium Aluminides," WRDC-TR-89-4145 Vol. I, Wright-Patterson Air Force Base, OH: WRDC/MLLN.

25. Lerch, B. 1991. "Matrix Plasticity in SiC/Ti-15-3 Composite," NASA TM-103760, Washington, DC: National Aeronautics and Space Administration.

26. Majumdar, B. S. and Newaz, G. M. 1992. "Inelastic Deformation of Metal Matrix Composites: Plasticity and Damage Mechanisms," *Philosophical Magazine A*, 66(2):187–212.

27. Johnson, W. S., Lubowinski, A. L., and Highsmith, A. L. 1990. "Mechanical Characterization of SCS6/Ti-15-3 Metal Matrix Composites at Room Temperature," *Thermal and Mechanical Behavior of Ceramic and Metal Matrix Composites, ASTM STP 1080*, Kennedy, J. M., Moeller, H. H. and Johnson, W. S., eds., Philadelphia, PA: American Society for Testing and Materials, pp. 193–218.

28. Watson, D. C. 1988. "Effect of Elevated Temperature Exposure on the Tensile Properties of SCS-6/Titanium Composite," AFWAL-TR-88-4170, Wright-Patterson Air Force Base, OH: AFWAL/MLSE.

29. Jeng, S. M., Yang, J.-M., and Yang, C. J. 1991. "Fracture Mechanics of Fiber-Reinforced Titanium Alloy Matrix Composites. Part III. Toughening Behavior," *Materials Science and Engineering*, A138:181–190.

30. Wawner, F. E. and Whatley, W. J. 1988. "The Effect of Elevated Temperature Exposure on the Properties of SiC/Ti-6Al-4V Composites," *Space Age Metal Technology, Proceedings of the 2nd International SAMPE Metals Conference*, August 2–4, 1988, Dayton, OH, pp. 470–479.

31. Russ, S. M. 1990. "Thermal Fatigue of Ti-24Al-11Nb/SCS-6," *Metallurgical Transactions A*, 21(A):1595–1602.

32. Revelos, W. C. and Smith, P. R. 1992. "Effect of Environment on the Thermal Fatigue Response of an SCS-6/Ti-24Al-11Nb Composite," *Metallurgical Transactions A*, 23(A):587–595.

33. Brindley, P. K., Bartolotta, P. A., and MacKay, R. A. 1989. "Thermal and Mechanical Fatigue of a SiC/Ti$_3$Al + Nb," HITEMP Review 1989, NASA

CP-10039, Washington, DC: National Aeronautics and Space Administration, pp. 52-1–52-14.

34. Revelos, W. C. and Roman, I. 1992. "Laminate Orientation and Thickness Effects on an SCS-6/Ti-24Al-11Nb Composite Under Thermal Fatigue," *Intermetallic Composites II,* Vol. 273, Miracle, D. B., Graves, J. A., and Anton, D. L., eds., Pittsburgh, PA: Materials Research Society, pp. 53–58.

35. Brindley, P. K. and Draper, S. L. 1993. "Failure Mechanisms of 0° and 90° SiC/Ti-24Al-11Nb Composites Under Various Loading Conditions," *Structural Intermetallics,* Darolia, R., Lewandowski, J. J., Liu, C. T. Martin, P. L., Miracle, D. B. and Nathal, M. V., eds., Warrendale, PA: The Minerals, Metals, & Materials Society, pp. 727–737.

36. Smith, P. R., Graves, J. A., and Rhodes, C. G. 1993. "Preliminary Mechanical Property Assessment of a SiC/Orthorhombic Titanium Aluminide Composite," *Structural Intermetallics,* Darolia, R., Lewandowski, J. J., Liu, C. T., Martin, P. L., Miracle, D. B., and Nathal, M. V., eds., Warrendale, PA: The Minerals, Metals & Materials Society, pp. 765–771.

37. Park, Y. H. and Marcus, H. L. 1983. "Influence of Interfacial Degradation and Environment on the Thermal and Fracture Fatigue Properties of Titanium-Matrix/Continuous SiC-Fiber Composites," *Mechanical Behavior of Metal Matrix Composites,* Hack, J. E. and Amateau, M. F., eds., Warrendale, PA: The Minerals, Metals & Materials Society, pp. 65–75.

38. MacKay, R. A. 1990. "Effect of Fiber Spacing on Interfacial Damage in a Metal Matrix Composite," *Scripta Metallurgica et Materialia,* 24:167–172.

39. Mall, S. and Ermer, P. G. 1991. "Thermal Fatigue Behavior of a Unidirectional SCS-6/Ti-15-3 Metal Matrix Composite," *J. of Composite Materials,* 25:1668–1686.

40. Cox, B. N., James, M. R., and Marshall, D. B. 1989. "Failure Mechanisms in Titanium Aluminide/SiC Composites," *Materials and Processing Move into the 90's,* Benson, S., Cook, T., Trewin, E. and Turner, R. M., eds., New York, NY: Elsevier Science Publishers B. V., pp. 313–320.

41. Revelos, W. C. 1993. "The Thermal Fatigue Response of an SCS-6/Ti-15Mo-3Al-2.6Nb-0.2Si (weight percent) Metal Matrix Composite," *Fatigue '93,* Vol. II, Bailon, J.-P. and Dickson, J. I., eds., United Kingdom: Engineering Materials Advisory Services, pp. 963–968.

42. Brindley, P. K. and Bartolotta, P. A. 1991. "Isothermal Fatigue Behavior of SiC/Ti-24Al-11Nb," HITEMP Review 1991, NASA CP-10082, Washington, DC: National Aeronautics and Space Administration, pp. 46-1–46-13.

43. Jeng, S. M., Alassouer, P. and Yang, J.-M. 1991. "Fracture Mechanisms of Fiber-Reinforced Titanium Alloy Matrix Composites, Part IV Low Cycle Fatigue," *Materials Science and Engineering,* A148:67–77.

44. Gabb, T. P., Gayda, J., and MacKay, R. A. 1990. "Isothermal and Nonisothermal Fatigue Behavior of a Metal Matrix Composite," *J. of Composite Materials,* 24:667–686.

45. Castelli, M. G. and Gayda, J. 1993. "An Overview of Elevated Temperature Damage Mechanisms and Fatigue Behavior of a Unidirectional SCS-6/Ti-15-3 Composite," *Reliability, Stress Analysis, and Failure Preventions,* New York, NY: American Society for Mechanical Engineers, pp. 213–221.

46. Bartolotta, P. A. and Verrilli, M. J. 1991. "Thermomechanical Fatigue Behavior of SiC/Ti-24Al-11Nb Under Different Environmental Conditions," HITEMP Review 1991, NASA CP-10082, Washington, DC: National Aeronautics and Space Administration, pp. 47-1–47-12.

47. Nicholas, T. and Russ, S. M. 1992. "Elevated Temperature Fatigue Behavior of SCS-6/Ti-24Al-11Nb," *Materials Science and Engineering,* A153:514–519.

48. Talreja, R. 1987. *Fatigue of Composite Materials,* Lancaster, PA: Technomic Publishing Co., Inc., pp. 25–40.

49. Majumdar, B. S. and Lerch, B. A. 1993. "Fatigue Mechanisms in a Ti-Based Fiber-Reinforced Metal Matrix Composite and Approaches to Life Prediction," *Titanium Metal Matrix Composites II,* WL-TR-93-4105, Smith, P. R. and Revelos, W. C., eds., Wright-Patterson Air Force Base, OH: Wright Laboratory, pp. 409–426.

50. Pollock, W. D. and Johnson, W. S. 1992. "Characterization of Unnotched SCS-6/Ti-15-3 Metal Matrix Composites at 650°C," *Composite Materials: Testing and Design, ASTM STP 1120,* Grimes, G. C., ed., Philadelphia, PA: American Society for Testing and Materials, pp. 175–191.

51. Castelli, M. G., Bartolotta, P. A., and Ellis, J. R. 1992. "Thermomechanical Testing of High Temperature Composites: Thermomechanical Fatigue (TMF) Behavior of SiC(SCS-6)/Ti-15-3," *Composite Materials: Testing and Design, ASTM STP 1120,* Grimes, G. C., ed., Philadelphia, PA: American Society for Testing and Materials, pp. 70–86.

52. Gabb, T. P., Gayda, J., and MacKay, R. A. 1991. "Nonisothermal Fatigue Degradation of a SiC/Ti Composite," *Advanced Composite Materials, Ceramic Transactions,* 19:527–534.

53. Gabb, T. P. and Gayda, J. 1992. "Isothermal and Nonisothermal Fatigue Damage/Failure Mechanisms in SiC/Ti-14Al-21Nb," *Titanium Matrix Composites,* WL-TR-92-4035, Smith, P. R. and Revelos, W. C., eds., Wright-Patterson Air Force Base, OH: Wright Laboratory, pp. 292–305.

54. Gayda, J., Gabb, T. P., and Freed, A. D. 1989. "The Isothermal Fatigue Behavior of a Unidirectional SiC/Ti Composite and the Ti Alloy Matrix," *Fundamental Relationships Between Microstructure and Mechanical Properties of Metal Matrix Composites,* Gungor, M. V. and Liaw, P. K., eds., Warrendale, PA: The Mining, Metallurgical, and Materials Society, pp. 497–514.

55. Harmon, D. M. and Saff, C. R. 1989. "Damage Initiation and Growth in Fiber Reinforced Metal Matrix Composites," *Metal Matrix Composites: Testing, Analysis, and Failure Modes, ASTM STP 1032,* Johnson, W. S., ed., Philadelphia, PA: American Society for Testing and Materials, pp. 237–250.

56. Gambone, M. L. and Bain, K. R. 1988. "Fractography of Titanium Aluminide Metal Matrix Composites," *Proceedings of the 2nd International SAMPE Metals Conference,* Aug. 2–4, 1988, Dayton, OH: Society for the Advancement of Materials and Process Engineering, pp. 487–497.

57. Johnson, W. S. 1992. "Damage Development in Titanium Metal-Matrix Composites Subjected to Cyclic Loading," *J. of Composites,* 24(3): 187–196.

58. Gabb, T. P. and Gayda, J. 1995. "Matrix Fatigue Cracking Mechanisms of α_2 TMC for Hypersonic Applications," *Life Prediction Methodology for Titanium Metal Matrix Composites, ASTM STP 1253,* Johnson, W. S., Larsen, J. M. and Cox, B. N., eds., Philadelphia, PA: American Society for Testing and Materials, pp. 107–126.

59. Kantzos, P., Telesman, J. and Ghosn, L. 1989. "Fatigue Crack Growth Behavior of SCS-6/Ti-15-3 Composite as Observed Through an SEM Loading Stage," HITEMP Review 1989, NASA CP-10039, Washington, DC: National Aeronautics and Space Administration, pp. 66-1–66-10.

60. Jira, J. R. and Larsen, J. M. 1993. "Crack Bridging Behavior in Unidirectional SCS-6/Ti-24Al-11Nb Composite," *Fatigue '93,* Vol. 2, Bailon, J.-P. and Dickson, I. J., eds., United Kingdom: Engineering Materials Advisory Services, pp. 1085–1090.

61. Ghosn, L. J., Telesman, J., and Kantzos, P. 1994. "Damage Tolerance Based Life Prediction Methodology in Titanium Matrix Composites," HITEMP Review 1994, NASA CP-10146, Vol. II, Washington, DC: National Aeronautics and Space Administration, pp. 39-1–39-10.

62. Collings, E. W. 1984. *The Physical Metallurgy of Titanium Alloys,* Metals Park, OH: American Society for Metals, pp. 56–57.

63. Brindley, W. J., Gabb, T. P., and Smith, J. W. 1995. "Embrittlement of the Surfaces and Crack Faces of TMC's During Fatigue," Orthorhombic Titanium Matrix Composites Workshop, WL-TR-95-4068, Smith, P. R., ed., Wright-Patterson Air Force Base, OH: Wright Laboratory, pp. 52–63.

64. Saitoh, Y. and Mino, K. 1993. "Embrittlement of Ti_3Al-10Nb-3V-1Mo Caused by Slight Oxidation," *Materials Transactions, JIM,* 34(4):393–395.

65. Kantzos, P., Bartolotta, P., Verrilli, M. J., and Ghosn, L. J. 1993. "The Effect of Fatigue Loading on the Interfacial Shear Properties of SCS-6/Ti-Based MMCs," *Proceedings of the American Society for Composites,* 8th Technical Conference, Oct. 19–21, 1993, Cleveland, OH, pp. 377–387.

66. Mall, S. and Portner, B. D. 1991. "Investigation of Fatigue Damage Mechanisms in SCS-6/Ti-15-3 Metal Matrix Composite at Elevated Temperature," *Mechanics of Composites at Elevated and Cryogenic Temperatures,* AMD-Vol. 118, New York, NY: American Society for Mechanical Engineers, pp. 239–249.

67. Gayda, J. and Gabb, T. P. 1992. "Isothermal Fatigue Behavior of a $[90]_8$ SiC/Ti-15-3 Composite at 426°C," *International J. of Fatigue,* 14:14–20.

68. Castelli, M. G. 1993. "Thermomechanical and Isothermal Fatigue Behavior of a $[90]_8$ Titanium Matrix Composite," *Proc. of the American Society for Composites,* 8th Technical Conference, October 19–21, 1993, Cleveland, OH, pp. 884–892.

69. Kantzos, P. and Telesman, J. 1989. "Fatigue Crack Growth Behavior of a SCS-6/Ti-15-3 Composite as Observed Through an SEM Loading Stage," HITEMP Review 1989, NASA CP-10039, Washington, DC: National Aeronautics and Space Administration, pp. 66-1–66-10.

70. Marshall, D. B., Shaw, M. C., James, M. R., Graves, J., Morris, W. L. and Porter, J. R. 1993. "Fatigue Resistant Ti_3Al Composites," WL-TR-93-4034, Wright-Patterson Air Force Base, OH: Wright Laboratory.

71. Castelli, M. G. 1994. "Isothermal Damage and Fatigue Behavior of SCS-6/Timetal 21S [0/90]$_s$ Composite at 650°C," NASA CR-195345, Washington, DC: National Aeronautics and Space Administration.

72. Castelli, M. G. 1995. "Characterization of Damage Progression in SCS-6/Timetal 21S [0]$_4$ Under Thermomechanical Fatigue Loading," *Life Prediction Methodology for Titanium Matrix Composites, ASTM STP 1253,* Johnson, W. S., Larsen, J. M. and Cox, B. N., eds., Philadelphia, PA: American Society for Testing and Materials.

73. Neu, R. W. and Nicholas, T. 1993. "Thermomechanical Fatigue of SCS-6/Timetal 21S Under Out-of-Phase Loading," *Thermomechanical Behavior of Advanced Structural Materials,* New York, NY: American Society for Mechanical Engineers, pp. 97–111.

74. Jeng, S. M. and Yang, J.-M. 1992. "Damage Mechanisms of SCS-6/Ti-6Al-4V Composites Under Thermal-Mechanical Fatigue," *Materials Science and Engineering,* A156:117–124.

75. Russ, S. M. Nicholas, T., Bates, M., and Mall, S. 1991. "Thermomechanical Fatigue of SCS-6/Ti-24Al-11Nb Metal Matrix Composite," *Failure Mechanisms in High Temperature Composite Materials,* AD-Vol.22/AMD-Vol. 122, New York, NY: American Society for Mechanical Engineers, pp. 37–43.

76. Gayda, J., Gabb, T. P., and Lerch, B. A. 1993. "Fatigue-Environment Interactions in a SiC/Ti-15-3 Composite," *International J. of Fatigue,* 15(1):41–45.

77. Bartolotta, P. A., Kantzos, P., Verrilli, M., and Dickerson, R. 1993. "TMF Damage Mechanisms of SCS-6/Ti-24Al-11Nb in Air and Argon Environments," HITEMP Review 1993, NASA CP-19117, Vol. II, Washington, DC: National Aeronautics and Space Administration, pp. 39-1–39-10.

78. Rangaswamy, P. and Jayaraman, N. 1994. "Residual Stresses in SCS-6/Ti-24Al-11Nb Composite. Part II. Finite Element Modeling," *J. of Composites Technology & Research, JCTRER,* 16(1):54–67.

79. Cerchiara, R. R., Meier, G. H., and Pettit, F. S. 1995. "Oxidation Studies on Ti-Al-Nb Neat and Composite Materials at Temperatures Between 500° and 900°C," *Orthorhombic Titanium Matrix Composites,* WL-TR-95-4068, Smith, P. R., ed., Wright-Patterson Air Force Base, OH: Wright Laboratory, pp. 15–40.

80. Brindley, W. J., Smialek, J. L., Smith, J. W., and Brady, M. P. 1995. "Environment Effects on Orthorhombic-Ti Matrix Materials," *Orthorhombic Titanium Matrix Composites,* WL-TR-95-4068, Smith, P. R., ed., Wright-Patterson Air Force Base, OH: Wright Laboratory, pp. 1–14.

81. Rosenberger, A. H. 1995. "Preliminary Assessment of the Environment Sensitivity of Ti-22Al-23Nb (a/o) Neat Laminates," *Orthorhombic Titanium Matrix Composites,* WL-TR-95-4068, Smith, P. R., ed., Wright-Patterson Air Force Base, OH: Wright Laboratory, pp. 41–51.

82. Neu, R. W. and Nicholas, T. 1993. "Effect of Laminate Orientation on the Thermomechanical Fatigue Behavior of a Titanium Matrix Composite," *J. of Composite Technology and Research, JCTRER,* 16(3):214–224.

83. Newaz, G. M. and Majumdar, B. S. 1995. "In-Phase Thermomechanical Fatigue Mechanisms in Unidirectional SCS-6/Ti-15-3 Metal Matrix Com-

posite," NASA Cr-195482, Washington, DC: National Aeronautics and Space Administration.

84. Neu, R. W. and Roman, I. 1993. "Acoustic Emission Monitoring of Damage in Metal Matrix Composites Subjected to Thermomechanical Fatigue," *Composite Science and Technology,* (52):1–8.

85. Nicholas, T. and Ahmad, J. 1994. "Modeling Fiber Breakage in a Metal Matrix Composite," *Composite Science and Technology,* 52:931–946.

86. Castelli, M. G. 1994. "Thermomechanical Fatigue Damage/Failure Mechanisms in SCS-6/Timetal 21S [0/90]$_s$ Composite," *Composites Engineering,* 4(9):931–946.

87. Russ, S. and Hansen, D. 1993. "Fatigue and Thermomechanical Fatigue of a SiC/Titanium [0/90]$_{2s}$ Composite," *Fatigue 93,* Bailon, J.-P. and Dickson, J. I., eds., UK: Engineering Materials Advisory Services, pp. 969–974.

88. Schubbe, J. J. and Mall, S. 1991. "Damage Mechanisms in a Cross-Ply Metal Matrix Composite Under Thermal-Mechanical Cycling," *Proc. of the International Conference on Composite Materials,* July 15–19, 1991, Honolulu, Hawaii.

89. Mirdamadi, M., Bakuckas, J. G. and Johnson, W. S. 1993. "Mechanisms of Strain Accumulation in Titanium Matrix Composites at Elevated Temperatures," *Mechanics of Composite Materials—Nonlinear Effects,* AMD Vol. 159, New York, NY: American Society for Mechanical Engineers, pp. 245–252.

90. Mirdamadi, M. and Johnson, W. S. 1994. "Prediction of Stress-Strain Response of SCS-6/Timetal-21S Subjected to a Hypersonic Flight Profile," NASA TM-109026, Washington, DC: National Aeronautics and Space Administration.

91. Aksoy, S. Z., Gayda, J. and Gabb, T. P. 1995. "Fatigue Behavior of [0]$_8$ SCS-6/Ti-6Al-4V Composite Subjected to High Temperature Turboshaft Design Cycles," *Thermo Mechanical Fatigue Behavior of Materials: Vol. II, ASTM STP 1263.* Philadelphia, PA: American Society for Testing and Materials.

92. Chatterjee, A., Gabb, T. P., and Gayda, J. 1995. "Simple Mission Profile Testing of Orthorhombic MMCs," *Orthorhombic Titanium Matrix Composites,* WL-TR-95-4068, Smith, P. R., ed., Wright-Patterson Air Force Base, OH: Wright Laboratory, pp. 180–200.

93. Gabb, T. P., Gayda, J., and Chatterjee, A. 1995. "Mission Cycle Behavior of Orthorhombic Titanium Matrix Composite," HITEMP Review 1995, NASA CP-10178, Vol. II, Washington, DC: National Aeronautics and Space Administration, pp. 25-1–25-11.

94. Gayda, J. and Gabb, T. P. 1993. "Fatigue Behavior of Clad Titanium Matrix Composite," *Proceedings of the American Society for Composites,* 8th Technical Conference, Oct. 19–21, 1993, Cleveland, OH, pp. 875–883.

95. Gayda, J., Gabb, T. P., and Aksoy, S. 1995. "Fatigue Behavior of a Clad SCS6/Ti-1100 Composite," HITEMP Review 1995, NASA CP-10178, Vol. II, Washington, D.C.: National Aeronautics and Space Administration, pp. 24-1–24-10.

96. Gambone, M. L. 1995. "SiC Fiber Strength After Consolidation and Heat-Treatment in Ti-22Al-23Nb Matrix Composite," *Orthorhombic Titanium Matrix Composites,* WL-TR-95-4068, Smith, P. R., ed., Wright-Patterson Air Force Base, OH: Wright Laboratory, pp. 341–352.

97. Smith, P. R. and Graves, J. A. 1995. "Tensile and Creep Properties of High Temperature Titanium Alloys in Neat Matrix Form," *Orthorhombic Titanium Matrix Composites,* WL-TR-95-4068, Smith, P. R., ed., Wright-Patterson Air Force Base, OH: Wright Laboratory, pp. 139–149.

6

Fatigue and Thermomechanical Fatigue Life Prediction

T. NICHOLAS

USAF Wright Laboratory
Wright-Patterson AFB, OH 45433

INTRODUCTION

THE INTEREST IN new high-performance aircraft and turbine engines over the last several years, which has brought attention to titanium matrix composites (TMCs) as a candidate for high-temperature structural applications, brings with it a number of questions related to design for fatigue. For fatigue life prediction in this material system, the role of thermal stresses induced by the large difference in coefficient of thermal expansion (CTE) between fiber and matrix must be considered when the material is subjected to thermal cycling. In applications where temperatures range from near or below room temperature to above 500 to 600°C, the thermally induced stress range in the fiber or matrix can exceed the mechanical stress range due to fatigue loading from zero applied stress to an applied stress near the yield strength of the composite. Thermal stresses and mechanical stresses, therefore, are important considerations for design. In many proposed applications, combinations of mechanical and thermal cycles, referred to as thermomechanical fatigue (TMF), may occur. TMF, therefore, must also be considered in the design process. In these cases, applied loading and temperatures are not sufficient to fully characterize the fatigue behavior or predict life of the composite. Phasing of the load and temperature excursions and the resulting stresses in the individual constituents, namely, the fiber, matrix, and the fiber-matrix interface, must also be considered. Therefore, micromechanics analysis, the determination of constituent stresses, must be employed in life prediction schemes to realistically represent the physical phenomena and mechanisms taking place. Added to these complexities are the environmental degradation of mechanical properties of the fiber or matrix with increasing exposure to elevated temperature and stress, the local inelastic deformation in the matrix, and the evolution of damage at the fiber-matrix interface. Thus, the ability to predict life, taking into account all of the above considerations, is a formidable task.

In this chapter, approaches that have been developed for life prediction in TMCs are reviewed and evaluated. The life prediction methodologies considered here are limited to the case of smooth bar fatigue, with no consideration being given to the problem of the propagation of a single dominant crack. This limits the applicability of the analyses to cases where damage is more or less uniformly distributed throughout the material. Furthermore, consideration is limited to total fatigue life where life is defined as either separation into two pieces, some measurable loss of load-carrying capability, or an unacceptable accumulation of strain. Analyses dealing with the loss of strength, the accumulation of strain, and the reduction of stiffness or other mechanical properties are not included in this chapter. Although some early investigations dealing with TMCs focused attention on damage accumulation due to thermal cycling, these works will not be covered here. It is believed that thermal cycling is no more than a subset of TMF; thus, fatigue life modeling will be restricted to discussion of isothermal fatigue (IF) and TMF. Fatigue under mission spectrum loading, which has received relatively little attention, will be addressed as an extension of life modeling under IF and TMF.

MECHANISMS OF FATIGUE

The complex microstructure of TMCs that includes titanium matrix regions, SiC fibers, and interfacial regions provides the opportunity for a number of mechanisms to affect the fatigue life of this class of materials. Although no attempt will be made to describe or list all of the possible mechanisms here, it will suffice to mention the most prominent, which form the basis of the concepts embedded in the various life prediction models. Fiber failure, whether through overload of single fibers or a bundle, fatigue through contact with the surrounding matrix and/or interface, or decrease of strength through a combined fatigue and environmental degradation process, is one of the primary mechanisms considered in fatigue life modeling. Matrix failure, primarily through multiple fatigue crack initiation and propagation, is another primary mechanism. Here, surface cracks that are subjected to the external environment and internal cracks from defects, such as a fiber-matrix interface, or a stress concentration due to the presence of off-axis fibers, can initiate the fatigue failure process. More complex fatigue mechanisms can involve combinations of the above, including combined mechanical and environmental interactions. Discussions and descriptions of the mechanisms leading to isothermal and nonisothermal fatigue in TMCs can be found in the literature [1,2,3] and are discussed in Chapter 5.

LIFE PREDICTION METHODOLOGIES

Stress in Fiber

The notion is widely held that for the behavior of a composite to be well characterized, the behavior of the composite constituents must be understood. In particular, the concept has been proposed that, for a TMC to fail due to fatigue, the fibers must fail. Thus, stress in the fiber should play an important role in fatigue life prediction. One of the first applications of this concept was by Johnson et al. [4] who showed good correlation between stress range in $0°$ fibers and cycles to failure for an SCS-6/Ti-15-3 composite. Noting that the stress range in the $0°$ fibers, $\Delta\sigma_f^0$, is related to strain range in the composite by:

$$\Delta\sigma_f^0 = E_f \Delta\varepsilon_c \qquad (1)$$

where E_f is the fiber modulus and $\Delta\varepsilon_c$ is the composite strain range, they used experimental measurements of stabilized strain range to correlate fatigue life with fiber stress range for four different layups of the Ti-15-3 composite at room temperature. The results, shown in Figure 1 which also shows the static strength at the first cycle, appear to correlate all of the data into a single curve. Pollock and Johnson [5] also concluded that fatigue life is controlled by the stress range in the $0°$ fiber for several layups of a Ti-15-3 composite at elevated temperature.

Under isothermal conditions, the stress in the $0°$ fiber can be obtained from the strain in the fiber that is the same as that in the composite, assuming that fibers have not been broken. The strain in the composite can be determined from either experimental measurements or from computations involving micromechanics analysis or rule-of-mixtures approaches, for example. Under nonisothermal conditions, however, the stress in the $0°$ fiber can be obtained computationally

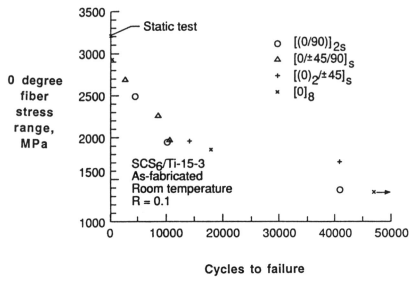

Cycles to failure

Figure 1. Cyclic stress range in 0° fiber versus number of cycles to laminate failure [4].

only through a micromechanics analysis, often requiring consideration of thermal residual stresses, viscoplastic flow in the matrix, and damage in the matrix (cracks), fibers (fiber breaks), or at the fiber-matrix interface (separation). To determine 0° fiber stresses experimentally involves a precise knowledge of the thermal stresses in the composite, which, in turn, requires an accurate micromechanics analysis and may require the incorporation of damage to achieve accurate results. Mirdamadi et al. [6] used a micromechanics analysis computer code VISCOPLY to determine the 0° fiber stress range to correlate IF and TMF life. A Ti-15-3 unidirectional composite was analyzed from data obtained under IP and OP TMF, IF, and under some nonisothermal conditions used at NASA Lewis. As shown in Figure 2, the 0° fiber stress range was able to consolidate data from in-phase (IP) and out-of-phase (OP) TMF on a single curve, IP and OP nonisothermal on a slightly different second curve, and isothermal data on much different curves, one for each temperature. Because the data cannot be consolidated using a single curve, they concluded that the fatigue strength of the 0° fiber was controlled by a combination of temperature, loading frequency, and time at temperature. However, for a given combination of these, they proposed that fatigue life was controlled by 0° fiber stress range. Some of the differences were attrib-

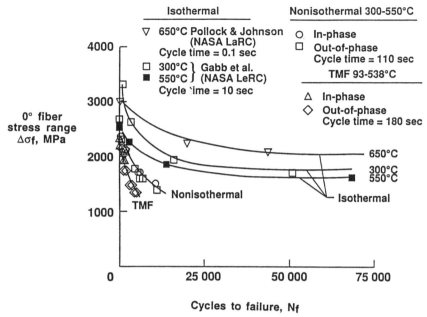

Figure 2. Stress range in 0° fiber as a function of cycles to failure [6].

uted to changes in fiber strength with time of exposure at elevated temperature.

The 0° fiber stress range has been used for more complicated loading conditions by Mirdamadi and Johnson [7]. They conducted experiments on fatigue life of SCS-6/TIMETAL®21S [0/90]$_{2S}$ under a generic hypersonic vehicle mission profile and under several mission profile segments to evaluate the most damaging portions of the mission. Fatigue lives were correlated with stresses in the 0° fibers as determined from analysis using VISCOPLY, which simulates damage in the form of fiber-matrix separation and matrix cracking to match experimental strain measurements. As in the prior works, correlation of fatigue life under a restricted range of conditions with stress range in the 0° fibers was achievable, but the correlation is not complete evidence that the stress in the fiber is the critical condition for fatigue failure, particularly because brittle materials, such as SiC, do not lend themselves well to a fatigue process.

Ward et al. [8,9] investigated the behavior of SCS-6/TIMETAL®21S [0/90]$_{2S}$ composite under room temperature fatigue at two different stress ratios, $R = 0.1$ and $R = 0.3$. On an applied stress range basis, data from the two values of R collapsed essentially onto a single curve as shown in Figure 3. It was pointed out that longer lives would normally be expected for lower R because of the lower mean stress at a given value of $\Delta\sigma$. They noted that Johnson et al. [4] had found that lives of SCS-6/Ti-15-3 with different laminate orientations, which contain at least some 0° plies, could be represented by the fiber stress range in those 0° plies, calculated as composite strain range multiplied by fiber modulus. Because the fiber was the same in both of these studies, the same life prediction methodology should apply if stress in the fiber is the governing factor in fatigue life prediction. However, they found that fatigue life for the Timetal composite is less than that for the Ti-15-3 composite for the same value of $\Delta\sigma$ in the 0° fiber as illustrated in Figure 3. They also pointed out that neat TIMETAL®21S has greater fatigue resistance than Ti-15-3. Possible explanations of these discrepancies included variability in fiber properties or fiber-matrix bond strength differences in the two composites. It may be added that the question of the validity of the use of $\Delta\sigma$ in the 0° fiber as the sole parameter for fatigue life prediction has to be raised.

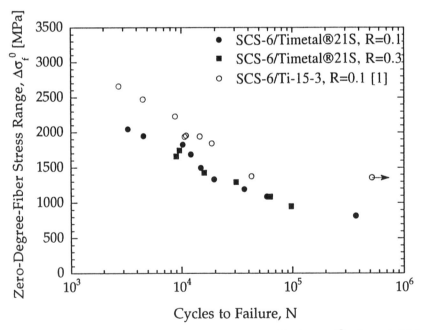

Figure 3. Comparison of fatigue lives between SCS-6/Timetal®21S and SCS-6/Ti-15-3 composites as a function of 0° fiber stress range at room temperature [9].

Another approach related to the use of stresses in the 0° fibers was developed by Gao and Zhao [10]. In that work, a model is developed for life prediction of a composite having 0° and off-axis plies. The load sharing between the plies is considered in the analysis after an assumed saturated state is reached. Both viscoplastic deformation and damage are considered in determining the stiffness reduction and strain accumulation in the off-axis plies. Although total life is the sum of the number of cycles to reach the saturated state plus the fatigue life of the 0° ply, the former is usually neglected. A simple fit to experimental data for the fatigue life of a 0° laminate based on maximum stress in that laminate is obtained using the equation:

$$\log |\sigma_{max}|_{0° \, laminate} = a \, (\log N)^2 + b \, (\log N) + \log X \tag{2}$$

where N is the number of cycles to failure, and X is ultimate tensile strength in a 0° laminate in the 0° direction. In this approach to fatigue life modeling, the determination of the stabilized value of accumulated strain and reduced stiffness of each lamina is required. Additionally, the model requires the fatigue data of 0° laminates. It is of interest to note that, as discussed later, Neu and Nicholas [11] were able to consolidate fatigue life data under in-phase and out-of-phase TMF for three different laminate orientations by normalizing the applied stress with respect to the ultimate tensile strength of the composite. Because the strength of the 0° laminate plays a major role in the strength and fatigue resistance of the composite, these life prediction approaches based on 0° laminate or 0° fiber stresses have a great deal of merit.

Strain in Matrix

Noting that stress range in the 0° fiber is related to strain range in the composite, see Equation (1), the strain range in the matrix is also identical to the strain range in the composite and may also be considered to be a criterion for fatigue life prediction under isothermal fatigue (IF). Majumdar and Newaz [12] conducted IP TMF and IF experiments on a quasi-isotropic SCS-6/Ti-15-3 composite. They used the 0° fiber stress range to consolidate data from each type of test, but, as in the investigations described previously, a different plot was obtained for each temperature or type of TMF test as shown in Figure 4. They noted that the same conclusions can be reached in comparing isothermal and TMF data using mechanical strain range as the correlating parameter.

The strain range approach for fatigue life prediction has been pursued by Halford et al. [13]. They proposed and outlined an approach for fatigue life prediction based on the computation of the steady-state strain-time profile in the matrix and the use of TMF fatigue data for the pure matrix material. The fatigue life of the matrix under TMF is predicted using the total strain version of strain range partitioning as described by Saltsman and Halford [14]. In this approach, the total strain range is decomposed into elastic and inelastic portions, thus:

$$\Delta\varepsilon_t = \Delta\varepsilon_{el} + \Delta\varepsilon_{in} \qquad (3)$$

where the elastic and inelastic portions are related to total fatigue life, N_f, in a power law fashion:

$$\Delta\varepsilon_{el} = B \ (N_f)^b \qquad (4)$$

$$\Delta\varepsilon_{in} = C \ (N_f)^c \qquad (5)$$

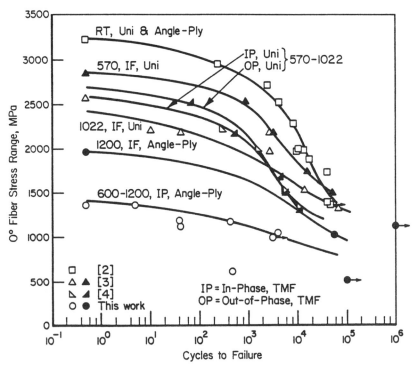

Figure 4. Consolidation of fatigue life data from IF and IP and OP TMF tests using 0° fiber stress as correlating parameter. Temperature is in °F [12].

It was noted that, in comparing fatigue lives based on strain range in the matrix of the composite with that of the pure matrix for several MMCs, the pure matrix material had considerably longer fatigue life than the composite for the same strain range. The results of this comparison for unidirectional SCS-6/Ti-24Al-11Nb at two different temperatures are shown in Figure 5. It can be seen that for equal lives, the strain range in the matrix had to be approximately double that in the composite. Some of the postulated reasons for this difference in-

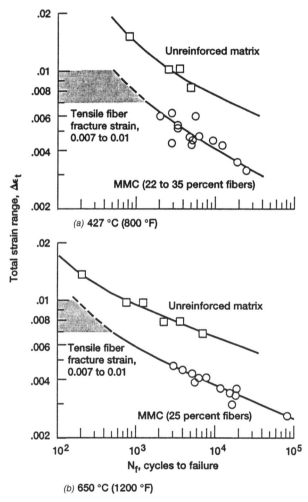

Figure 5. Comparison of fatigue resistance of SCS-6/Ti-24Al-11Nb composite with Ti-24Al-11Nb unreinforced matrix material: (a) 427°C and (b) 650°C [13].

cluded the differences in mean stress or strain ratio, load transfer to the fibers in the composite through stress relaxation in the matrix and subsequent fractures of the fibers, and possible early microcrack initiation in the composite matrix at fiber-matrix interfaces due to local stress concentrations and environmental influences.

The methodology developed in [13] for prediction of fatigue life of a unidirectional composite was applied to the case of a ring under low cycle fatigue [15]. The application involved cycling with superimposed hold times under constant stress at a temperature of 800°F (427°C) of an SCS-6/Ti-15-3 unidirectional composite. Although no experimental data were obtained at the time of publication, the fatigue life prediction analysis was conducted. It was noted that the volume of composited material in the rings was nearly 600 times greater than that in a single test coupon and that statistics of the probability of defects being present might influence the results.

Method of Universal Slopes

Methods used for low cycle fatigue life prediction of metals have been employed on TMCs. Bartolotta [16] employed a simple equation similar to the method of Universal Slopes used for monolithic metals by Manson [17] to correlate fatigue life with tensile properties (ultimate strength, ductility, and modulus). The basic form of the equation employs tensile properties similar to the Universal Slopes equation which, for life prediction, has the form:

$$N_f = A \left(\frac{\sigma_{ult}}{E} \right)^\alpha (\varepsilon_f)^\beta (\varepsilon_{max})^\gamma \tag{6}$$

where the composite tensile properties are used in place of constituent properties: ultimate tensile strength (σ_{ult}), modulus (E), and fracture strain (ε_f). N_f is the fatigue life, whereas ε_{max} is the maximum applied strain, A is a constant, and α, β, and γ are exponents. The method was applied to isothermal fatigue data on an SCS-6/Ti-24Al-11Nb unidirectional composite to correlate lives at 425 and 815°C. The data are shown in Figure 6 plotted as maximum strain against life (log-log) and produce a good correlation with Equation (6). It was found that they could also correlate data at intermediate temperatures and for cross-ply material. However, data for a quasi-isotropic composite could not be correlated. In their summary, they acknowledged that the method cannot be used to account for mean stress effects, TMF, or fully reversed loading.

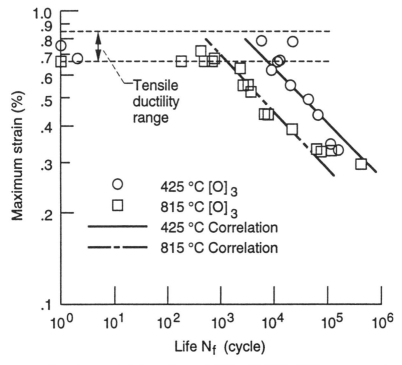

Figure 6. Correlation of fatigue lives of SCS-6/Ti-24Al-11Nb using maximum strain from method of Universal Slopes [16].

Radhakrishan and Bartolotta [18] extended the work of Bartolotta [16] to include a matrix-dominated failure term and a fiber term. The matrix term, in turn, is comprised of two terms, one representing matrix plasticity and one matrix elasticity. The two terms are given as:

$$\varepsilon_{max} = A_f \, (N_f)^{\alpha_f} \tag{7}$$

$$\varepsilon_{max} = A_m \, \frac{\sigma_{ult}}{E} \, (N_f)^{\alpha_m} + B_m \, (N_f)^{\beta_m} \tag{8}$$

where Equation (7) represents the fiber-dominated life, whereas Equation (8) represents fatigue life dominated by the matrix. The first term in Equation (8) is based on the Universal Slopes method, similar to Equation (6), where σ_{ult} and E now represent the ultimate tensile strength and modulus of the matrix material, respectively. The three terms provide a method of representing the three regions in a

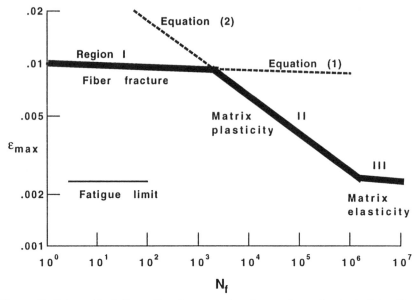

Figure 7. Schematic fatigue life diagram showing regions dominated by fiber fracture, matrix plasticity, and matrix elasticity as proposed by Talreja [19].

Talreja diagram [19] as shown schematically in Figure 7. The mode of failure associated with each region is indicated in the figure. Using maximum strain as the correlating parameter for each term, the model represented by Equations (7) and (8) was able to correlate isothermal data for several composites at several different temperatures.

Semiempirical Methods

Because of the complexity of the general life prediction problem, which may involve differences in frequency, temperature, stress ratio, and nonisothermal conditions, semiempirical methods have been developed to correlate large bodies of fatigue data. Such an approach has been taken by Tong and Chamis [20] who made modifications to the METCAN code [21] so that it can be calibrated to simulate TMF based on experimental data. They used the code to correlate in-phase and out-of-phase fatigue life data on a unidirectional SCS-6/Ti-24Al-11Nb composite over two different temperature ranges in air and under TMF from 425 to 815°C in argon. In their approach, upon which METCAN is based, a multifactor interaction relationship, including an effect of oxidation, was employed. A similar approach is taken by Rabzak et al. [22] where they develop a capability for accurate isothermal life prediction of SCS-6/Ti-24Al-11Nb at room and elevated (650°C) temperature using MMLT, which is an outgrowth of METCAN. The data correlated well with the predictions, although oxidation was not specifically considered in this approach as it had been in the previous work of Tong and Chamis [20]. In the life prediction scheme, use was made of maximum elastic and inelastic cyclic strains to develop the fatigue life model. Recently, the METCAN code has demonstrated the capability of predicting fatigue lives under isothermal and TMF conditions for a unidirectional TMC [23]. The authors, however, point out that it is necessary to have the proper data available for accurate life prediction, particularly the final values for various factors in the equations used. These values may not be available especially for new materials or for long-term effects.

S-N Curves

Life prediction under TMF conditions can be accomplished by plotting experimental data as *S-N* curves for each of the experimental conditions considered. Fatigue life can then be related to the maximum applied stress for the particular fatigue profile. Such an approach has been used by Hart and Mall [24] for an SCS-6/Ti-15-3 quasi-isotropic composite under IP and OP TMF and IF at T_{max} of the TMF cycle. A plot of the data from these three test conditions is shown in Figure 8. The regions of behavior on the *S-N* plot, delineated primarily by the crossover of the IP and OP TMF curves, were related to governing mechanisms identified through fractography. In addition, micromechanical stresses, determined from the METCAN program [25], were used to validate the mechanisms and define the critical constituent stresses in each region. No attempt was made, however, to quantify the observed behavior in terms of a model relating fatigue life to micromechanical stresses. In an earlier study, Mall and Schubbe [26] investigated IP and OP TMF of SCS-6/Ti-15-3 [0/90]$_{2S}$ and found what was first termed the crossover behavior as shown in Figure 9 for their data. In a companion study by Mall and Portner [27], results from isothermal fatigue tests at different frequencies at the maximum temperature of the TMF tests showed that, at low stresses, IF produced longer lives than either TMF condition. At high stresses, however, isothermal lives at the same frequency as TMF produced shorter lives than either TMF condition. The crossover of IP and OP TMF curves had been observed to occur at the stress corresponding

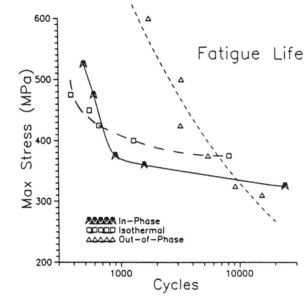

Figure 8. Fatigue life versus maximum applied stress for quasi-isotropic TMC under TMF from 149–427°C and IF at 427°C [24].

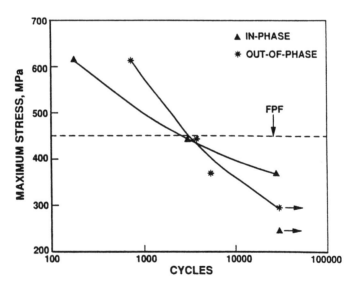

Figure 9. Fatigue lives from IP and OP TMF tests on a cross-ply SCS-6/Ti-15-3 composite (149–427°C) [26].

222

to the second point of nonlinearity in the high-temperature stress-strain curve of the composite [26,28] as shown in Figure 9. The nonlinearity is found at a stress where first ply failure (FPF) occurs. Results from the study by Hart and Mall [24], however, did not exhibit such a correlation of crossover stress with FPF stress. Strain accumulation for both minimum and maximum strain was observed in both IP TMF and IF tests. This was attributed to matrix creep, an observation that was also made by Mall et al. [28]. In all of these observations, the governing mechanisms and the relative positions of *S-N* curves are dependent on the material, the layup, and the specific temperature range used in the experiments. Frequency is also an influencing factor, especially when comparing IF results with those obtained under TMF, because the latter is limited to low frequencies experimentally because of practical considerations. The concept of a crossover point and the physical basis for its existence is discussed later.

Damage Parameter

The Smith-Watson-Topper (SWT) fatigue stress damage parameter, based on calculated matrix stresses, has been applied by Gravett [29] for predicting fatigue life of unidirectional SCS-6/Ti-15-3 composites under isothermal fatigue. This approach is valid for conditions when fatigue is matrix stress dominated. The correlating parameter has the form:

$$\sigma_{m, eff} = \sqrt{\sigma_{m, max} \frac{\Delta\varepsilon}{2} E_m} \qquad (9)$$

where $\sigma_{m, max}$ is the maximum matrix stress, $\Delta\varepsilon$ is the strain range (in the matrix or composite), and E_m is the modulus of the matrix material. This parameter constitutes an empirical relationship that relates failure at a given life under nonzero mean stress in the matrix to failure at the same life under zero cyclic mean stress. Life prediction was based on a combination of a concentric cylinder micromechanical model and the SWT stress parameter, Equation (9). The life prediction was found to be valid for any load-temperature cycle and for different fiber volume fractions.

The applicability of $\Delta\sigma$ in the 0° fiber as a correlating parameter for fatigue life prediction was evaluated by Mirdamadi and Johnson [7] in experiments on fatigue life of SCS-6/TIMETAL®21S [0/90]$_{2S}$ under several hypersonic vehicle mission profile segments. Fatigue lives were correlated with stresses in the 0° fibers as determined from micromechanics analysis. In addition to the use of fiber stress range, correlations for the various profiles investigated included using maximum fiber stress and a Smith-Watson-Topper (SWT) effective stress parameter. The SWT parameter [30], which takes into account stress ratio effects, provides an effective strain for correlating fatigue lives in the form:

$$\Delta\varepsilon_{eff} = \left(\frac{\varepsilon_{max}\Delta\varepsilon}{2}\right)^{1/2} \tag{10}$$

For assumed elastic behavior in the fiber, the parameter can be written as an effective stress in the form:

$$\Delta\sigma_{eff} = \left(\frac{\sigma_{max}\Delta\sigma}{2}\right)^{1/2} \tag{11}$$

where strain and stress range and maximum strain and stress in these two equations refer to the values in the 0° fibers in the composite. The SWT parameter collapsed the limited experimental data into the narrowest band and, thus, appeared to be the best of the three correlating parameters evaluated.

The SWT approach is a method for considering a combination of stress range and maximum stress in a single parameter. An alternate method of accounting for a combination of stress range and maximum stress was introduced by Mall et al. [28] in a fiber failure term in a life fraction model described in a subsequent section:

$$N_f = 10^{N_0}\left(1 - \frac{\sigma_{max}(1 - R)^m}{\sigma^*}\right) \tag{12}$$

Here, σ_{max} and R both refer to the stresses in the fiber. The exponent, m, which can take a value between 0 and 1, can represent any condition between stress range ($m = 1$) and maximum stress ($m = 0$) being the correlating fiber stress parameter. In comparison, the SWT parameter is equivalent to the case where $m = 0.5$ and is less general.

Multiple-Mechanism-Based Modeling

From the use of stress range or strain range in a composite as a criterion for isothermal fatigue failure and noting that failure can be attributed to either matrix or fiber stresses, it becomes apparent that two (or more) different mechanisms may be involved in composite fatigue, depending on variables, such as temperature, frequency, or stress ratio. Under nonisothermal fatigue, additional complications arise, which point the way toward consideration of the constituent stresses in life prediction. Comparing in-phase and out-of-phase TMF, e.g., maximum stresses can occur at either the maximum or minimum temperature of the TMF cycle, respectively. Furthermore, the stress ranges are considerably different for each of the constituents for the two TMF cases because of the thermal stresses induced during in-phase or out-of-phase cycling. These observations, combined with extensive life prediction analysis of a broad database, led Neu and Nicholas [31] to the conclusion that a single parameter cannot be used to correlate fatigue life data obtained under the most general conditions of isothermal fatigue (IF) and TMF.

In their investigation, Neu and Nicholas [31] pointed out that a single parameter cannot be used in correlating fatigue data from a variety of IF and TMF conditions because the mechanisms are different for each type of test, frequency, stress ratio, temperature, etc. In their work, several parameters and models to correlate the cycles to failure of a unidirectional metal matrix composite (SCS-6/TIMETAL®21S) undergoing thermal and mechanical loading were examined. Three different cycle types were considered: OP TMF, IP TMF, and IF. Using a large body of fatigue data on the unidirectional composite covering maximum temperatures up to 815°C, a single parameter based on either the fiber or matrix behavior was shown to be unable to correlate the cycles to failure of all the data. Examples of these are presented in Figures 10 and 11 where fatigue life is correlated with maximum fiber stress and matrix stress range, respectively. These parameters are used in the linear life fraction model, discussed later. The figures show that either parameter is unable to correlate data from all three types of tests evaluated, IP and OP TMF from 150 to 650°C, and IF at 650°C. In the case of the maximum fiber stress (Figure 10), OP TMF data are not correlated well, whereas for the matrix stress range (Figure 11), IP TMF and IF data are not well correlated. Neu and Nicholas [31] presented two prediction methods that assume that life may be dependent on at least two fatigue damage mechanisms and, therefore, must consist of at least two terms. The first method, the linear life fraction model (LLFM), showed that by using the response of the constituents, the life of these different cycle types was better correlated using two simple empirical relationships: one describing the fatigue damage in the matrix and the other fiber-dominated damage. The second method, the dominant damage model (DDM), is more complex but additionally brings in the effect of the environment. The

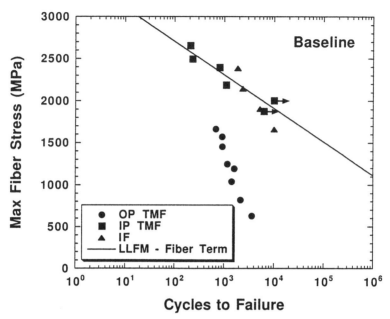

Figure 10. Fatigue life as function of maximum fiber stress showing correlation with fiber term in LLFM [31].

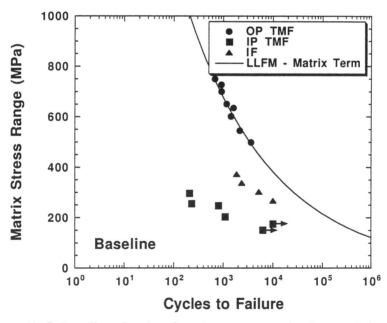

Figure 11. Fatigue life as function of matrix stress range showing correlation with matrix term in LLFM [31].

226

latter method improved the predictions of the effects of the maximum temperature, temperature range, and frequency especially under OP TMF and IF. Figure 12 illustrates the capability of the DDM to correlate TMF data at different temperatures and fiber volume fractions (V_f) while, at the same time, it demonstrates the sensitivity of IP TMF life to V_f. The predictions versus experimental lives are shown in Figure 13 for the DDM approach, and similar correlation capability was demonstrated by the LLFM. Both methods are described in the following section. The steady-state response of the constituents was determined using a one-dimensional micromechanics model with viscoplasticity, FIDEP [32]. The residual stresses due to the CTE mismatch between the fiber and matrix during processing were included in the analysis. Another feature of the modeling approach was the use of micromechanical stresses from the tenth cycle, after the majority of initial stress redistribution had taken place due to creep in the matrix. Furthermore, the FIDEP calculations used a viscoplastic model incorporating damage as developed by Neu et al. [33]. From this study, it was concluded that a general life prediction model for

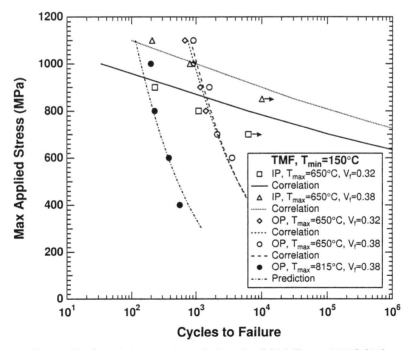

Figure 12. Correlations and predictions for DDM; T_{min} = 150°C [31].

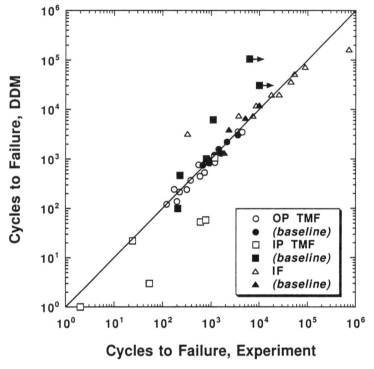

Figure 13. Comparisons of data and predictions using DDM [31].

MMCs that is capable of predicting the cycles to failure for different cycle types requires at least two terms. The justification was that at least two distinct fatigue damage mechanisms occur, depending on the cycle type. None of the single correlating parameters, which either describe the matrix or fiber behavior, were able to correlate fatigue life for all cycle types within a factor of 10.

A comparison of the capability of both single parameter and multi-parameter models to predict fatigue lives under a broad combination of isothermal and TMF conditions is presented by Calcaterra et al. [34]. It is shown there that, for data obtained under limited conditions, any number of models can provide decent correlations between predicted and experimentally observed lives. For predicting fatigue lives under the broad range of conditions available, on the other hand, all of the models suffered some weaknesses. It should be noted, also, that the capability of any model to correlate a body of experimental data is influenced by the combinations of conditions under which the data were obtained. Thus, a database that did not contain OP TMF conditions might allow a model that did not handle matrix-dominated fatigue well to represent the database adequately. Similarly, data obtained under tests where creep in the matrix dominates might influence one model more than another. Additionally, the number of tests conducted under a specific set of conditions, as a fraction of the total number of tests, can have an influence in assessing the correlating capability of a given model. The following sections describe some of the more general models for handling a broad database using more than one correlating parameter to account for the complex material behavior and multiple mechanisms governing that behavior.

Dominant Damage Model

The framework for developing a mechanism-based life prediction model for metal matrix composites is described by Neu [35]. For SCS-6/Ti-24Al-11Nb (atomic percent) $[0]_8$, three dominant damage mechanisms were identified: (1) matrix fatigue damage, (2) surface-initiated environmental damage, and (3) fiber-dominated damage. Damage expressions were developed for each mechanism along with a method for determining the constants. The damage is summed to obtain the total life. The model is capable of making predictions for a wide range of histories, including isothermal fatigue (IF) at different frequencies and stress ratios, thermomechanical fatigue (TMF) under in-phase and out-of-phase cycling conditions, thermal cycling at constant stress, and stress holds at either maximum or minimum stress.

The life prediction model sums the damage contributions from fatigue of matrix, surface-initiated environmental damage in the matrix, and fiber-dominated damage using the linear damage summation concept:

$$\frac{1}{N_f} = \frac{1}{N_f^{fat}} + \frac{1}{N_f^{env}} + \frac{1}{N_f^{fib}} \tag{13}$$

where N_f represents the cycles to failure, and superscripts in the three right-hand terms represent fatigue, environment, and fiber, respectively. A low cycle fatigue strain-life relation incorporating mean stress effects, based on stresses and strains in the matrix material, is used to represent N_f^{fat}:

$$\frac{\Delta \varepsilon_m^{(m)}}{2} = \frac{\sigma_f' - \sigma_m^{(m)}}{E^{(m)}} \left(2 N_f^{fat} \right)^b + \varepsilon_f' \left(2 N_f^{fat} \right)^c \tag{14}$$

where $\sigma_m^{(m)}$ is the mean stress, $\Delta\varepsilon_m^{(m)}$ is the mechanical strain range in the matrix, and $E^{(m)}$ is the matrix material modulus at room temperature. The remaining parameters are obtained from low cycle fatigue tests on the matrix material at room temperature as described in Neu [35]. The micromechanical stresses are computed using the concentric cylinder model, FIDEP.

The environmental term is derived assuming that environmentally enhanced fatigue damage is controlled by the rate of oxidation of the matrix. Computing the critical oxide thickness when environment-dominated growth ends and the crack grows faster than the oxide growth rate, the environmental damage per cycle can be determined:

$$\frac{1}{N_f^{env}} = \left[\frac{C_{crit}}{\Phi_{ox}D_{eff}}\right]^{-1/\beta} 2^{a/\beta} t_c^{(1-a/\beta)} \left(\Delta\varepsilon_m^{(m)}\right)^{(2+a)/\beta} \tag{15}$$

where t_c is the period of a cycle, Φ_{ox} is the effective phasing of the cycle and is a function of the ratio of the thermal strain rate, $\dot{\varepsilon}_{th}^{(m)}$, and mechanical strain rate, $\dot{\varepsilon}_m^{(m)}$, of the matrix:

$$\Phi_{ox} = \frac{1}{t_c} \int_0^{t_c} \exp\left[-\frac{1}{2}\left(\frac{\left(\dot{\varepsilon}_{th}^{(m)}/\dot{\varepsilon}_m^{(m)}\right) + 2}{\xi_{ox}}\right)^2\right] dt \tag{16}$$

where ξ_{ox} is an empirical constant that controls the severity of the phasing effect between OP TMF and IF. D_{eff} in Equation (15) is an effective oxidation constant described by an Arhennius-type expression averaged over a cycle:

$$D_{eff} = \frac{1}{t_c} \int_0^{t_c} D_0 \exp\left(-\frac{Q_{ox}}{RT}\right) dt \tag{17}$$

where D_0 is the diffusion coefficient, R is the gas constant, and Q_{ox} is the apparent activation energy for diffusion. The remaining parameters in Equation (15) and the procedures for determining their values are described in detail in the paper by Neu [35].

Finally, a fiber fatigue damage term is developed based on the concept of a stress-assisted environmental degradation of the carbon layers of the SCS-6 fibers and fracture of the weakened fibers, the resulting expression being:

$$\frac{1}{N_f^{fib}} = A_{fib} \Phi_{fib} \exp\left(-\frac{Q_{fib}}{RT}\right)\left[\frac{\sigma_z^{(f)}}{\sigma_T}\right]^m dt \tag{18}$$

where the kinetics of the environmental attack are embedded in the expression and the term Φ_{fib} is used to account for phasing between stress in the fiber and temperature. This expression utilizes the axial fiber stress, $\sigma_z^{(f)}$ normalized with respect to the ultimate tensile strength of the fiber, σ_T. The expression for the phasing factor Φ_{fib} has a form similar to Equation (16).

The TMF life can then be determined by summing three distinct dominant damage mechanisms that are described by expressions involving the constituent stress and strains, temperature, time, and parameters describing the kinetics of damage. Both the surface-initiated environmental and fiber-dominated damage terms are dependent on temperature through Arhennius-type laws as seen in Equations (17) and (18).

Successful predictions of TMF life for a variety of stress and temperature histories were made. The model demonstrated that the life of in-phase cycling was controlled by fiber-dominated damage, whereas out-of-phase cycling was controlled by surface-initiated environmental matrix cracking. The model was able to predict the frequency, stress ratio, and hold time effects under IF conditions, where damage can be a combination of surface-initiated environmental and fiber-dominated damage. Successful predictions were also made at a lower temperature (150°C) where damage accumulation from surface-initiated environmental and fiber-dominated damage mechanisms is negligible. The predictions of the fatigue life and the dominant damage mode using this model generally agreed well with the experimentally determined lives and the modes of damage accumulation as evidenced through fractographic observations. Examples of the capability of the model to match experimental data at different frequencies and stress ratios and under TMF conditions are presented in Figures 14 and 15, respectively.

Life Fraction Model

The method that has been used by the author and co-workers over the last several years for consolidating data from a variety of isothermal and TMF tests is the life fraction model. The concept of linear and nonlinear life fraction models has been applied in a series of investigations [1,28,36–39]. In the modeling, damage is considered to accumulate simultaneously due to independent mechanisms, each represented in the model by a life fraction. When the sum of the life fractions from each mechanism equals one, failure is predicted. In practice, one damage mode typically dominates in an individual test,

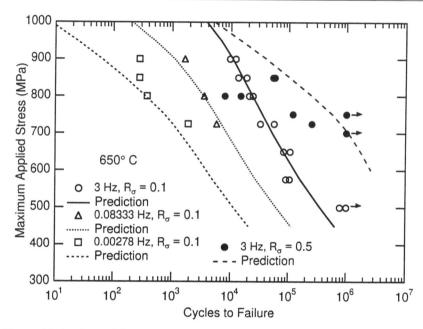

Figure 14. Isothermal fatigue life predictions from dominant damage model and experiments for SCS-6/Ti-24Al-11Nb [0]$_8$ [35].

so that the life fraction concept becomes more of a discrimination test of the dominant mechanism, and life is then governed by the related term. This concept is also embedded in the dominant damage model, Equation (13), in the previous section. In developing these models, consideration was given to the fact that fatigue under isothermal conditions can be dominated by cycle-dependent mechanisms at high frequencies or time-dependent and environmental mechanisms at low frequencies [36–38]. TMF, on the other hand, is typically conducted at a single frequency, and the modeling must consider mechanisms dominated by the stresses in the fiber or matrix, dependent on the stress and temperature history. Thus, a fiber-dominated term and a matrix-dominated term are used in the damage summation law [1,28,38,39]. For the most general loading conditions, which involve variations in test type (e.g., isothermal versus TMF), load or temperature frequency, and maximum temperature, consideration of all of the above mechanisms is required. Thus, cycle-dependent and time-dependent behavior and fiber- and matrix-dominated modes must be addressed in the model development.

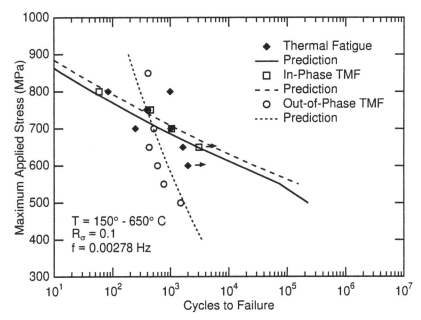

Figure 15. TMF life predictions from dominant damage model and experiments for SCS-6/Ti-24Al-11Nb [0]$_8$ [35].

A review of the procedures developed by the author and his colleagues over the last several years for predicting elevated temperature fatigue life of MMCs is presented by Nicholas [40]. In that study, modeling approaches involving concepts of both linear and nonlinear summation of damage from cycle-dependent and time-dependent mechanisms are summarized. The analyses utilize the micromechanical stresses in the constituents as parameters in the various versions of life fraction models. Modeling has been applied to IF at different frequencies and temperatures, and TMF under both in-phase and out-of-phase loading conditions at different temperature ranges and maximum temperatures. Experimental data were used as the basis for determining the parameters embedded in the models. The numerical results, in turn, provided insight into the dominant mechanisms controlling fatigue life under a given condition. The capability to correlate experimental data from a wide variety of test conditions for several versions of a damage summation model was demonstrated in that study. An example of the ability of the model to consolidate data on unidirectional SCS-6/Ti-24Al-11Nb at three different stress levels

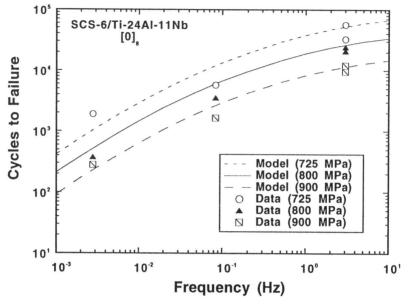

Figure 16. Comparison of model predictions with experimental data as a function of frequency at 650°C [36].

over a broad frequency range is illustrated in Figure 16. The same model was adapted to 0.003 Hz cycling with superimposed hold times at 725 MPa up to 5000 sec and produced good correlation with experimental data on the same material as illustrated in Figure 17. In general, the model was found to have the capability to capture the general trends in the experimental data over a wide range of conditions.

An outline of the life fraction model follows. Over a period of years, a nonlinear damage summation model was developed in a form in which time-dependent damage has a square root dependence:

$$\frac{N}{N_c} + \left(\frac{N}{N_t}\right)^{0.5} = 1 \tag{19}$$

where N_c represents the cycles to failure due to cyclic damage and N_t is the cycles to failure due to time-dependent damage mechanisms. In this type of model, damage is considered to accumulate simultaneously due to independent mechanisms, each represented in the model by a life fraction. When the sum of the damage induced by all mechanisms equals one, failure is predicted. In practice, one damage mode

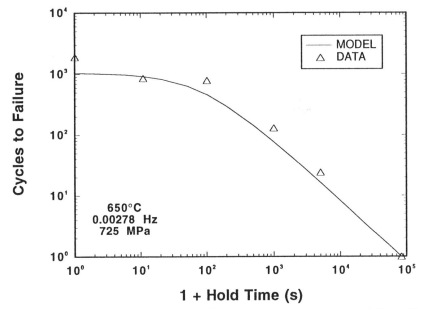

Figure 17. Effect of superimposed hold time at maximum load on fatigue life [36].

typically dominates in an individual test, so that the life fraction concept becomes more of a discrimination test of the dominant mechanism and life is then governed by the related term. To correlate both isothermal and TMF data, the cyclic damage term in Equation (19) is decomposed into fiber and matrix contributions, similar to the approach used by Russ et al. [39] for TMF at a single frequency:

$$\frac{1}{N_c} = \frac{1}{N_f} + \frac{1}{N_m} \qquad (20)$$

where the fiber- and matrix-dominated terms are given in terms of fiber and matrix micromechanical stresses, respectively:

$$N_f = 10^{N_0 \left(1 - \frac{\sigma_{f,\,max}\,(1-R_f)^p}{\sigma^*}\right)} \qquad 21)$$

$$N_m = B \, (\Delta\sigma_m)^{-n} \qquad (22)$$

where subscripts f and m refer to stresses in the fiber and matrix, respectively, and N_0, σ^*, B, n, and p are empirical constants. The time-dependent term (N_t) in Equation (19) is obtained by integrating the incremental damage using time to failure, t_c, as a function of stress:

$$t_c = A\, \sigma^{-m} \tag{23}$$

which produces the following expression:

$$\frac{1}{N_t} = \frac{1}{f\,\sigma_{max}\,(1-R)} \int_{R\,\sigma_{max}}^{\sigma_{max}} \frac{d\sigma}{A\,\sigma^{-m}} \tag{24}$$

Carrying out the integration for a triangular wave form results in the following term for use in Equation (19):

$$N_t = \frac{Af\,(m+1)\,(1-R)}{\sigma_{max}^{\,m}(1-R^{m+1})} \tag{25}$$

where σ is applied stress, σ_{max} is its maximum value, f is cycle frequency, R is stress ratio, and A and m are empirical constants.

For the time-dependent term under TMF conditions, the integration of Equation (24) is carried out using Equation (23), but A is now a function of temperature. For simplicity of integration over a TMF cycle and for lack of experimental data, the variation of parameter A with temperature was taken in the form:

$$A(T) = \frac{A(T_{max} - T_{min})}{T - T_{min}} \tag{26}$$

Reasonably good correlation was obtained for this model with all of the 650°C isothermal data at different frequencies and the TMF data as illustrated in Figure 18 [38]. In this modeling, the micromechanical stresses were obtained from a concentric cylinder model FIDEP for the [0] ply using a viscoplastic model for the matrix, but the [90] ply was represented by an elastic-plastic model fit to FEM results. The micromechanical modeling for this approach used stresses obtained after the first complete cycle of loading.

Modifications to the model for representing data at various frequencies, stress ratios, temperatures, and temperature range for TMF utilized the nonlinear life fraction model, Equation (19), which has cycle-dependent and time-dependent terms. The cycle-dependent term is obtained from a fiber-dominated and a matrix-dominated term, Equation (20), but the computations were based on microme-

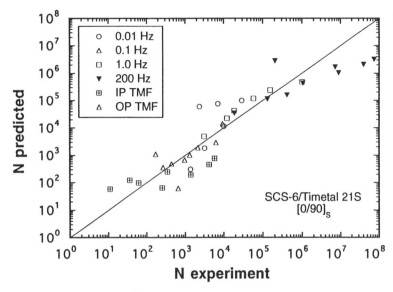

Figure 18. Correlation of life fraction model predictions with experiments for cross-ply composite at a temperature of 650°C [38].

chanical stresses in the tenth cycle. It was found through a number of computations that stress equilibrium is generally approached by 10 complete cycles during which creep and stress redistribution take place. For modeling purposes, the results were observed to be more consistent when the saturation stresses in the fiber and matrix were used than when the first cycle stresses were used. Equations (21) and (22) were used for the fiber and matrix stress terms. As before, the time-dependent or environmental term is based on integration of creep damage based on applied stress level and time, Equation (24), and time-dependent behavior is based on a time to failure as a function of stress, Equation (23).

Temperature dependence of the fiber stress term, the matrix stress term, and the time-dependent or environmental term was introduced in the following manner:

$$\sigma^* = \sigma_0^* - C_1\, T_{max} \tag{27}$$

$$B = B_1\left(1 + \frac{B_2}{T_{abs}}\right) \tag{28}$$

$$A = \frac{A_1}{exp\left(\dfrac{-A_2}{T_{abs}}\right)} \tag{29}$$

where σ^*, B, and A are now the temperature-dependent terms in Equations (21), (22), and (23), respectively, and T_{abs} is absolute temperature ($T_{abs} = T^\circ C + 273$). The remaining parameters, σ_0, C_1, B_1, B_2, A_1, and A_2 are empirical constants. These expressions are used in characterizing the isothermal fatigue behavior at any temperature.

Data from the complete set of fatigue experiments performed on SCS-6/TIMETAL®21S [0]$_4$ were utilized to determine the constants in this model based on a least squares error minimization scheme. The data, including test conditions and cycles to failure, are summarized in Neu and Nicholas [31]. A total of 48 experimental data points were used in the calibration of the model and included isothermal data at frequencies from 0.01 to 200 Hz and isothermal and TMF maximum temperatures up to 815°C. The majority of the data were from isothermal tests at 650°C and TMF tests from 150 to 650°C at a frequency of 0.006 Hz. Because each data point required a computation of the micromechanical stresses at the appropriate conditions and the computations were carried out for 10 complete cycles and involve viscoplastic matrix behavior, no predictions were made for other than experimental conditions. The correlation of the model with the experimental values of cycles to failure is best presented in a plot of computed cycles versus experimental values, Figure 19 [40].

Application to Other Material Layups

Although the modeling described up to this point represented the data sets adequately, it can be noted that TMF tests were generally conducted at a single frequency, which, in turn, was also used for the isothermal tests. The expansion of the experimental database to conditions involving TMF at different frequencies and different maximum temperatures and temperature ranges, such as those reported by Neu and Nicholas [41], pointed out the necessity for expanding the modeling to include frequency and temperature effects into a single model. Furthermore, the modeling was limited to [0] and [0/90] composite architectures in which analytical tools were available to compute micromechanical stresses. Although little progress has been made in model development to address life prediction of alternate architectures, some experimental data provide guidance on how the fatigue life of a quasi-isotropic layup can be approximated. Neu and

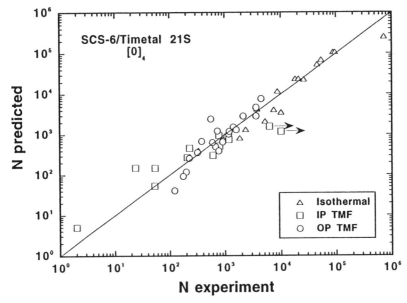

Figure 19. Correlation of life fraction model predictions with experiments for unidirectional composite over a range of temperatures and frequencies [40].

Nicholas [11] conducted TMF experiments on SCS-6/TIMETAL®21S with three different laminate orientations: $[0]_4$, $[0/90]_S$, and $[0/\pm45/90]_S$. Both IP and OP stress-controlled tests were conducted under a temperature cycle of 150 to 650°C and a stress ratio of 0.1. It was found that the fatigue lives for these three different orientations could be consolidated within a factor of 3 by normalizing the maximum applied stress (σ_{max}) by the ultimate tensile strength at the σ_{max} temperature of the TMF cycle. The TMF data on a normalized stress plot are presented in Figure 20. It can be seen that the data correlate quite well, given the amount of scatter generally observed in TMF tests on this material system. No other attempt appears to have been made to predict TMF life in the quasi-isotropic, $[0/\pm45/90]_S$, layup in this class of materials.

Roush et al. [42] have also investigated isothermal and TMF behavior of materials with alternate architectures. In their investigation, they evaluated SCS-6/Ti-15-3 in a $[\pm45]_{2S}$ laminate orientation at a single frequency (0.02 Hz) under IP and OP TMF (149 to 427°C) and IF at 427°C. The fatigue lives obtained from these tests, plotted as a function of maximum applied stress, are presented in Figure 21. It

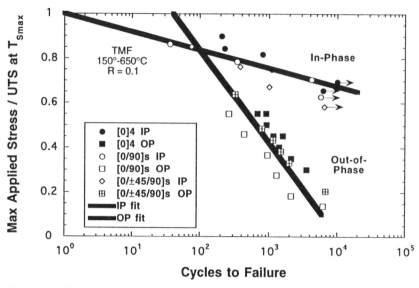

Figure 20. Cycles to failure against normalized stress for three lay-ups of SCS-6/Timetal®21S under IP and OP TMF [11].

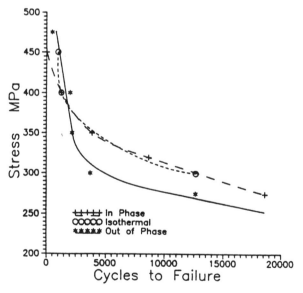

Figure 21. Applied stress versus cycles to failure for angle-ply TMC [42].

240

can be seen that at high-stress levels, above 400 MPa, all three test conditions produced similar lives, although IP TMF was the most detrimental condition. Below 400 MPa, IP TMF and IF produced similar lives, whereas OP TMF produced lives that were generally shorter by a factor of 2. For all conditions tested, fiber-matrix interface failure was the initial mode of damage. Micromechanics analysis performed using the METCAN code [25] confirmed the high interfacial stresses, the presence of large shear stresses, and high-fiber stresses under all conditions tested. The analysis, however, produced only qualitative results because nonlinear matrix material properties were not available. Therefore, elastic properties had to be assumed and, furthermore, the interface was modeled as rigid and fiber breakage was not considered. Additionally, significant fiber rotation occurred during the tests, which was not taken into account in the METCAN analysis. These tests were also characterized by large amounts of fiber breakage and large strain accumulation because of the lack of 0° fibers to constrain the composite in the axial loading direction.

STRAIN-CONTROL AND REVERSED-LOADING TESTS

Interest in titanium matrix composites has been spurred by interest in their potential use in hypersonic aircraft where thin panels could form a part of the fuselage. Because of the interest in thin sections and the relatively high cost of the material, thick sections have not been fabricated to any great extent. The practical consequences of these considerations is that most of the testing of TMCs has been conducted under load control and has been confined largely to tension-only conditions. There is interest in the use of TMCs in aeroengine components where thick sections and compressive loading would be important considerations. Therefore, there have been several investigations to evaluate the reversed loading behavior and strain-controlled testing. The interest in strain control derives partially from experience and practice in LCF of metals where, at a notch or stress concentration in a component, the material undergoes local plastic deformation in an otherwise elastically constrained field. Under this condition, the material sees a more or less constant strain range and will often undergo compression due to reversed plasticity, even though the far-field loading is tensile. Furthermore, stress-controlled testing in metals will normally produce unconstrained strain ratchetting, although no such ratchetting should occur in TMCs because of the constraint in the 0° fibers. The compressive loading under strain control and the use of thin specimens has hindered the testing of TMCs under reversed loading and strain-controlled conditions. The addition of antibuckling guides to overcome this problem can present other constraint problems, such as friction, which have deterred many investigators from pursuing this aspect of fatigue behavior. One of the first reported investigations of strain-controlled fatigue in a TMC is that of El-Soudani and Gambone [43] who tested SCS-6/Ti-6Al-4V in 0° and 90° orientations at room and elevated temperature. Using a strain ratio of 0.1, there was no indication of compressive stresses developing during the test even though a gradual decrease in maximum and minimum stress during the tests was reported. It appears from the data, however, that stress redistribution during the first few cycles may have been allowed to occur before the steady-state ($R_\varepsilon = 0.1$) condition was applied, thereby preventing compressive stresses from developing.

To compare fatigue lives under strain control versus load control, Lerch and Halford [44] conducted a series of tests on unidirectional SCS-6/Ti-15-3 under both conditions using stress and strain ratios of 0 and −1. Denoting the stress- or strain-controlled condition with subscripts σ and ε, respectively, they observed that, in all tests, the stress or strain changed in the early part of the fatigue test but remained relatively constant thereafter for most of the test. Because these tests were conducted at an elevated temperature of 427°C, creep or stress relaxation in the matrix was the obvious cause of the initial redistribution of stresses in the composite. The fatigue life diagrams of life as a function of either total strain range (Figure 22) or total stress range (Figure 23) showed that the tests conducted under $R_\sigma = 0$ produced shorter lives than those conducted under $R_\varepsilon = 0$ by a factor of approximately 2 to 3. The data from the tests conducted under $R_\sigma = -1$ and $R_\varepsilon = -1$ produced nominally equivalent lives, and these were considerably longer than those under either $R_\sigma = 0$ or $R_\varepsilon = 0$ when plotted against total strain range or stress range. Similar lives under stress or strain control under fully reversed loading conditions have also been observed by Sanders et al. [45] on a cross-ply SCS-6/Ti-15-3 composite. The lack of difference in lives for the fully reversed loading conditions in stress or strain control is probably due to the fact that the resultant strain or stress ranges are still close to being fully reversed. Although no micromechanical analyses were conducted to evaluate the constituent stresses in this investigation, the experimental data clearly show that the composite is subjected to a positive R_ε when subjected to $R_\sigma = 0$ or to a negative R_σ when subjected to $R_\varepsilon = 0$ because of the initial creep or stress relaxation in the matrix over the first few cycles. Thus, in comparing fatigue lives on a stress or strain range basis, the strain or stress ratio must be taken into account to make a valid comparison. In the data of Lerch and Halford [44] shown in Figures 22 and 23, values of approximately $R_\varepsilon = 0.28$ and $R_\sigma = -0.24$ represent the steady-state conditions under stress or strain control, respectively, for an applied $R = 0$. In plotting the data, it is not surprising to find that, for a given stress or strain range, the lives are longest at lowest values of R and shortest at highest values of R. In Figure 22, e.g., the three curves of progressively longer lives correspond to values of R_ε of 0.28, 0, and -1, respectively. Similarly, in Figure 23, progressively increasing lives for a given stress range correspond to values of R_σ of 0, -0.24, and -1, respectively. This serves to illustrate why stress ratio or mean stress must be taken into account when representing fatigue lives on a plot of applied stress or strain range against cycles to failure. If the stress or strain range does not stabilize, as in the case of low-frequency fatigue where fiber fracture occurs during the test [44,46], then the prior comments are not valid because a single value of stress or strain range has no particular meaning and an "average" value cannot represent a phenomenon that occurs throughout the test. This is especially true when testing composites that have no 0° fibers, as in the tests of Lerch et al. [47], where a $[\pm 30]_{8S}$ SiC-Ti-15-

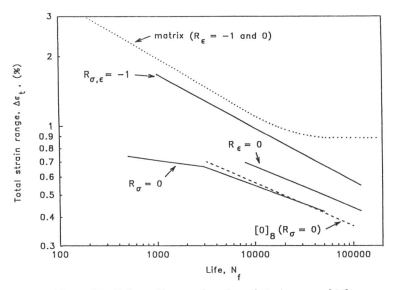

Figure 22. Fatigue life as a function of strain range [44].

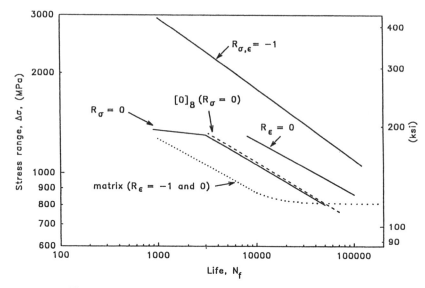

Figure 23. Fatigue life as a function of stress range [44].

3 composite showed significant strain ratchetting under zero tension load-controlled testing at 427°C. Although strain range remained fairly constant in those tests, the mean strain increased from less than 0.5 to beyond 4% during the test. However, under fully reversed strain-controlled testing using a 32-ply material to prevent buckling, the amount of strain ratchetting was very small.

A hybrid method of strain-controlled testing was previously used by Bartolotta [16] and subsequently by Sanders and Mall [48] where $R_\varepsilon = 0$ conditions are modified to prevent the specimen from going into compression. In this case, the maximum strain is kept constant, but the minimum strain is continually increased to keep the minimum stress at zero. In such a test, the strain range is generally decreasing throughout the test. Although Bartolotta concluded that the lives in SCS-6/Ti-24Al-11Nb at $R = 0$ were no different under stress- or strain-controlled conditions, the use of hybrid strain control probably accounts for this similarity. Lerch and Halford [44] found that hybrid strain-control tests at $R_\varepsilon = 0$ produced equivalent lives to those obtained under $R_\sigma = 0$ testing and concluded, further, that such hybrid strain-control tests should not be used to predict response under true strain-controlled conditions. Results similar to those of Bartolotta [16] were obtained by Sanders and Mall [49] on unidirectional SCS-6/Ti-

15-3 at 427°C when they compared fatigue lives on a strain range (at half-life) basis from their tests using the hybrid strain-control mode with available data obtained under stress control. Although the lives were comparable from both test types, no comments were offered regarding the potential comparison of strain-controlled testing with the hybrid strain-control mode. Equivalent lives under hybrid strain control and stress control were also observed by them for the same material in the 90° laminate orientation.

Fully reversed loading tests have been conducted by Boyum and Mall [50] on SCS-6/Ti-15-3 $[0/90]_{2S}$ laminates at room temperature. Using a short gauge length specimen of 25 mm on this 8-ply material, they were able to conduct the tests at a frequency of 10 Hz without any antibuckling guides. Results under tension-compression (TC) loading at $R_\sigma = -1$, when compared with those obtained under tension-tension (TT) loading under stress control at $R_\sigma = 0.1$, showed that TC produces shorter lives than TT when compared on a maximum stress basis (Figure 24), whereas TT produces shorter lives than TC when compared on a stress range basis (Figure 25). The differences can be explained on the basis of mean stress in the TT tests, which is detrimental to fatigue life, a conclusion similar to that of Lerch et al. [47] from tests on a [±30] laminate tested under strain-controlled conditions. The study of fatigue under fully reversed cycling was extended by Boyum and Mall [51] to elevated temperatures where a unique antibuckling guide was employed. Similar experiments were subsequently conducted by Sanders et al. [45] using a similar antibuckling guide. In [51], the room temperature tests of a previous study [50] were duplicated, and the results validated those of prior tests in which no guide was used. Comparison of elevated temperature data with room temperature data under TC cycling showed that longer fatigue lives are obtained at elevated temperature for all stress levels, although the differences narrow as stresses are increased. The longer lives at elevated temperature were attributed to improved matrix ductility. At lower stresses, where the greatest differences in lives were observed, there was considerably more strain accumulation and matrix cracking at room temperature. No micromechanical stress calculations were reported for this investigation.

Fully reversed fatigue testing at $R_\sigma = -1$ has been conducted by Majumdar and Newaz [52] using an antibuckling guide in a chamber filled with inert gas. Their investigation included tests with three different fiber volume fractions. In comparing data obtained at different values of R, they observe that strain range is better than stress range in correlating data at different R and different fiber volume fractions

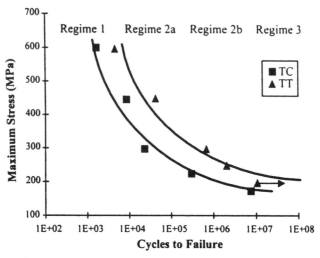

Figure 24. S-N curves on maximum stress basis at room temperature [50].

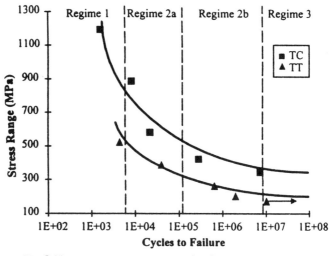

Figure 25. S-N curves on stress-range basis at room temperature [50].

246

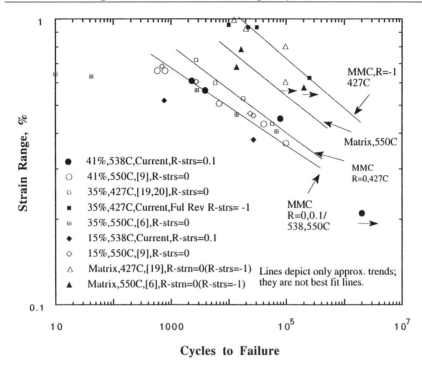

Figure 26. Fatigue life of SCS-6/Ti-15-3 at elevated temperature plotted on a strain range basis [52].

as illustrated in Figures 26 and 27, respectively. However, neither provides a method for correlating the fully reversed loading data with data obtained at values of R near zero. They point out that data at different R could be collapsed by accounting for mean stress as is done often for monolithic metals, but no such attempt was made in their work.

INTERPRETING MODELING RESULTS

In addition to providing a method for predicting fatigue lives, mechanism-based life fraction or multiple mechanism-based models can provide additional information regarding the mechanisms governing the fatigue process. In particular, Neu [53] has used numerical results from a dominant damage model [35] to map out regions in load and temperature space where each of three mechanisms is dominant. The task is accomplished numerically by predicting fatigue lives from the

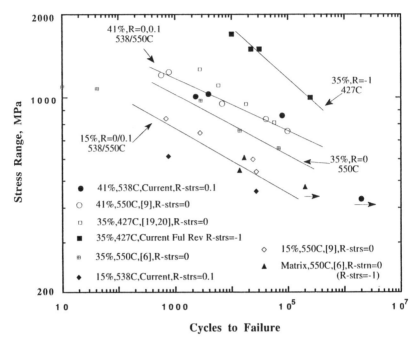

Figure 27. Fatigue life of SCS-6/Ti-15-3 at elevated temperature plotted on a stress range basis [52].

three-term model for each of a series of closely spaced loading and temperature conditions and then noting which term produces the lowest fatigue life. The results are then plotted in the form of a mechanism map as illustrated for OP and IP TMF in unidirectional SCS-6/TIMETAL®21S in Figures 28 and 29, respectively. These plots are for a value of ΔT of 500°C and show, at a glance, which conditions are dominated by which mechanism. Both curves show, e.g., that at low applied stresses and high maximum temperatures, failure is dominated by a combined matrix fatigue and environment mechanism. This region is one of essentially pure thermal cycling. The maps also serve to explain why IP TMF (Figure 29) can produce equal lives at both low and high applied stresses at high maximum temperatures. At high applied stresses, the failure is fiber dominated due to the high-fiber stresses resulting from the combined contribution of thermal and mechanical stresses at maximum load. At low applied stress, on the other hand, thermal stresses in the matrix dominate.

Similar observations on dominant mechanisms, without the use of maps, were made by Nicholas and Johnson [46] by examining the in-

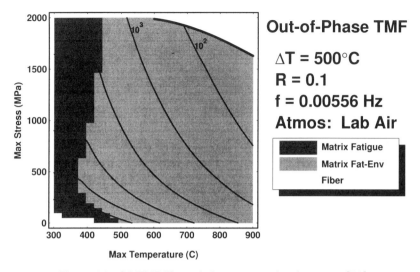

Figure 28. OP TMF life and damage mechanism map [53].

dividual terms in a life fraction model and noting which terms predicted the shortest lives. Such observations, combined with a reexamination of experimental data, led them to the conclusion that creep, in-phase TMF, hold time fatigue loading, and low frequency IF were all governed by a time-dependent mechanism. Under certain condi-

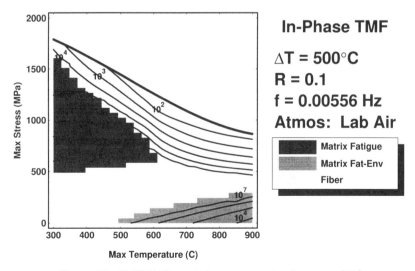

Figure 29. IP TMF life and damage mechanism map [53].

tions, the mechanism was simply the relaxation of stress in the matrix, which, in turn, produces fiber stresses above the strength of the weakest fibers and the subsequent failure of a bundle of fibers. Results from the four types of tests are shown in Figure 30 in which time to failure is plotted against maximum stress. Using an effective time, which accounts for the actual time during a cycle that the material is subjected to high stress at high temperature, when relaxation occurs, the data are seen to fall in a single band. The results indicate that there was essentially no fatigue involved in the cases studied, including the in-phase TM"F" tests, low frequency isothermal "fatigue," and hold time "fatigue" tests. The apparent lack of trend of decreasing life with increasing stress would appear to demonstrate that the data are indicative of a statistical process more than a deterministic one. Further, the time-dependent nature of the tests conducted implies that failure occurs only if the applied stresses are within a narrow band so that the degradation of strength with continued time of exposure is not a dominant issue. The data seem to represent the equivalent of Region I of a Talreja diagram [19], which is governed by the statistical failure of fibers due to fiber stresses that are sufficiently high to cause initiation and propagation of failure of a bundle of fibers. These are just an example of cases where life modeling, based on understanding of mechanisms, produces results that can be referred to the mechanisms on which the model was developed.

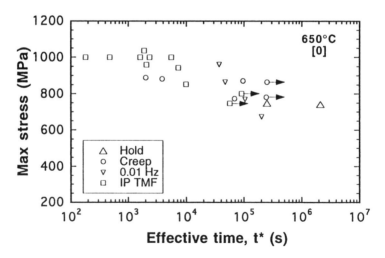

Figure 30. Stress-effective time plot for all tests at 650°C [46].

As another example, Johnson et al. [54] evaluated fatigue life under generic hypersonic vehicle spectrum loading, where it was noted that holding a significant load at temperature was detrimental to the fatigue life. For the cross-ply material used in that investigation, SCS-6/Ti-15-3 [0/90]$_{2S}$, and at the loads and temperatures investigated, the material accumulated strain in a manner characteristic of classical creep behavior. However, for this material, the strain accumulation was due primarily to internal damage in the form of fiber-matrix debonding and matrix cracking and not due to deformation associated with matrix creep or because of an accumulation of 0° fiber breaks.

An understanding of mechanisms coupled with numerical results from fatigue life modeling can also be used to explain experimental results, such as the commonly observed "crossover behavior" in in-phase and out-of-phase TMF plots of cycles to failure against maximum applied stress. Although attempts have been made to relate the crossover stress to the points of nonlinearity in the high-temperature isothermal stress-strain curve of the composite [26,28], other data show no such correlation [24,27,42]. The lack of correlation is especially true in the cases where off-axis plies are used in the composite architecture [24,42]. As noted previously, analysis of data has shown that in-phase TMF is governed primarily by stress relaxation in the matrix and subsequent failure of the fiber bundle. Furthermore, this mechanism occurs over a narrow range of applied stresses and, thus, the IP TMF data produce relatively flat *S-N* curves. Conversely, OP TMF produces curves that are relatively steep and is associated with a matrix fatigue mechanism. Because the thermal stresses provide a major contribution to the matrix fatigue under many test conditions reported in the literature, the results are relatively insensitive to the applied mechanical loads. The net result, therefore, is a "crossover" of a nearly flat in-phase curve with a very steep out-of-phase curve, and this occurs at a stress level that represents the value in the in-phase tests. This value, in turn, is related to fiber bundle strengths and relaxed matrix strengths. It could, therefore, be related to stress-strain behavior where fiber breakage first becomes apparent. An example of such behavior is illustrated in Figure 31 where IP and OP TMF data are shown along with isothermal data at the maximum and minimum temperatures of the TMF cycle. For the conditions used in these experiments, the high-temperature IF data produce longer lives than the IP TMF data, and the low-temperature IF data produce significantly longer lives than the OP TMF data. The ordering of the data is dependent on specific conditions that compete or add the thermal stresses in fiber and matrix with the associated mechanical stresses.

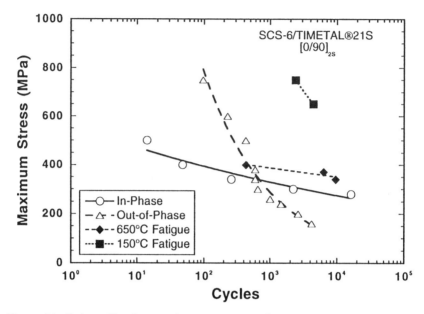

Figure 31. Fatigue life diagram for 150 and 650°C IF and 150–650°C TMF [40].

A different ordering of the curves can be produced under different conditions as illustrated previously, e.g., in Figure 8. The flatness and steepness of the IP and OP TMF curves can vary with material and layup in addition to test conditions as illustrated in Figures 8, 9, and 15. This makes it risky to draw general conclusions on failure modes from one material and set of loading conditions when predicting the behavior of a second material or loading condition.

As more experiments and analyses are conducted, conclusions drawn from different investigations start to fall into consistent patterns. For example, the observation that fatigue life is controlled by the stress range in the 0° fiber was reported by Pollock and Johnson [5] for SCS-6/Ti-15-3. Nicholas and Johnson [46], on the other hand, concluded that only low frequency IF and IP TMF tests were controlled by the stresses in the fibers, which resulted from stress relaxation in the matrix. Thus, time-dependent creep in the matrix has to be considered. They observed that an average stress of approximately 2000 MPa in the fiber is sufficient to initiate catastrophic failure in a composite with SCS-6 fibers. In the analysis by Mirdamadi et al. [6] of the composite tested by Pollock and Johnson [5], the calculated 0°

fiber stress range was about 2000 MPa with $R = 0.07$ for the fiber. Although they conclude that $0°$ fiber stress range controls life, they note that this is "within a given set of parameters (i.e., temperature, loading frequency, and time at temperature)." An analysis of a large body of IP TMF data over a wide range of conditions from several layups of SCS-6/TIMETAL®21S shows that a maximum fiber stress of slightly more than 2000 MPa correlates fatigue lives from 10 to 10^4 cycles to failure [34]. The authors note that the best correlation between fiber stress and fatigue life would be a horizontal line that, although not useful for predicting lives, could be a useful tool for design by establishing an upper limit for fiber stress. Other investigations have led to the conclusion that hold time at temperature is a very significant contributor to fatigue life under mission spectrum loading [55,56], which seems to corroborate the interpretation of Nicholas and Johnson [46] that one mode of failure involves fiber stresses reaching a critical value after stress relaxation in the matrix has taken place.

Fiber Breakage

A number of studies have been performed in which fiber breakage is identified as one of the modes of failure in TMCs that contain fibers in the $0°$ orientation. These studies range from direct observations of fiber breakage, indirect observations, and inferences drawn from experimental observations of strain accumulation. For example, Russ et al. [1] were one of the first to observe continuous strain accumulation under IP TMF and reasoned that the strains might be the result of matrix stress relaxation. In subsequent work, the breaking of fibers was also considered to be necessary to obtain such large strains. An example of strain accumulation under IP TMF and IF, both at a low frequency of 0.0056 Hz, is illustrated in Figures 32 and 33, respectively, for a cross-ply SCS-6/TIMETAL®21S composite. The important feature that both figures illustrate is that maximum and minimum strain accumulate at the same rate, consistent with a creep-type phenomenon. OP TMF, on the other hand, usually shows accumulation of maximum strain only, whereas minimum strain remains constant, implying a damage mechanism, such as matrix cracking which produces a decrease in specimen stiffness [1]. Strain accumulation under IP TMF has also been observed in Ti-15-3 composites in a [±45] orientation [42], a quasi-isotropic layup [24], and in the TIMETAL®21S composite in three orientations: [0], [0/90], and quasi-isotropic [11]. The presence of $0°$ fibers in any layup would seem to rule out the possibility of continued strain accumulation without breakage of fibers. A

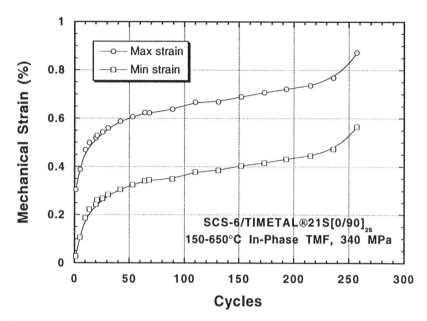

Figure 32. Maximum and minimum strain history for IP TMF, 340 MPa, 150–650°C [1].

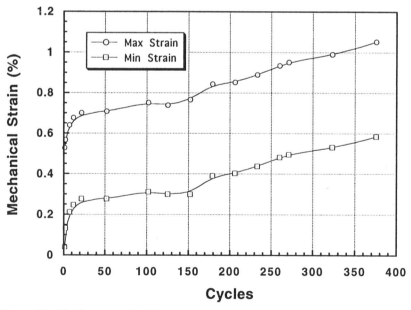

Figure 33. Maximum and minimum strain history for IF at 650°C, 400 MPa [1].

254

strain accumulation mechanism due to fiber breakage, rather than creep, was proposed by Mall et al. [28] based on observations under both IP TMF and low-frequency IF, although no direct evidence of breaking fibers was presented. Gabb et al. [2] compared results from sequential and simultaneous TMF tests on two SiC reinforced titanium matrix materials, Ti-24Al-11Nb and Ti-15-3, and noted that matrix relaxation induces load shedding from the matrix to the fibers in both systems under IP TMF. Fiber cracking under IP TMF has been reported in the literature by Castelli and Gayda [57] along with strain accumulation. A related observation was made by Bartolotta and Verrilli [58] who also observed that there was little or no matrix cracking under IP TMF. It was also observed that, although there was a strong environmental influence on the fatigue lives under OP TMF, due to the environmental susceptibility of the matrix at high temperatures, no conclusive evidence of environmental influences under IP TMF was obtained. This observation indicates that the fiber strength, which governs the IP TMF behavior, is not affected by environment. The most direct observations of cracked fibers, obtained from polished sections, have been reported by Majumdar and Newaz [52], Castelli and Gayda [57], Lerch and Halford [44], and Neu and Roman [59], among others. The first two studies report such findings under IP TMF, whereas Lerch and Halford [44] make their observations under low-frequency isothermal conditions. Majumdar and Newaz [52] also found fiber fractures under 0.01 Hz IF under high stresses along with experimental observations of strain ratchetting. Neu and Roman [59] observed numerous fiber fractures on polished cross sections but recorded only a few fractures with acoustic emission monitoring. It is also of interest to note that many closely spaced fiber cracks have been observed and reported by Brindley et al. [60] on specimens that have undergone tensile testing. To add more confusion to the picture, Nicholas and Ahmad [61] presented an analysis of fiber breakage, which attempted to explain how closely spaced fiber fractures in almost every fiber could still result in both significant load carrying capability of the composite and minimal stiffness reduction, consistent with reported experimental observations.

As noted earlier, Nicholas and Johnson [46] relate failure under IP TMF to the progressive failure of a bundle of fibers as the stress in the matrix relaxes over time due to stress at temperature. Such a phenomenon can occur under pure creep or under low-frequency fatigue, provided that matrix cracking does not occur first due to the stress range in the matrix, early cracking as a result of the stress concentration produced by off-axis plies, or environmental degradation of the fatigue resistance of the matrix near the surface. All of these phenomena would have to be considered in the development of a robust model for fatigue life prediction. These conditions, notwithstanding, the failure mode and its modeling due to breakage of a bundle of fibers is fairly straightforward, although the statistical aspects could provide mathematical complications. The most complicating feature of the modeling of fatigue life due to the progressive breakage of fibers as a result of increasing stress levels from matrix load shedding is determining the number of fibers that are broken and the shear lag response that ultimately shares the load with adjacent fibers. Although several studies have shown polished cross sections containing multiple fiber breaks [44,52,59], the polishing technique has not been questioned because control samples, either untested or tested under OP TMF where fiber breaks are not expected, show no indications of broken fibers when polished under identical conditions. Only recently has it been pointed out that the residual stress state in a composite that has been subjected to IP TMF or low-frequency IF may result in fibers in tension at room temperature [62], depending on the material, test temperature, stress level, frequency, and number of cycles. These tensile residual stresses in the fibers are hypothesized to cause cracking in the brittle SiC fibers when subjected to mechanical polishing. This has been verified in a systematic series of tests in which the matrix of the other half of a polished specimen, that showed extensive fiber cracking after IP TMF, was electrochemically dissolved, and only a very few broken fibers were found [62]. This does not rule out fiber breakage as a mechanism of failure. Rather, it points out that the number of breaks may be significantly less than that revealed by polishing of cross sections, and these numbers may be only a small fraction of the total number of fibers in the cross section except when the point of instability and significant strain increase is reached near the end of a test. It is of interest to point out that Majumdar and Newaz [52] observed, from metallographically polished sections, that periodic fiber fractures occurred under IP TMF at stress levels as low as 300 MPa in a SCS-6/Ti-15-3 composite. On the other hand, interrupted experiments, in which fibers were extracted by leaching away the matrix, appeared to indicate that there was no significant fiber strength degradation due to TMF cycling.

The tensile residual stresses that arise under IP TMF conditions are apparent from micromechanical calculations that, in turn, were used as the basis of several different life prediction methodologies as noted previously. For example, plots of fiber stress after 10 TMF cycles in a cross-ply SCS-6/TIMETAL®21S composite (Figure 34) clearly

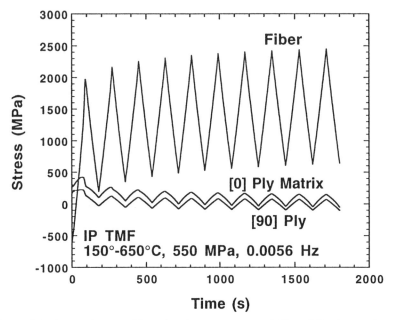

Figure 34. Stress-time profiles showing response of fiber, [0] ply matrix, and [90] ply over 10 cycles to IP TMF during 180 s cycle, σ_{max} = 550 MPa, 150–650°C [63].

show the existence of high tensile stresses, about 700 MPa, at minimum load and temperature (150°C) under in phase TMF [63]. These stresses relax by about 100 MPa [33] when the temperature is lowered to room temperature, but the residual stress is still tensile and is approximately 600 MPa for the case shown in the work by Coker et al. [63]. On the other hand, OP TMF computations in the same study show that minimum fiber stress, which occurs at a maximum temperature of 650°C, are much less than 500 MPa, which is the amount of compression developed when cooling from 650°C to room temperature. Thus, specimens tested under OP TMF will have compressive residual stresses in the fibers when cooled to room temperature and polished. The same is true for untested specimens because consolidation from the processing temperature results in fibers in compression [33]. Thus, it is not surprising to expect that brittle silicon fibers, which are in residual tension, will develop cracks when polished after testing under IP TMF. For the case of IF, the residual stresses in the fiber after testing and then cooling to room temperature can be either tensile or compressive, depending on the material and test conditions.

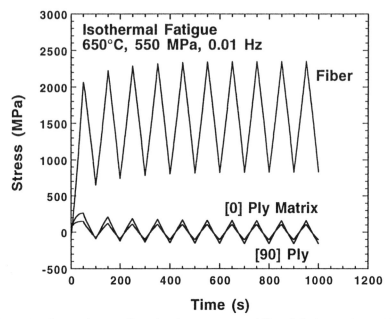

Figure 35. Stress-time profiles showing response of fiber, [0] ply matrix, and [90] ply over 10 cycles to 0.01 Hz isothermal fatigue, σ_{max} = 550 MPa, 650°C [63].

For the case computed by Coker et al. [63], at 550 MPa maximum stress, 0.01 Hz, and 650°C, the fiber stress at minimum load (Figure 35) is about 800 MPa, which will reduce to about 300 MPa at room temperature. For this example, the fiber is still in tension, and this may help explain why some IF tested specimens are found to have many broken fibers when examined after polishing.

Similar observations pointing to fibers being in tension after IP TMF testing can be drawn based on micromechanical calculations by Neu and Roman [59]. For a unidirectional SCS-6/TIMETAL®21S composite, tensile residual stresses are apparent by noting the stresses in the fiber at minimum load and temperature (150°C) after IP TMF cycling at a maximum stress of 1000 MPa and maximum temperature of 650°C. The polished cross sections show numerous broken fibers, with the breaks spaced at only a few fiber diameters. In that study, acoustic emission indicated only a few fiber breakage events. It is of interest to note their observation: "Broken fibers were found in both in-phase tests after testing . . . however, more fiber fractures than the number of AE events indicating fiber fractures were found." This is further support for the hypothesis that fibers fracture during IP TMF, but the number of fractures appears to be far less than indicated on polished cross sections where fibers may crack due to polishing under a state of residual tension.

Comparisons of IP TMF and Isothermal Fatigue

Majumdar and Newaz [52] observed that, in comparing IP TMF behavior with that under isothermal conditions at the maximum temperature of the TMF cycle, the "net effect is a significantly shorter life under in-phase TMF compared with IF in the fiber dominated region of fatigue." They did not point out, however, that the two sets of tests were run at different frequencies. In fact, the isothermal tests were conducted at 1 Hz, whereas the TMF tests were run at much lower frequencies. If frequency is considered in this comparison, the same conclusion may be drawn, but the differences are less significant. Data of this type, where time at stress and temperature affects the stress relaxation behavior of the matrix, should be compared on a time basis and not strictly on a cycle count basis. There are several comparisons of IF data with IP TMF data reported in the literature. Of significance in these comparisons are several factors, which include the matrix material, composite layup, frequency, stress level, and temperature (or temperature range in TMF). Thus, general statements are difficult to make when discussing differences between IF and IP TMF life. One of the experimental observations that contributes to the making of a valid comparison is the amount of strain ratchetting, which, in general, is attributed to the creep behavior of the matrix and the subsequent breaking of fibers due to the high stresses developed.

The comparisons between IF and IP TMF life of Majumdar and Newaz [52], while made primarily at different frequencies, were also made for a limited number of cases under identical frequencies. These still show that for the material studied, unidirectional SCS-6/Ti-15-3, the IP TMF lives are shorter than the IF lives for a given applied maximum stress. A temperature range of 300 to 538°C was reported in that study. A similar comparison was made by Castelli and Gayda [57] on the same material, which led them to conclude that "cyclic lives determined under TMF conditions were greatly reduced from those obtained under comparable isothermal conditions." It should be noted that, in their case, the temperature range for TMF tests was 93 to 538°C. Although these comparisons were made at different frequencies or cycle times (3 min TMF vs. 3 sec isothermal), the conclusions are not altered when the data are compared on a time instead of a cycle basis at high stresses. The differences, however, are slight. At lower stresses, on the other hand, the isothermal lives are lower when compared on a time basis. They also observed significantly more strain ratchetting under IP TMF than under isothermal conditions. The increased ratchetting under TMF was attributed to "the increasing elastic modulus and yield strength of the matrix (due to the decreasing temperature) during unload. Here, the "cold" nominally elastic unload reduces the degree of strain reversal, in turn, enhancing the tensile strain ratchetting effects associated with the matrix load shedding." Although no micromechanical computations are presented to support this hypothesis, recent results on a TIMETAL®21S composite tend to support this explanation of fully reversed plasticity in the matrix under IF with no such reversal under IP TMF [64]. Thus, the strain ratchetting under IF is not larger because of creep at high temperature, but reversed loading retards the net strain accumulation in this material at the temperatures at which it was tested (550°C).

Comparisons between IF and IP TMF life for another material, unidirectional SCS-6/Ti-24Al-11Nb composite, can be found in the works by Bartolotta and Verrilli [58] and Gabb et al. [2]. Here, over a TMF temperature range of 215 to 815°C, IP TMF lives were found to be significantly lower than isothermal LCF lives. Although the cycle times were different for the two test types (180 vs. 10 sec), comparison of the data on a time basis still shows the TMF lives to be considerably shorter at high-stress levels. At lower stress levels, the fatigue life data tend to converge when compared on a time basis, whereas at even lower stresses, the data tend to cross over. In making a similar comparison on unidirectional SCS-6/Ti-15-3, Gabb et al. [2] show only a small difference between TMF (180-sec cycle) and IF (10-sec cycle) lives at high stresses, and this difference disappears when comparisons are made on a time basis. As in the other cases noted above, the curves tend to converge at lower stresses on a cycle basis, thus IF lives are shorter than IP TMF lives at low stress levels.

Comparisons can be made for a third material, SCS-6/TIMETAL® 21S, under IF and IP TMF testing at identical frequencies and maximum temperatures. Nicholas [40] compares the data of Mall et al. [28] on a cross-ply composite to show that the IF and IP TMF curves are nearly parallel and very close to one another, although the isothermal lives are somewhat higher as shown in Figure 31. The isothermal tests at low temperature produce much longer lives than the OP TMF tests, but the curves are nearly parallel. For the same material, but in a unidirectional layup, Neu and Nicholas [31] compare IF and IP TMF data on two plots, Figures 12 and 36. Comparing the IP TMF data at a volume fraction of 0.38 in Figure 12 with 650°C IF data at $V_f = 0.37$ in Figure 36 shows that the IP TMF lives are slightly shorter for a high stress of 1000 MPa, but IF lives are shorter at lower stress levels. Here, the comparisons are made at slightly different frequencies, 0.0067 Hz for TMF and 0.01 Hz for IF. Finally, micromechanical modeling supported by experiments can be used to compare behavior in this same material in two orientations. Neu et al. [33] compare stress-strain loops and strain accumulation under the frequencies used experimentally in their works, 0.0067 Hz for TMF and 0.01 Hz for IF. For a unidirectional composite, the amount of strain ratchetting is much greater under IP TMF than IF. However, there is significant hysteresis on the first cycle in IF, indicating that significant stress redistribution takes place between fiber and matrix, which leads to fully reversed loading in the matrix in subsequent cycles. For the cross-ply material, ratchetting takes place under IP TMF and IF, but IF shows large hysteresis loops on all cycles, whereas

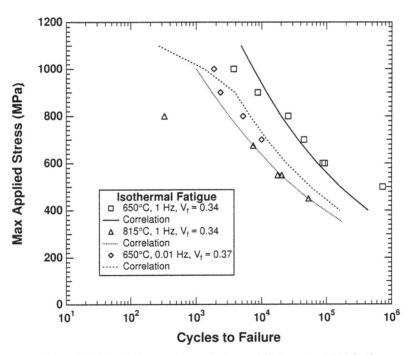

Figure 36. Correlations and predictions of IF data for DDM [31].

TMF cycles show less hysteresis. What is not accounted for in these calculations, however, is the strain accumulation which may be due to the breakage of fibers as discussed in the previous section. Finally, one can use micromechanical calculations and fatigue life modeling to explain differences in behavior from one material to another under similar conditions. In comparing IF at the maximum temperature of a TMF cycle to IP TMF, the stress-strain behavior of the matrix material provides useful information. Figure 37 presents such a plot for unidirectional SCS-6/TIMETAL®21S based on micromechanical computations using the modeling described above [63]. It is clear that the hysteresis loop of the matrix material under isothermal fatigue is much larger than that obtained under IP TMF. What cannot be determined from the calculations alone is whether the hysteresis loop is large enough to produce matrix fatigue before the strain ratchetting and associated stress redistribution causes fiber fracture. In the case presented, the maximum fiber stress is 2477 MPa under IF and 2579 MPa under IP TMF. Such high fiber stresses usually lead to fiber fracture, most likely before matrix fatigue occurs in this material. For comparison, similar curves are shown in Figure 38 for unidirectional SCS-6/Ti-6Al-4V under typical loading conditions for that material. Although there is very little hysteresis in the matrix under IF or IP

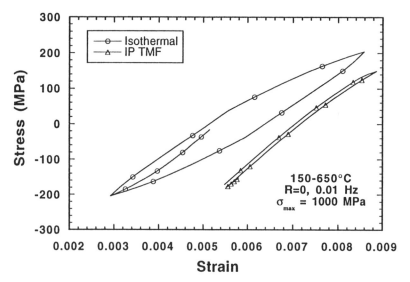

Figure 37. Matrix stress-strain behavior for unidirectional SCS-6/Timetal®21S after 10 cycles.

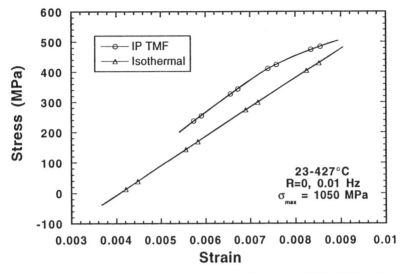

Figure 38. Matrix stress-strain behavior for unidirectional SCS-6/Ti-6Al-4V after 10 cycles.

TMF, the maximum applied stress is higher than in the Timetal composite. Furthermore, the maximum fiber stresses are lower: 2104 and 2063 for IF and IP TMF, respectively. One should expect more of a tendency for matrix fatigue to occur than fiber breakage in this composite compared with the Timetal composite for the conditions evaluated. What these and other observations demonstrate are factors that not only must be considered in life prediction modeling but show up in the numerical analyses that form the basis of such modeling. It is clear from the above comparisons that different mechanisms are dominant, depending on the material and orientation. Furthermore, the stress distribution between fiber and matrix is both complicated and changing with number of cycles when damage, creep, or plasticity are present under cyclic loading. Reversed plasticity or creep, e.g., is illustrated in Figure 37 for a matrix material that is constrained by the 0° fiber. In general, creep in the matrix, leading to increased stress in the fibers, fully reversed plasticity in the matrix, which retards strain accumulation, and matrix fatigue, which causes cracking in the matrix, to be the dominant life limiting factor, all must be considered in fatigue life modeling, along with any associated environmental effects. The dominance of these mechanisms in various regions of a Talreja diagram [19] are pointed out and discussed in detail by Majum-

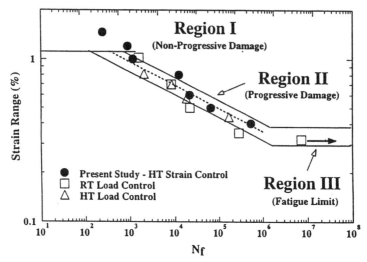

Figure 39. Fatigue life curve for SCS-6/Ti-15-3 [0/90]$_{2S}$ under fully reversed loading as a function of strain range [45].

dar and Newaz [52], Sanders and Mall [49] and Sanders et al. [45], among others, and have been incorporated into fatigue life modeling by Radhakrishan and Bartolotta [18], e.g., and made the basis for mechanism maps by Neu [53]. The Talreja diagram, illustrated schematically in Figure 7, has both the characteristic shape and

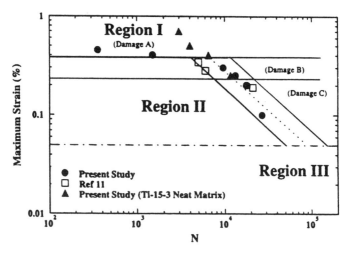

Figure 40. Fatigue life curve for SCS-6/Ti-15-3 [90]$_8$ under hybrid strain controlled loading as a function of maximum strain [65].

mechanistic basis that represents isothermal TMC fatigue life data for materials containing 0° fibers when plotted as a function of maximum strain as illustrated by Figures 6 and 39, e.g., in this chapter. The three regions of a Talreja diagram have also been observed in a transversely loaded unidirectional composite by El-Soudani and Gambone [43] and Sanders and Mall [65], among others, as shown in Figure 40. In this case, although the behavior is dominated by the matrix material, regions of rapid fracture, matrix fatigue, and infinite life are observed just as in composites containing 0° fibers. The strain-controlled mode of testing had to be used to prevent strain ratchetting for the transversely loaded material [26]. It is of interest to note that the data in the fatigue region match data from neat material tested under strain-controlled fatigue, indicating that matrix fatigue governs the

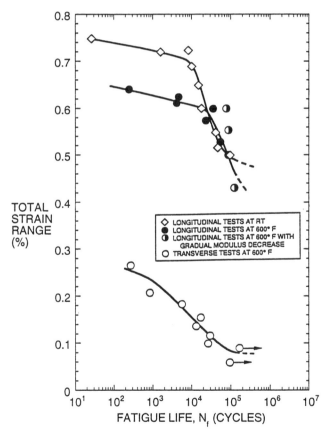

Figure 41. Comparison of fatigue behavior of [0] and [90] orientations of SCS-6/Ti-6Al-4V at room and elevated temperature [43].

behavior in this region. It should also be pointed out that failure does not always occur under strain-controlled testing, so an additional criterion of 75% reduction in maximum stress was used for both the composite and neat material tests. Finally, it should be noted that micromechanical stresses or strains should be considered when comparing data rather than applied stresses or strains because of the large thermal residual stresses present in the TMCs at room temperature. Comparisons of data that form a Talreja diagram, as illustrated in Figure 41, show different applied strains for Region I at room and elevated temperatures. Yet the 0° fibers are in compression at room temperature, so a larger applied strain is necessary to reach the fiber strain limit at room temperature than at elevated temperature. A figure such as this could easily be misleading if one were to conclude, incorrectly, that the strain to failure of the fiber is greater at room temperature.

DISCUSSION AND SUMMARY

Methods have been developed to predict, or at least correlate, fatigue life of TMCs with one or more parameters. For isothermal conditions, a single parameter is sometimes sufficient to correlate data obtained over a range of applied stresses. It may not be important, in this case, to relate the parameter to any particular governing mechanism. For example, strain range in the composite is related directly to stress range in the fiber and stress range in the matrix material. When mean stresses and loading frequencies affect the fatigue life, the single-parameter approach may break down, particularly if the wrong mechanism is assumed in choosing the correlating parameter. This is particularly true when nonisothermal loading conditions, such as IP or OP TMF, are considered. For any but the simplest loading conditions and geometries, it becomes readily apparent that the governing mechanisms are related to stress or strain states in the constituents. Micromechanics analysis, therefore, becomes an important ingredient in developing a life prediction methodology. In addition to this, environmental effects play a significant role at higher temperature so that time has to enter into the modeling explicitly. Several approaches have been developed to take most of these factors into account and appear to have the capability to consolidate data obtained over a wide variety of conditions. The use of stabilized stress or strain states, obtained after a few loading cycles, appears to improve the predictive capabilities by accounting for stress redistribution due to creep, damage, or stress relaxation over the first few cycles. Little or no work appears to have been done in developing a predictive capability for fatigue life under spectrum loading or for composite layups other than unidirectional or cross-ply. Finally, a full appreciation of the role of time-dependent behavior in the matrix material has only recently been realized in life prediction studies.

REFERENCES

1. Russ, S. M., T. Nicholas, D. G. Hanson and S. Mall. 1994. "Isothermal and Thermomechanical Fatigue of Cross-Ply SCS-6/β21-S," *Science and Engineering of Composite Materials,* 3, pp. 177–189.
2. Gabb, T. P., J. Gayda, P. A. Bartolotta and M. G. Castelli. 1993. "A Review of Thermomechanical Fatigue Damage Mechanisms in Titanium and Titanium Aluminide Matrix Composites," *Int. J. Fatigue,* 15, pp. 413–422.
3. Jeng, S. M., J.-M. Yang and S. Aksoy. 1992. "Damage Mechanisms of SCS-6/Ti-6Al-4V Composites under Thermal-Mechanical Fatigue," *Mat. Sci. Engr.,* A156, pp. 117–124.
4. Johnson, W. S., S. J. Lubowinski and A. L. Highsmith. 1990. "Mechanical Characterization of Unnotched SCS6/Ti-15-3 Metal Matrix Composites at Room Temperature," *Thermal and Mechanical Behavior of Metal Matrix and Ceramic Matrix Composites, ASTM STP 1080,* J. M. Kennedy, H. H. Moeller and W. S. Johnson, Eds., American Society for Testing and Materials, Philadelphia, pp. 193–218.
5. Pollock, W. D. and W. S. Johnson. 1992. "Characterization of Unnotched SCS-6/Ti-15-3 Metal Matrix Composites at 650°C," *Composite Materials: Testing and Design (Tenth Volume), ASTM STP 1120,* Glenn C. Grimes, Ed., American Society for Testing and Materials, Philadelphia, pp. 175–191.
6. Mirdamadi, M., W. S. Johnson, Y. A. Bahei-El-Din and M. G. Castelli. 1993. "Analysis of Thermomechanical Fatigue of Unidirectional Titanium Metal Matrix Composites," *Composite Materials: Fatigue and Fracture, Fourth Volume, ASTM STP 1156,* W. W. Stinchcomb and N. E. Ashbaugh, Eds., American Society for Testing and Materials, Philadelphia, pp. 591–607.
7. Mirdamadi, M. and W. S. Johnson. 1996. "Modeling and Life Prediction Methodology for Titanium Matrix Composites Subjected to Mission Profiles," *Life Prediction Methodology for Titanium Matrix Composites, ASTM STP 1253,* W. S. Johnson, J. M. Larsen and B. N. Cox, Eds., American Society for Testing and Materials, Philadelphia, pp. 573–594.
8. Ward, G. T., D. J. Herrmann and B. M. Hillberry. 1993. "Stress-Life Behavior of Unnotched SCS-6/Ti-β21S Cross-Ply Metal Matrix Composites at Room Temperature," *FATIGUE 93,* Volume II, J.-P. Bailon and J. I. Dickson, Eds., EMAS, pp. 1067–1072.
9. Ward, G. T., D. J. Herrmann and B. M. Hillberry. 1995. "Fatigue-Life Behavior and Matrix Fatigue Crack Spacing in Unnotched SCS-

6/Timetal®21S Metal Matrix Composites," *Journal of Composites Technology and Research, JCTRER,* 17(3), pp. 205–211.

10. Gao, Z. and H. Zhao. 1995. "Life Predictions of Metal Matrix Composites under Isothermal and Nonisothermal Fatigue," *Journal of Composite Materials,* 29, pp. 1142–1168.

11. Neu, R. W. and T. Nicholas. 1994. "Effect of Laminate Orientation on the Thermomechanical Fatigue Behavior of a Titanium Matrix Composite," *Journal of Composites Technology and Research, JCTRER,* 16, pp. 214–224.

12. Majumdar, B. S. and G. M. Newaz. 1991. "Thermomechanical Fatigue of a Quasi-Isotropic Metal Matrix Composite," *Composite Materials: Fatigue and Fracture (Third Volume), ASTM STP 1110,* T. K. O'Brien, Ed., American Society for Testing and Materials, Philadelphia, pp. 732–752.

13. Halford, G. R., B. A. Lerch, J. F. Saltsman and V. K. Arya. 1993. "Proposed Framework for Thermomechanical Fatigue (TMF) Life Prediction of Metal Matrix Composites (MMCs)," *Thermomechanical Fatigue Behavior of Materials, ASTM STP 1186,* H. Sehitoglu, Ed., American Society for Testing and Materials, Philadelphia, pp. 176–194.

14. Saltsman, J. F. and G. R. Halford. 1988. "An Update of the Total-Strain Version of Strainrange Partitioning," *Low Cycle Fatigue, ASTM STP 942,* H. D. Solomon, G. R. Halford, L. R. Kaisand and B. N. Leis, Eds., American Society for Testing and Materials, Philadelphia, pp. 329–341.

15. Halford, G. R. 1993. "MMC Ring Fatigue and Fracture Life Prediction: An Engineering Model," *Reliability, Stress Analysis, and Failure Prevention,* DE-Vol. 55, American Society of Mechanical Engineers, New York, pp. 307–315.

16. Bartolotta, P. A. 1991. "Fatigue Life Prediction of an Intermetallic Matrix Composite at Elevated Temperatures," *Failure Mechanisms in High Temperature Composite Materials,* AD-Vol. 22/AMD-Vol. 122, G. K. Haritos, G. Newaz and S. Mall, Eds., American Society of Mechanical Engineers, New York, pp. 45–54.

17. Manson, S. S. 1965. "Fatigue: A Complex Subject-Some Simple Approximations," *Exp. Mech.,* 5, pp. 193–226

18. Radhakrishan, V. M. and P. A. Bartolotta. 1993. "Considerations Concerning Fatigue Life of Metal Matrix Composites," NASA Tech Memo 106144, Lewis Research Center, Cleveland, OH.

19. Talreja, R. 1981. "Fatigue of Composite Materials: Damage Mechanisms and Fatigue Life Diagrams," *Proceedings of the Royal Society of London,* Vol. A378, pp. 461–475.

20. Tong, M. T. and C. C. Chamis. 1993. "Simulation of Thermomechanical Fatigue Behavior of SCS6/Ti-24Al-11Nb Composite," HITEMP Review 1993, NASA Conference Publication 19117, Vol II, pp. 40-1–40-10.

21. Chamis, C. C., P. L. N. Murthy and D. A. Hopkins. 1988. "Computational Simulation of High Temperature Metal Matrix Composites Cyclic Behavior," NASA TM-102115, Lewis Research Center, Cleveland, OH.

22. Rabzak, C., D. A. Saravanos and C. C. Chamis. 1993. "Optimal Synthesis of SCS6/Ti-24Al-11Nb Composites for Improved Fatigue Behavior,"

HITEMP Review 1993, NASA Conference Publication 19117, Vol II, pp. 41-1–41-11.

23. Tong, M. T., S. N. Singhal, C. C. Chamis and P. L. N. Murthy. 1996. "Simulation of Fatigue Behavior of High Temperature Metal Matrix Composites," *Life Prediction Methodology for Titanium Matrix Composites, ASTM STP 1253*, W. S. Johnson, J. M. Larsen and B. N. Cox, Eds., American Society for Testing and Materials, Philadelphia, pp. 540–551.

24. Hart, K. A. and S. Mall. 1993. "Response of a Quasi-Isotropic Metal Matrix Composite under Thermomechanical Fatigue," *Thermomechanical Behavior of Advanced Structural Materials*, AD-Vol. 34, AMD-Vol. 173, W. F. Jones, Ed., American Society of Mechanical Engineers, New York, pp. 47–60.

25. Hopkins, D. A. and C. C. Chamis. 1988. "A Unique Set of Micromechanics Equations for High Temperature Metal Matrix Composites," *Testing Technology of Metal Matrix Composites, ASTM STP 964*, P. R. Giovanni and N. R. Adsit, Eds., American Society for Testing and Materials, Philadelphia, pp. 176–196.

26. Mall, S. and J. J. Schubbe. 1994. "Thermomechanical Fatigue Behavior of a Cross-Ply SCS-6/Ti-15-3 Metal Matrix Composite," *Composites Science and Technology*, 50, pp. 49–57.

27. Mall, S. and B. Portner. 1992. "Characterization of Fatigue Behavior in Cross-Ply Laminate of SCS-6/Ti-15-3 Metal Matrix Composite at Elevated Temperature," *ASME J. Eng. Mat. Tech.*, 114, pp. 409–415.

28. Mall, S., D. G. Hanson, T. Nicholas, and S. M. Russ. 1992. "Thermomechanical Fatigue Behavior of a Cross-Ply SCS-6/β21-S Metal Matrix Composite," *Constitutive Behavior of High Temperature Composites*, MD-Vol. 40, B. S. Majumdar, G. M. Newaz and S. Mall, Eds., American Society of Mechanical Engineers, New York, pp. 91–106.

29. Gravett, P. W. 1993. "A Life Prediction Method for Matrix Dominated Fatigue Failures of SCS-6/Ti-15-3 MMC," *Reliability, Stress Analysis, and Failure Prevention*, DE-Vol. 55, American Society of Mechanical Engineers, New York, pp. 223–230.

30. Smith, K. N., R. Watson and T. H. Topper. 1970. "A Stress-Strain Function for the Fatigue of Metals," *Journal of Materials*, 5, pp. 767–778.

31. Neu, R. W. and T. Nicholas. 1996. "Methodologies for Predicting the Thermomechanical Fatigue Life of Unidirectional Metal Matrix Composites," *Advances in Fatigue Lifetime Predictive Techniques: Third Volume, ASTM STP 1292*, M. R. Mitchell and R. W. Landgraf, Eds., American Society for Testing and Materials, Philadelphia, pp. 1–23.

32. Coker, D., N. E. Ashbaugh and T. Nicholas. 1993. "Analysis of the Thermomechanical Fatigue Behavior of [0] and [0/90] SCS-6/TIMETAL®21S Composites," *Thermomechanical Behavior of Advanced Structural Materials*, AD-Vol. 34, AMD-Vol. 173, W. F. Jones, Ed., American Society of Mechanical Engineers, New York, pp. 1–16.

33. Neu, R. W., D. Coker and T. Nicholas. 1996. "Cyclic Behavior of Unidirectional and Cross-Ply Titanium Matrix Composites," *International Journal of Plasticity*, 12, pp. 361–385.

34. Calcaterra, J. R., W. S. Johnson and R. W. Neu. 1997. "A Comparison of Life Prediction Methodologies for Titanium Matrix Composites Subjected to Thermomechanical Fatigue," *13th Symposium on Composite Materials: Testing and Design*, ASTM STP 1242 American Society for Testing and Materials, Philadelphia (accepted for publication).

35. Neu, R. W. 1993. "A Mechanistic-Based Thermomechanical Fatigue Life Prediction Model for Metal Matrix Composites," *Fatigue and Fracture of Engineering Materials and Structures*, 16(8), pp. 811–828.

36. Nicholas, T. and S. M. Russ. 1992. "Elevated Temperature Fatigue Behavior of SCS-6/Ti-24Al-11Nb," *Mat. Sci. Eng.*, A153, pp. 514–519.

37. Nicholas, T., S. M. Russ, N. Schehl and A. Cheney. 1993. "Frequency and Stress Ratio Effects on Fatigue of Unidirectional SCS-6/Ti-24Al-11Nb Composite at 650°C," *FATIGUE 93*, Volume II, J.-P. Bailon and J. I. Dickson, Eds., EMAS, pp. 995–1000.

38. Nicholas, T., S. M. Russ, R. W. Neu and N. Schehl. 1996. "Life Prediction of a [0/90] Metal Matrix Composite under Isothermal and Thermomechanical Fatigue," *Life Prediction Methodology for Titanium Matrix Composites, ASTM STP 1253*, W. S. Johnson, J. M. Larsen and B. N. Cox, Eds., American Society for Testing and Materials, Philadelphia, pp. 595–617.

39. Russ, S. M., T. Nicholas, M. Bates and S. Mall. 1991. "Thermomechanical Fatigue of SCS-6/Ti-24Al-11Nb Metal Matrix Composite," *Failure Mechanisms in High Temperature Composite Materials*, AD-Vol. 22/AMD-Vol. 122, G. K. Haritos, G. Newaz and S. Mall, Eds., American Society of Mechanical Engineers, New York, pp. 37–43.

40. Nicholas, T. 1995. "An Approach to Fatigue Life Modeling in Titanium Matrix Composites," *Mat. Sci. Eng.*, A200, pp. 29–37.

41. Neu, R. W. and T. Nicholas. 1993. "Thermomechanical Fatigue of SCS-6/TIMETAL®21S under Out-of-Phase Loading," *Thermomechanical Behavior of Advanced Structural Materials*, W. F. Jones, Ed., AD-Vol. 34/AMD-Vol. 173, ASME, New York, pp. 97–111.

42. Roush, J. T., S. Mall and W. H. Vaught. 1994. "Thermo-Mechanical Fatigue Behavior of an Angle-Ply SCS-6/Ti-15-3 Metal-Matrix Composite," *Composites Science and Technology*, 52, pp. 47–59.

43. El-Soudani, S. M. and M. L. Gambone. 1989. "Strain-Controlled Fatigue Testing of SCS-6/Ti-6Al-4V Metal Matrix Composite," *Fundamental Relationships Between Microstructures and Mechanical Properties of Metal Matrix Composites*, P. K. Liaw and M. N. Gungor, Eds., The Metallurgical Society of AIME, Warrendale, PA, pp. 669–704.

44. Lerch, B. and G. Halford. 1995. "Effects of Control Mode and R-ratio on the Fatigue Behavior of a Metal Matrix Composite," *Mat. Sci. Eng.*, A200, pp. 47–54.

45. Sanders, B. P., S. Mall and L. B. Dennis. 1995. "Isothermal Fatigue Behavior of a Cross-Ply SCS-6/Ti-15-3 Metal Matrix Composite," *Fatigue and Fracture at Elevated Temperatures*, AD-Vol. 50, A. Nagar and S. Mall, Eds., American Society of Mechanical Engineers, New York, pp. 91–100.

46. Nicholas, T. and D. A. Johnson. 1996. "Time- and Cycle-Dependent Aspects of Thermal and Mechanical Fatigue in a Titanium Matrix Compos-

ite," *Thermo-Mechanical Fatigue Behavior of Materials, ASTM STP 1263,* M. J. Verrilli and M. G. Castelli, Eds., American Society for Testing and Materials, Philadelphia, pp. 331–351.

47. Lerch, B. A., M. J. Verrilli and G. R. Halford. 1993. "Fully-Reversed Fatigue of a MMC," *Proceedings of the American Society for Composites: Eighth Technical Conference,* Technomic Publishing Co., Inc., Lancaster PA, pp. 388–396.

48. Sanders, B. and S. Mall. 1994. "Longitudinal Fatigue Response of a Metal Matrix Composite under Strain Controlled Mode at Elevated Temperature," *Journal of Composites Technology and Research, JCTRER,* 16, pp. 304–313.

49. Sanders, B. and S. Mall. 1995. "Isothermal Fatigue Behavior of a Titanium Matrix Composite under a Hybrid Strain Controlled Loading Condition," *Mat. Sci. Eng.,* A200, pp. 130–139.

50. Boyum, E. A. and S. Mall. 1995. "Fatigue Behavior of a Cross-Ply Titanium Matrix Composite under Tension-Tension and Tension-Compression Cycling," *Mat. Sci. Eng.,* A200, pp. 1–11.

51. Boyum, E. A. and S. Mall. 1994. "Fatigue Response of a Titanium Matrix Composite under Tension-Compression Cycling," *Proceedings of the American Society for Composites: Ninth Technical Conference,* Technomic Publishing Co., Inc., Lancaster PA, pp. 523–529.

52. Majumdar, B. S. and G. M. Newaz. 1995. "Constituent Damage Mechanisms in MMCs under Fatigue Loading, and Their Effects on Fatigue Life," *Mat. Sci. Eng.,* A200, pp. 114–129.

53. Neu, R. W. 1996. "Thermomechanical Fatigue Damage Mechanism Maps for Metal Matrix Composites," *Thermo-Mechanical Fatigue Behavior of Materials, ASTM STP 1263,* M. J. Verrilli and M. G. Castelli, Eds., American Society for Testing and Materials, Philadelphia, pp. 280–298.

54. Johnson, W. S., M. Mirdamadi and J. G. Bakuckas, Jr. 1996. "Damage Accumulation in Titanium Matrix Composites under Generic Hypersonic Vehicle Flight Simulation and Sustained Loads," *Thermo-Mechanical Fatigue Behavior of Materials, ASTM STP 1263,* M. J. Verrilli and M. G. Castelli, Eds., American Society for Testing and Materials, Philadelphia, pp. 252–265.

55. Mirdamadi, M. and W. S. Johnson. 1993. "Fatigue of $[0/90]_{2S}$ SCS-6/Ti-15-3 Composite under Generic Hypersonic Vehicle Flight Simulation," *FATIGUE 93,* Volume II, J.-P. Bailon and J. I. Dickson, Eds., EMAS, pp. 951–956.

56. Mirdamadi, M., J. G. Bakuckas and W. S. Johnson. 1993. "Mechanisms of Strain Accumulation in Titanium Matrix Composites at Elevated Temperatures," *Mechanics of Composite Materials—Nonlinear Effects,* AMD-Vol. 159, American Society of Mechanical Engineers, New York, pp. 245–252.

57. Castelli, M. G. and J. Gayda. 1993. "An Overview of Elevated Temperature Damage Mechanisms and Fatigue Behavior of a Unidirectional SCS-6/Ti-15-3 Composite," *Reliability, Stress Analysis, and Failure Prevention,* DE-Vol. 55, ASME, New York, pp. 213–221.

58. Bartolotta, P. A. and M. J. Verrilli. 1993. "Thermomechanical Fatigue Behavior of SiC/Ti-24Al-11Nb in Air and Argon Environments," *Composite Materials: Testing and Design, ASTM STP 1206,* E. T. Componesch, Jr., Ed., American Society for Testing and Materials, Philadelphia, pp. 190–201.

59. Neu, R. W. and I. Roman. 1994. "Acoustic Emission Monitoring of Damage in Metal Matrix Composites Subjected to Thermomechanical Fatigue," *Composites Science and Technology,* 52, pp. 1–8.

60. Brindley, P. K., S. L. Draper, J. I. Eldridge, M. V. Nathal and S. M. Arnold. 1992. "The Effect of Temperature on the Deformation and Fracture of SiC/Ti-24Al-11Nb," *Metall. Trans.,* 23A, pp. 2527–2548.

61. Nicholas, T. and J. Ahmad. 1994. "Modeling Fiber Breakage in a Metal Matrix Composite," *Composites Science and Technology,* 52, pp. 29–38.

62. Nicholas, T., M. G. Castelli and M. L. Gambone. 1997. "Fiber Breakage in Metal Matrix Composites—Reality or Artifact?," *Scripta Materialia,* 36, pp.585–592.

63. Coker, D., R. W. Neu and T. Nicholas. 1997. "Analysis of the Thermoviscoplastic Behavior of [0/90] SCS-6/Timetal®21S Composites," *Thermo-Mechanical Fatigue Behavior of Materials, ASTM STP 1263,* M. J. Verrilli and M. G. Castelli, Eds., American Society for Testing and Materials, Philadelphia, pp. 213–235.

64. Kroupa, J. L. and M. Bartsch. 1996. Wright Laboratory, unpublished data.

65. Sanders, B. P. and S. Mall. 1996. "Transverse Fatigue Response of a Metal Matrix Composite under Strain-Controlled Mode at Elevated Temperature: Part I—Experiments," *Journal of Composites Technology and Research, JCTRER,* 18, pp. 15–21.

7

Creep Behavior

J.-M. YANG

Department of Materials Science and Engineering
University of California
Los Angeles, CA 90095-1505

INTRODUCTION

TITANIUM MATRIX COMPOSITES (TMC) are currently being developed to meet the increasing requirements of high-speed aerospace vehicles and advanced propulsion systems [1–6]. Reinforcing titanium alloys with continuous silicon carbide fibers can substantially improve the stiffness-to-weight and strength-to-weight capabilities over those of conventional titanium and nickel-based alloys. Furthermore, these properties can be retained to much higher temperatures in the composites because the SiC fibers exhibit little degradation in properties to temperatures as high as 900°C. In addition to the strength and stiffness, the TMCs reinforced with SiC fibers are expected to provide significant improvement in fatigue and creep performance [7,8]. Potential aerospace applications include hoop-reinforced turbine engine compressor components, such as bladed rings and disks, axially loaded struts, and actuator rods for engines and airframes.

Despite the potentially attractive properties of TMCs for high temperature structural applications, much less effort has been directed toward understanding the creep behavior and deformation mechanisms in these materials. Because the stiffness, strength, and creep resistance of the CVD-SiC fibers are significantly higher than those of the titanium alloys, it is expected that the fibers will carry the majority of load during creep loading [8]. This will effectively refrain the matrix from creeping. Thus, the creep deformation mechanisms in the TMCs may be quite different from those typically observed in the titanium alloys. It is generally recognized that the factors that will affect the creep behavior of the TMCs include the strength and strength

273

distribution of the SiC fibers, composition and creep properties of the titanium alloys, nature and properties of the fiber-matrix interface, fiber orientation, residual stress, temperature, stress, and environment. This chapter summarizes the current understanding of the creep behavior and deformation mechanisms in fiber-reinforced titanium matrix composites. The results from experimental investigation that provide insight into the creep response and microstructural damage mode of several fiber-reinforced titanium-based composites will be discussed. The micromechanical models proposed for predicting the creep strain and rupture life for TMCs will also be discussed.

CREEP BEHAVIOR AND DEFORMATION MECHANISMS OF TMCS

In this section, the experimental investigation to characterize the creep response and microstructural damage mode of several SiC (SCS-6) fiber-reinforced titanium-based composites will be discussed. The typical creep properties of several titanium alloys are plotted in Figure 1 using the Larson-Miller parameter for time to 0.2% creep strain [9]. Apparently, the creep properties of the titanium alloys are sensitive to the variations of alloy compositions and microstructure. In general, the α-phase Ti exhibits better creep resistance and strength than the β-phase Ti. Therefore, most titanium alloys developed for high-temperature applications contain a higher percentage of α-phase. The excellent creep resistance of the titanium aluminide intermetallic alloys, such as Ti-25Al-10Nb-3V-1Mo and Ti-24Al-11Nb, has been attributed to the ordered α_2-phase, which is more resistant to both diffusional and dislocation creep.

Tensile Creep of Unidirectional SCS-6/Ti-6Al-4V Composite

The tensile creep behavior of unidirectional SCS-6/Ti-6Al-4V composites was extensively studied by Schwenker [9] at temperatures ranging from 427 to 650°C and stresses ranging from 621 to 1380 MPa. Figure 2 shows the creep curves as a function of temperature and applied stress [9]. The results indicate that, upon loading, the composite exhibits a primary or transient stage in which the creep rate decreases gradually with time. At low stress the creep rate continues to decrease to a near-zero value, and strain approaches some limiting value. This response is apparently governed by a mechanism whereby the matrix relaxes and stress is redistributed into the fibers

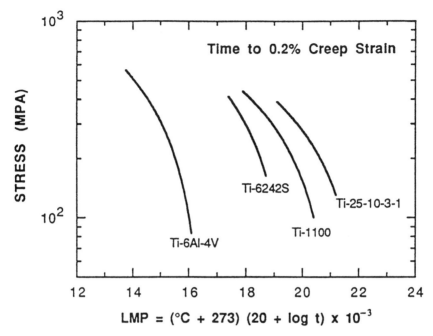

Figure 1. Larson-Miller plot of time to 0.2% creep strain for various titanium alloys [9].

[10]. However, at higher applied stresses, the creep rate decreases to essentially a constant value rather than continuously decreasing to near zero. At certain testing conditions, the constant creep region extends over hundreds of hours, analogous to the steady-state or secondary creep stage observed in the monolithic matrix alloys. The minimum creep rates as a function of applied stress at two different temperatures were plotted in Figure 3 [9]. The stress exponents obtained from a simple power law creep for the SCS-6/Ti-6-4 composites tested at 427 and 538°C were found to be 18.6 and 16.0, respectively.

The effect of temperature on the creep behavior of the SCS-6/Ti-6Al-4V composite is plotted in Figure 4 [9]. It is obvious that increasing the temperature has the same effect as increasing the stress. At 427°C, the composite exhibits asymptotic creep behavior. As the temperature is increased to 538°C, a constant creep region was observed. Above 538°C, the creep rate and rupture life also become increasingly sensitive to temperature. The transient creep response in the composite also depends on temperature. Comparison of creep response at

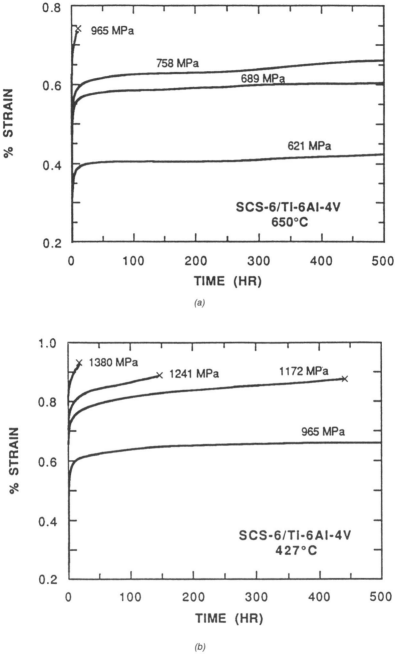

Figure 2. The longitudinal tensile creep curves as a function of temperatures and applied stresses for SCS-6/Ti-6Al-4V composites: (a) 427°C, (b) 638°C, and (c) 650°C.

276

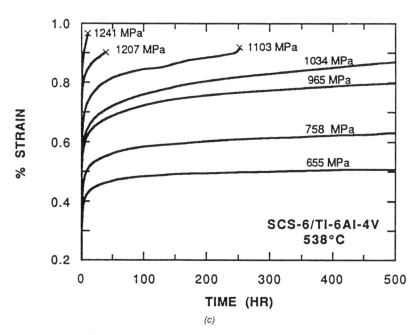

Figure 2. (cont.) The longitudinal tensile creep curves as a function of temperatures and applied stresses for SCS-6/Ti-6Al-4V composites: (a) 427°C, (b) 638°C, and (c) 650°C.

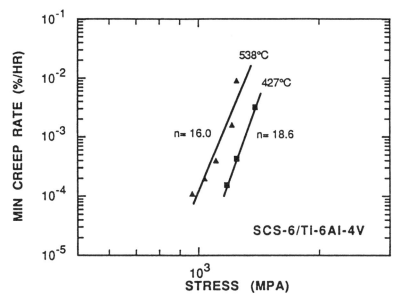

Figure 3. The minimum creep rates plotted against applied stresses for SCS-6/Ti-6Al-4V composites.

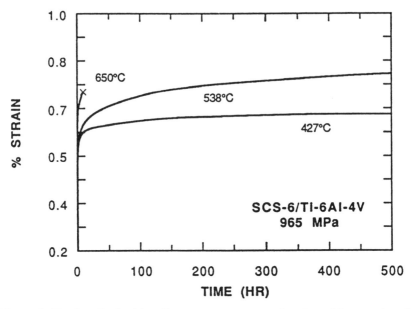

Figure 4. The longitudinal tensile creep curves as a function of temperature for SCS-6/Ti-6Al-4V composites.

427 and 650°C shows that the creep rate in the transient region decreases more rapidly at higher temperatures.

Schwenker [9] used in situ acoustic emission monitoring coupled with microstructural analysis to study the microstructural damage during the secondary creep stage. Two primary modes of damage were identified as being responsible: fiber fracture and oxidation damage to the carbon coating on the SCS-6 fibers. Fiber fracture occurred primarily along the outer surfaces of the composite and, in particular, along the machined surface edges [Figure 5(a)]. Acoustic emission monitoring shows that fiber fracture occurred at a nearly constant frequency throughout the secondary creep stage [Figure 5(b)]. The frequency of fiber fracture increased with applied stress. The additional strain resulting from fiber fracture appears to be controlled by a stress redistribution mechanism. Under creep loading conditions, interfacial shear stresses gradually relax and limit the stress that can be transmitted to the broken fibers. Over time this stress must be redistributed to the undamaged fibers. These fibers, in turn, elastically expand with creeping matrix to yield the time-dependent strain. The confinement of fiber failure to a region along the outer surface sug-

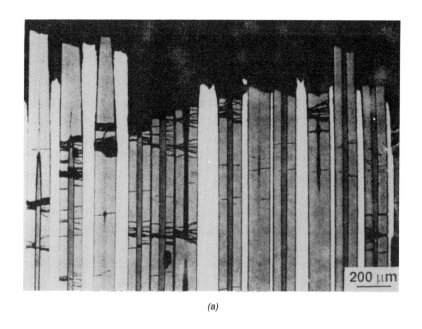

(a)

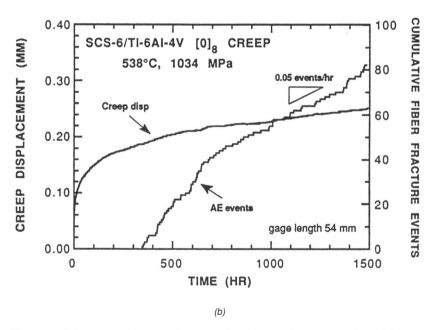

(b)

Figure 5. (a) Fiber fracture in the crept SCS-6/Ti-6Al-4V composites, (b) creep displacement curve and cumulative AE events during creep of SCS-6/Ti-6Al-4V composite at 538°C/1034 MPa.

279

gests that environment may play an important role in reducing fiber strength. The carbon-rich coating on the SCS-6 fibers are susceptible to oxidation damage at creep temperature of 538°C and above. The oxidation damage proceeds down the interface and is accelerated by fiber fracture. The absence of carbon coating is expected to limit severely the transfer of load to broken fibers. The penetration of the interfacial oxidation damage into the machined edges of the composite, which have exposed fibers, was observed to be accelerated by poor fiber spacing. In this case, oxidation can rapidly progress from fiber to fiber by a combination of pipeline oxidation down the interfaces and short-circuit diffusion between fibers at points where they are in contact. Improved fiber distribution and bonding are expected to reduce the rate of damage penetration.

The transverse tensile creep behavior of the SCS-6/Ti-6-4 composites was investigated by Wright [11] at 438°C with stresses ranging from 105 to 210 MPa. Extensive fiber-matrix debonding and matrix cracking was found to be the primary creep damage mechanisms. Matrix cracks advanced along the debonded fiber-matrix interface and propagated intergranually into the matrix before rupturing by dimpled shear as shown in Figure 6.

Tensile Creep Behavior of SCS-6/Ti-25Al-10Nb-3Mo-1V Composite

The tensile creep behavior of the SCS-5/Ti-25-10 composite was also studied by Schwenker [9] at temperatures ranging from 538 to 760°C and stresses ranging from 414 to 1034 MPa. The creep curves of the composite as a function of applied stresses at 650°C are shown in Figure 7(a) [9]. At higher applied stresses, the composite exhibited a constant or secondary creep stage similar to that observed in the SCS-6/Ti-6-4 composite. Temperature has the same effect as that observed in the SCS-6/Ti-6-4 composite, i.e., at higher temperatures, the composite exhibited a constant creep response at lower stresses. Both creep rates and rupture lives are sensitive to stress and temperature above this transition. The results show that the SCS-6/Ti-25-10 composite exhibits a more pronounced tertiary creep stage than in the SCS-6/Ti-6-4 composite. The minimum creep rates as a function of applied stress at two different temperatures were plotted in Figure 7(b) [9]. The stress exponents obtained from a simple power law creep for the SCS-6/Ti-25Al-10Nb-3V-1Mo composites tested at 650 and 760°C were found to be 8.45 and 7.49, respectively. Microstructural examination

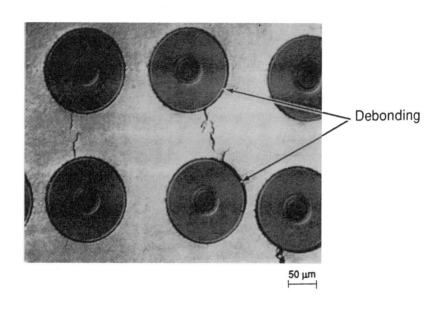

Debonding

50 μm

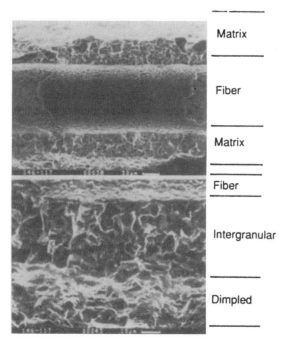

Matrix

Fiber

Matrix

Fiber

Intergranular

Dimpled

Figure 6. Microstructure of the crept SCS-6/Ti-6Al-4V composite showing fiber/matrix debonding and matrix cracking.

281

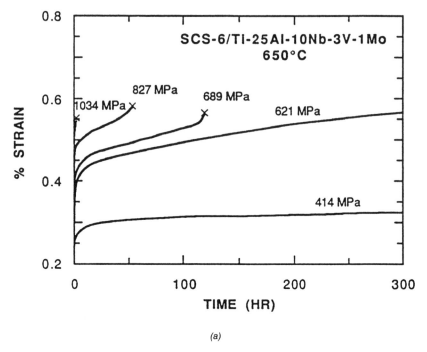

(a)

Figure 7. (a) *The longitudinal tensile creep for SCS-6/Ti-25Al-10Nb-3V-1Mo composites at 650°C; (b) minimum creep rates plotted against applied stresses.*

reveals that the creep damage mechanisms in the secondary creep stage are similar to those observed in the SCS-6/Ti-6-4 composite.

Tensile Creep Behavior of SCS-6/Ti-1100 Composite

The tensile creep behavior of the SCS-5/Ti-1100 composite was also studied by Schwenker [9] and Gambone [12] at temperatures ranging from 538 to 760°C and stresses ranging from 414 to 1034 MPa. The creep curves of the composite as a function of applied stresses at 650°C are shown in Figure 8(a). The effect of heat treatment on the creep behavior and microstructural damage mode of the SCS-6/Ti-1100 composite was also investigated by Gambone [12]. Results show that the mean fiber strength after heat treatment at 950°C for 430 hours decreased substantially from 4177 to 1456 MPa, and the Weibull's modulus also decreased from 8.98 to 5.88 [Figure 8(b)]. The fiber strength degradation was attributed to the consumption of the

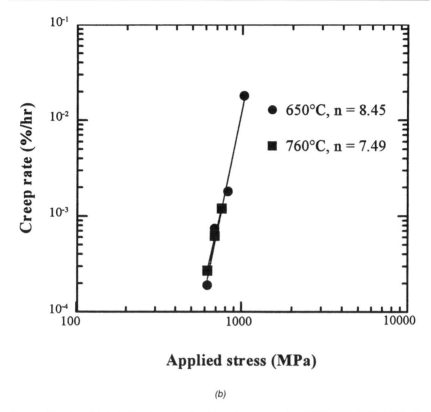

Applied stress (MPa)

(b)

Figure 7. (cont.) (a) *The longitudinal tensile creep for SCS-6/Ti-25Al-10Nb-3V-1Mo composites at 650°C; (b) minimum creep rates plotted against applied stresses.*

C-rich coating layer on the SCS-6 fiber by reaction between the fiber and matrix. Surface flaws were created in areas where the coating was breached and the SiC fiber was attacked by the interfacial reaction. Heat treatment also embrittled the Ti-1100 matrix. Therefore, during creep loading, intergranular matrix cracks developed rapidly in the heat-treated composite and grew by environmentally assisted creep crack growth. Both fiber strength degradation and matrix embrittlement reduced the creep life of the heat-treated composite.

Schwenker [9] also compared the creep properties of SCS-6/Ti-6-4, SCS-6/Ti-25-10, and SCS-6/Ti-1100 composites. Figure 9(a) shows the creep behavior of these three composites as a function of applied stress at 538°C. To illustrate the difference in creep strength, time-to-rupture data for these three composites were plotted in Figure 9(b)

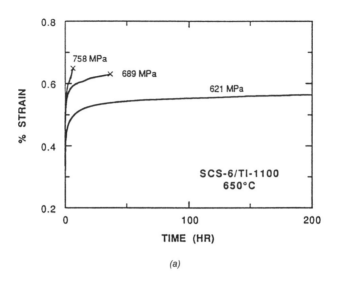

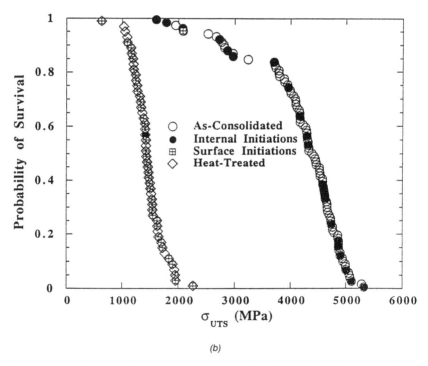

Figure 8. (a) The longitudinal tensile creep for SCS-6/Ti-1100 composites at 650°C; (b) effect of heat treatment at 950°C/430 hours on strength distribution of SCS-6 fiber.

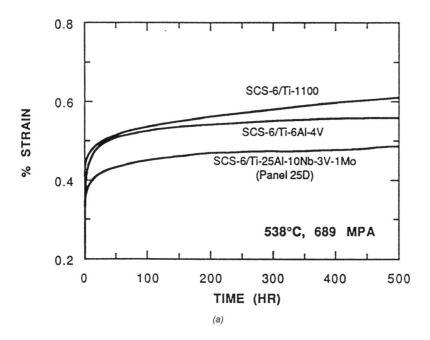

(a)

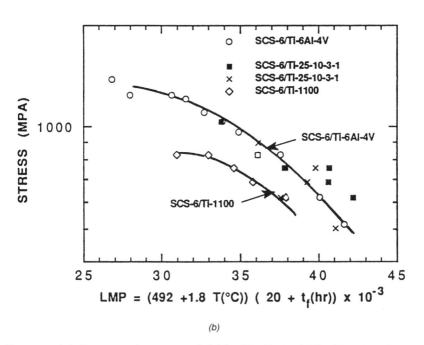

(b)

Figure 9. (a) Creep performance of SCS-6/Ti-6Al-4V, SCS-6/Ti-25Al-10Nb-3V-1Mo and SCS-6/Ti-1100 composites at 538 and 650°C; (b) Larson-Miller plot for SCS-6/Ti-6Al-4V, SCS-6/Ti-25Al-10Nb-3V-1Mo and SCS-6/Ti-1100 composites.

285

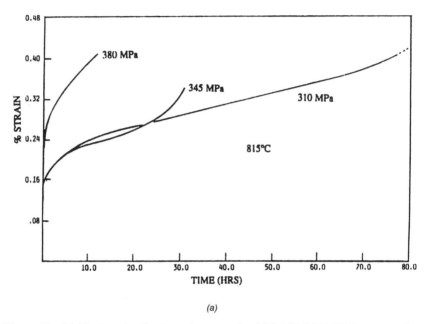

(a)

Figure 10. (a) The longitudinal tensile creep for SCS-6/Ti-24Al-11Nb composites at 815°C; (b) the longitudinal tensile creep for SCS-6/Ti-24Al-11Nb composites at 345 MPa.

using Larson-Miller parameter. The plot shows that the SCS-6/Ti-6-4 and SCS-6/Ti-25-10 composites exhibit the highest creep strength. The SCS-6/Ti-1100 composite has the least creep resistance, exhibited higher secondary creep rates, and had a shorter rupture life than the other composites. This can be primarily attributed to low matrix ductility and brittle matrix cracking. Embrittlement of the Ti-1100 matrix is possibly due to the precipitation of titanium silicides during composite processing. During creep exposure, brittle matrix cracking was observed to initiate along the surface edges of the composite and propagate into the fiber-matrix interface. Such cracking significantly accelerated the rate of environmental attack of the fiber-matrix interfaces during creep exposure. The results indicate that a matrix must provide resistance to cracking and environmental embrittlement to achieve optimum creep performance. Gambone [12] proposed to control the matrix microstructure by heat treating the composite above the β-transus temperature following consolidation to improve the cracking resistance of the matrix and extend the creep life of the composite without significantly damaging the fibers or interface.

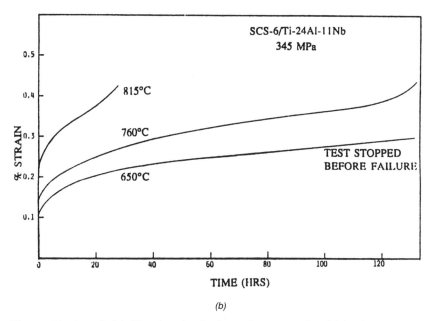

(b)

Figure 10. (cont.) (a) The longitudinal tensile creep for SCS-6/Ti-24Al-11Nb composites at 815°C; (b) the longitudinal tensile creep for SCS-6/Ti-24Al-11Nb composites at 345 MPa.

Tensile Creep Behavior of SCS-6/Ti-24Al-11Nb Composite

The tensile creep behavior of the SCS-6/Ti-24Al-11Nb composite was investigated by Khobaib [13,14] and Gambone [15] over temperature ranges from 650 to 815°C and stresses ranging from 310 to 650 MPa. The creep curves as a function of stress and temperature are plotted in Figure 10(a). At 815°C and 310 MPa, the composite exhibit three distinct stages, an early region of primary creep extending to about 25 hours followed by a linear region. The tertiary region is a short region consisting of approximately the final 7% of the total life. The stress exponent obtained from the plot of minimum creep rates versus applied stress was found to be nearly 16.6 [13]. The effect of temperature on the creep behavior of the composite at 345 MPa is shown in Figure 10(b). Obviously, the time-to-failure was found to increase with decreasing stress and temperature. The effect of fiber orientation on the creep behavior of the SCS-6/Ti-24Al-10Nb composite was also studied. Figure 11(a) shows the creep curves of the unidirectional, [0/90] laminated composites, and unreinforced matrix at 815°C. The differences

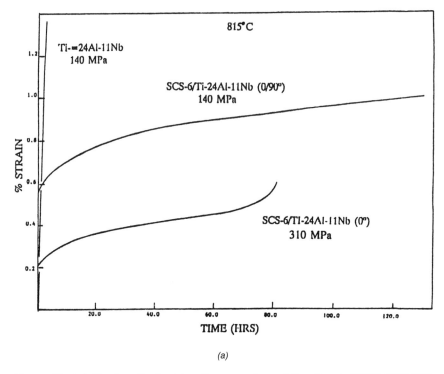

(a)

Figure 11. (a) The longitudinal tensile creep for unidirectional, [0/90] SCS-6/Ti-24Al-11Nb composites and unreinforced matrix alloy at 815°C; (b) Larson-Miller plot for various titanium alloys and composites.

in creep behavior becomes more apparent on a Larson-Miller plot shown in Figure 11(b). The creep plot of [0/90] laminate falls to the right of the titanium aluminide matrix plot, showing little improved creep resistance over Ti-24Al-11Nb [13]. The stress dependence of creep rupture life at various temperatures for the SCS-6/Ti-24Al-10Nb composite along the longitudinal and transverse directions is shown in Figure 12 [15,16]. As expected, the transverse properties are significantly lower than the longitudinal creep properties.

Khobaib [13] also examined the fracture surfaces of the crept composites to determine the damage and failure mechanisms. In most cases, multiple cracks were found to initiate in the unidirectional composite. The crack initiated at surface flaws, either from a broken fiber, embrittled matrix, molybdenum cross weave, or some other surface defects. The ruptured surface shows two distinct zones of

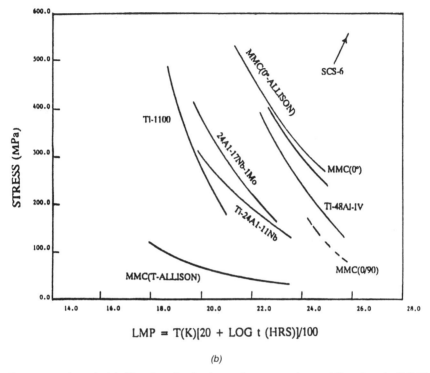

STRESS (MPa)

600.0

500.0

400.0

300.0

200.0

100.0

0.0

14.0 16.0 18.0 20.0 22.0 24.0 26.0 28.0

TI-1100

MMC(0°-ALLISON)

SCS-6

24AI-17Nb-1Mo

MMC(0°)

TI-48AI-IV

Ti-24AI-11Nb

MMC(0/90)

MMC(T-ALLISON)

$$LMP = T(K)[20 + LOG \ t \ (HRS)]/100$$

(b)

Figure 11. (cont.) *(a) The longitudinal tensile creep for unidirectional, [0/90] SCS-6/Ti-24Al-11Nb composites and unreinforced matrix alloy at 815°C; (b) Larson-Miller plot for various titanium alloys and composites.*

failure: a relatively flat area in the creep zone [Figure 13(a)] followed by a jagged fracture zone that exhibits extensive fiber pull out associated with final failure [Figure 13(b)]. Apparently, the defects at the edge or surface provide an easy path for diffusion of oxygen, leading to severe degradation of the fiber. The process of fiber damage and cracking continues until the reduced section of the composite cannot sustain the applied load, causing catastrophic failure. The fracture surface of the [0/90] laminate tested at 815°C and 140 Mpa is shown in Figures 14(a) and 14(b). The matrix shows high density of cracks [Figure 14(a)]. Some of the cracks appear to initiate from the interfacial reaction zone of the 0° fiber and later connected to the 90° fiber or vice versa. The surface of the 90° fibers appears to be severely degraded due to environmental attack [Figure 14(b)]. Appar-

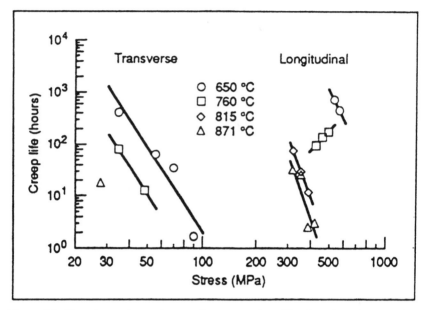

Figure 12. The stress dependence of creep rupture life at various temperatures for the SCS-6/Ti-24Al-11Nb composite along the longitudinal and transverse directions.

ently, the two ends of the 90° fibers at open edges provide easy access for oxygen. The carbon core of the SCS-6 fiber was also found to be severely damaged, resulting in pores and cracks that are indications of severe stress-assisted environmental degradation and provide crack initiation site.

Tensile and Flexural Creep Behavior of SCS-6/Ti-6242 Composite

The tensile creep behavior of the SACS-6/Ti-6242 composite was studied by Jeng and Yang at 600°C [18]. The minimum creep rates plotted as a function of stress are shown in Figure 15. The stress exponent was found to be 17.5, which is similar to that reported in the SCS-6/Ti-6-4 composite shown in Figure 3. The flexural creep behavior of the SCS-6/Ti-6242 composite was investigated by Jeng et al. [20] at stresses ranging from 700 to 1300 MPa and temperatures between 550 and 700°C. Microstructural examination of the crept composite at the secondary creep stage reveals that multiple fiber fracture was the primary microstructural damage mode [Figure 16(a)].

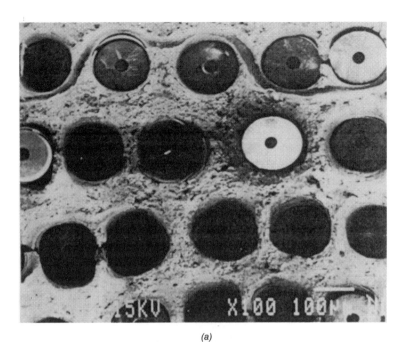

(a)

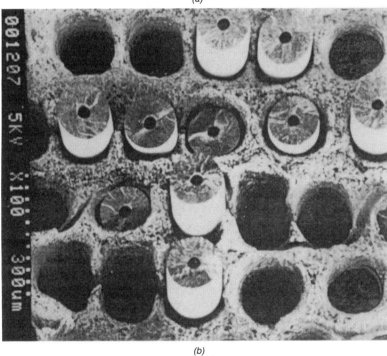

(b)

Figure 13. Fracture surface of the crept unidirectional SCS-6/Ti-24Al-11Nb composite: (a) flat feature in early stage of fracture and (b) extensive fiber pull out associated with overload fracture.

291

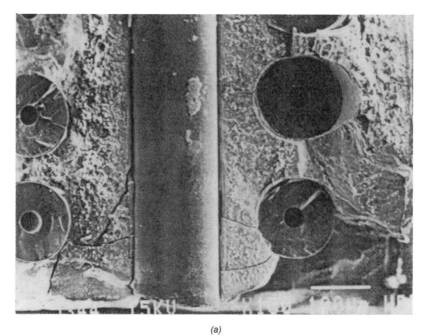

(a)

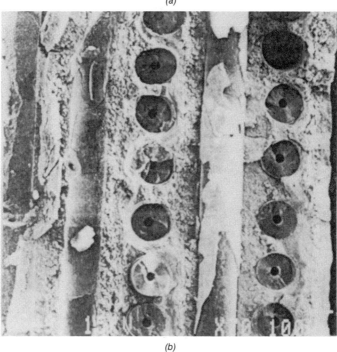

(b)

Figure 14. Fracture surface of the crept [0/90] SCS-6/Ti-24Al-11Nb composite: (a) high density of matrix cracks and (b) degradation of C-rich coating layer.

292

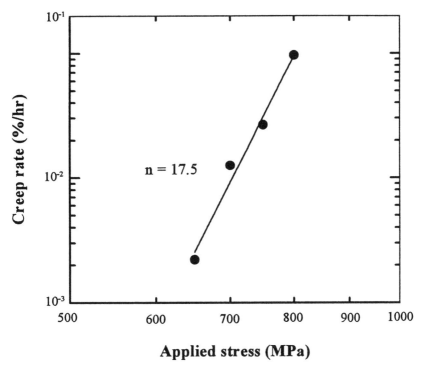

Applied stress (MPa)

Figure 15. *The minimum longitudinal tensile creep rates plotted against applied stresses for SCS-6/Ti-6242 composites at 600°C.*

However, prior to the onset of tertiary creep stage, fiber fracture, cracking of the interfacial reaction layer, and matrix intergranular cracking that extended from the broken fiber ends were observed as shown in Figures 16(b) and 16(c). The onset of tertiary creep is induced by significant fiber fracture, leading to the reduction of its load-bearing capability [Figure 16(d)].

Creep Cracking of SCS-6/Ti-25Al-10Nb-3Mo-1V Composite

Chiu et al. [19] studied the effect of fiber coating on the creep cracking behavior of a notched SCS-6/Ti-25-10 composite. The longitudinal creep curves of the notched SCS-6/Ti-25-10 and Ag/Ta-coated SCS-6/Ti-25-10 composites tested at 700°C under an initial stress intensity factor of 15 MPa $m^{1/2}$ are shown in Figure 17(a). The duplex Ag/Ta coating has been shown to be effective in controlling the interfacial reactions and in suppressing the matrix cracking near the interface

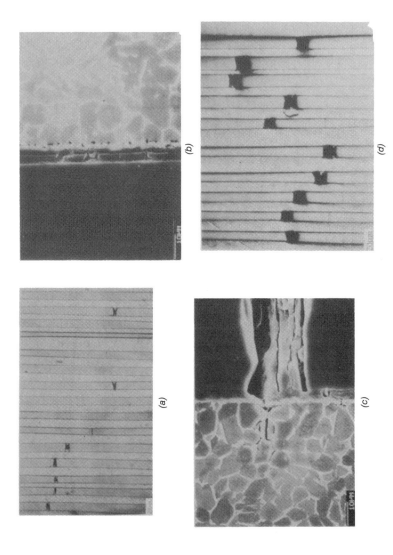

Figure 16. Microstructure of the crept SCS-6/Ti-6242 composites (a) fiber frracture, (b) cracking of the interfacial reaction layer, (c) matrix cracking entending from broken fiber ends, and (d) extensive fiber fracture at tertiary creep stage.

294

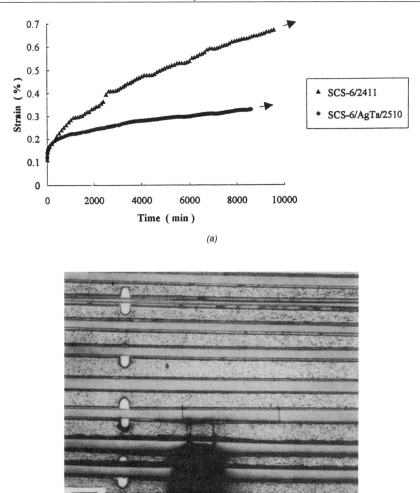

(a)

(b)

Figure 17. (a) The creep curves of the notched SCS-6/Ti-25Al-10Nb-3V-1Mo and Ag/Ta-coated SCS-6/Ti-25Al-10Nb-3V-1Mo composites at 700°C (K_i = 15 Mpa $m^{1/2}$); (b) the damage zone at the notch tip.

during thermomechanical loading [21]. The interfacial shear strength of the Ag/Ta-coatedSCS-6/Ti-25Al-10Nb-3V-1Mo composite (140 MPa) [21] has been found to be much higher than that of the uncoated composites (115 Mpa) [22]. The creep rate of the uncoated compsites is significantly higher than that of the coated composite. Microstructural examination of the crept composite reveals the existence of a damage zone near the crack tip for both coated and uncoated composites. Multiple fiber fracture, interfacial debonding, and matrix cracking originating from the broken fiber ends were found in the damage zone as shown in Figure 17(b). Furthermore, examination of the interfaces at several fibers away from the crack tip showed that the fiber-matrix interface in the coated composite remained intact, whereas the interface in the uncoated composite showed evidence of considerable debonding and oxidation. Therefore, the higher creep rate in the uncoated composite is thought to be due to insufficient load transfer to the fibers, and consequently, more severe matrix plastic flow. When tested at a stress intensity factor of 20 MPa $m^{1/2}$, the uncoated composite failed shortly after the load was applied, whereas the coated composite lasted for over 30 hours before final failure. At such a high stress intensity level, the first few fibers ahead of the notch tip fractured upon loading. The ingression of oxygen through the broken fiber ends and notch tip embrittled the interface and matrix. As time elapsed, the oxygen-induced embrittlement accelerated fiber rupture, leading to the onset of catastrophic failure. It is evident that the Ag/Ta-coated composite is more resistant to creep cracking than the uncoated one. The improvement is attributed to higher interfacial shear strength, which facilitates more efficient load transfer, and the coating, which retards the oxidation of both the matrix and interface.

Summary Remarks

The minimum longitudinal tensile creep rates versus applied stresses for several unidirectional SCS-6 fiber-reinforced titanium-based composites are plotted in Figure 18. The stress exponents obtained from the slope of the creep rates versus stresses plot are listed in Table 1. It is obvious that the stress exponents for these composites are very similar, indicating that the dominant creep deformation mechanisms are similar. It is also noted that the stress exponents in the composites are substantially higher than those of the unreinforced matrix alloys. Based on the microstructural observation, the sequence of creep deformation in the composite along the fiber direction can be summarized as follows. Upon loading, the load carried by the matrix will be continuously transferred to the fiber. As a result, substantial stress relaxation in the matrix occurs due to a restraining effect from the fibers. The highly stressed fibers then rupture where high stresses and fiber flaws coincide. At the rupture sites, the elastic

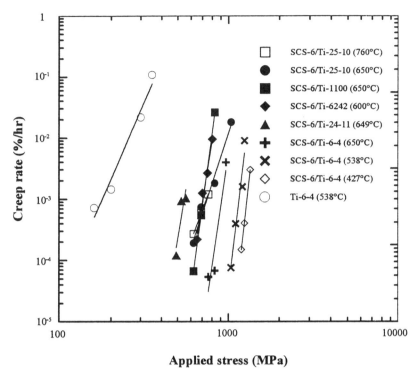

Figure 18. The minimum longitudinal tensile creep rates versus applied stresses for several SCS-6 fiber-reinforced titanium-based composites.

load released by the fiber is transferred back to the matrix, which sustains further creep deformation. Creep deformation of the matrix continues, causing local stress redistribution until intact fiber segments adjacent to or further along the ruptured fiber supply a constraining

Table 1. The stress exponent for TMCs.

Materials	Stress Exponent	Temperature (°C)	Stress (MPa)
SCS-6/Ti-6-4	16	538°C	965-1241
SCS-6/Ti-6-4	18.6	427°C	1172-1380
SCS-6/Ti-1100	17	650°C	621-758
SCS-6/Ti-6242	17.5	600°	650-800
SCS-6/Ti-24-11	16.6	815°C	310-380
SCS-6/Ti-25-10	8.45	650°C	621-1034
Ti-6-4	7.1	538°C	200-400

force, allowing the stress in the matrix to again relax. A constant creep rate is reached when a dynamic equilibrium is established between the rate of load transfer from the matrix to the fiber and subsequent fiber rupture and load transfer from the fiber to the matrix. As creep deformation proceeds further, this process propagates through the whole composite, resulting in multiple fiber fracture. The rupture of the composite is due to the accumulation of microstructural damage, which includes fiber fracture, environmental-assisted matrix cracking, fiber strength degradation, and interface degradation.

The incorporation of reinforcing fibers clearly impart substantial creep resistance to the composite along the fiber direction. However, the unidirectional TMC also exhibits a highly anisotropic creep behavior. To provide a measure of the long-term high-temperature capability of TMCs, Larsen et al. [3] used the Larson-Miller plot to correlate the stress rupture lives of TMC laminates with different fiber orientations and a range of testing conditions. The density-corrected Larson-Miller plot for SCS-6/TIMETAL®21S, SCS-6/Ti-24Al-11Nb and the comparison materials are shown in Figure 19. It is obvious that the [0] and [90] laminates exhibited the best and worst creep proper-

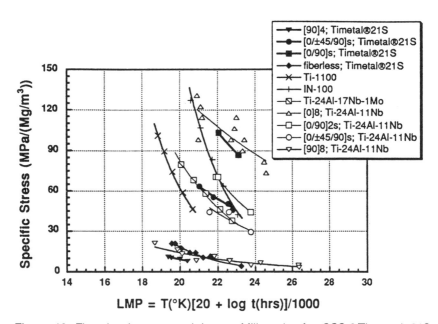

Figure 19. The density-corrected Larson-Miller plot for SCS-6/Timental 21S, SCS-6/Ti-24Al-11Nb and the comparison materials.

ties. The creep capability of the [0/90] and [0/45/90] laminates fall in the intermediate range. In summary, the creep response in the TMCs along the fiber direction is dominated by the reinforcing fibers. As a result, the creep behavior of the composite is sensitive to the applied stress levels and fiber orientation. The environmentally induced damage in the fiber and matrix and the microstructural instability of the titanium matrix and fiber-matrix interface also play an important role in limiting the creep performance of TMCs.

CREEP MODELING OF TITANIUM MATRIX COMPOSITES

Several micromechanical models have been developed for predicting the creep behavior of continuous fiber-reinforced metal matrix composites. In this section, the merits and limitations of these models will be discussed. Analytical models for predicting the creep behavior of discontinuous fiber-reinforced metal matrix composites will not be discussed. A comprehensive discussion of creep models for both discontinuous and continuous fiber-reinforced ceramic matrix composites can be found in Reference [22].

Creep Model for an Undamaged Composite

McLean [10,23] had proposed a model to describe the longitudinal creep deformation of an unidirectional fiber-reinforced metal matrix composite. This model is based on "continuous damage mechanics" for a composite consisting of an elastic fiber embedded in a power law creeping matrix. The fiber-matrix interface bonding is assumed to be perfect. The overall strain rate of the composite can be described by the following governing equations:

$$\dot{\varepsilon} = \alpha_1 \alpha_2 \dot{\varepsilon}_m (1 - S_1)^m \qquad (1)$$

where

$$\alpha_1 = \left[1 + \frac{E_f V_f}{E_m (1 - V_f)} \right]^{-1}$$
$$\alpha_2 = (1 - V_f)^{-m}$$

where $\dot{\varepsilon}_m$ represents the creep rate of the matrix carrying all of the applied stress σ_o, E_f and E_m are Young's modulus of the fiber and matrix, repsectively, V_f is the volume fraction of the fibers, m is the exponent constant of the matrix, and S_1 represents a normalized value for the stresses carried by the fiber:

$$S_1 = \frac{\sigma_f V_f}{\sigma_o} \tag{2}$$

Meanwhile, the S_1 can be described by:

$$\dot{S}_1 = H_1 \dot{\varepsilon} \tag{3}$$

where

$$H_1 = \frac{V_f E_f}{\sigma_0}$$

For the composite under constant applied stress σ_o, the initial conditions for Equations (1) and (3) are:

$$\varepsilon(0) = \frac{\sigma_0}{E_f V_f + E_m(1 - V_f)}$$
$$S_1(0) = H_1 \varepsilon(0) \tag{4}$$

The creep deformation of the fiber-reinforced metal matrix composites without microstructural damage can be predicted by the governing equations and initial conditions [Equation (4)]. Physically, S_1 can be treated as a dimensionless internal stress (stress carried by the fibers) that modifies the matrix creep behavior. The constant, H, is a strain-hardening coefficient that controls increments of the internal stress. If the fibers do not fail, the internal stress S_1 would gradually increase as the deformation accumulates, which, in turn, would reduce the stress carried by the matrix. When the fibers fully support the applied stress ($S_1 = 1$), the composite will not deform as function of loading time. However, if damages were induced during processing and/or creep loading, the above equations developed by McLean are inadequate to describe the creep behavior of the composites. McLean [24] modified his model later to incorporate the fact of fiber fracture by treating the composite as one with many short fibers inside. Yet this needs not be true if the break density is not very high. Although Goto and McLean [25] have tried to accommodate an "interface" phase with creep behavior different from other constituents to simulate the steady-state creep of composites, the model still overlooks some important phenomena, such as load transfer, stress recovery, etc [5]. To accurately predict the creep behavior, the following two damage-induced characteristics need to be taken into consideration. The first consideration is to change the internal stress S_1 and its increasing rate, $\dot{S}_1$, resulting from fiber breakage and microvoids formation between reaction layer-matrix interface. To determine the effects of these two damage mechanisms on S_1, several aspects must be considered. These include the interfacial shear strength between the fiber and matrix, stress recovery at the broken fibers, fiber modulus degradation, and the statistical distribution of fiber strength. The second consideration comes from the modification of the matrix creep characteristics, $\dot{\varepsilon}_m$, resulting from the stress concentration and matrix cracking near the broken fiber ends and environmental-assisted matrix cracking.

Creep Models for Composites with Fiber Fracture and Interface Debonding

To assess the effect of fiber fracture on the creep behavior of a composite, it is necessary to understand how fiber fracture affects the load-carrying capability of the broken and surviving fibers. Kelly and Davies [26] proposed a theory concerning the reloading of a broken fiber by analyzing the mechanics of load transfer in a discontinuously reinforced composite. Recently, Curtin [27] developed an expression for the "characteristic length" of a fiber in a single-filament composite by assuming that the fiber strength distribution is defined by Weibull statistics rather than a unique fracture strength. The characteristic length δ_c is related to the region in which stress recovery occurs. Du and McMeeking [28] utilize Curtin's concepts to modify McLean's model to study the effect of fiber fracture and the subsequent stress relaxation in the broken fibers on creep response. The average fiber stress, which includes the stress in both broken and unbroken fibers over a gauge length of $2L_f$, where L_f is the distance required to reload the fiber to a stress of εE_f, is derived as:

$$\bar{\sigma}_f = (1 - \phi)\varepsilon E_f + \frac{\phi}{2}\varepsilon E_f \qquad (5)$$

where $\phi = \dfrac{2L_f}{L_o}\left(\dfrac{\varepsilon E_f}{\sigma_o}\right)^m$ is the cumulative number of fiber fracture that will occur in a fiber. Then, the normalized stress in the matrix can be obtained by substituting Equation (5) into the rule of mixture ($\sigma = V_f \bar{\sigma}_f + V_m \sigma_m$):

$$\frac{\sigma_m}{S_c} = \frac{1}{V_m}\left(\frac{\sigma}{S_c} - V_f \frac{\varepsilon E_f}{S_c}\left(1 - \frac{1}{2}\left(\frac{\varepsilon E_f}{S_c}\right)^{m+1}\right)\right) \tag{6}$$

where S_c is a normalized constant defined as $\left(\dfrac{2\sigma_o^m \tau_m L_o}{d}\right)^{1/m+1}$. Differ-

entiating equation (6) and substituting it into McLean's power law representation:

$$\dot{\varepsilon} = \dot{\varepsilon}_m = \frac{\dot{\sigma}_m}{E_m} + B\sigma_m^n \tag{7}$$

Finally, a differential equation for the strain rate can be obtained:

$$\frac{\dot{\varepsilon} E_f}{S_c} = \frac{B\sigma_c^{n-1}E_f\left\{\dfrac{1}{V_m}\left\{\dfrac{\sigma}{S_c} - V_f \dfrac{\varepsilon E_f}{S_c}\left(1 - \dfrac{1}{2}\left(\dfrac{\varepsilon E_f}{S_c}\right)^{m+!}\right)\right\}\right\}^n}{1 + \dfrac{E_f}{E_m}\dfrac{V_f}{V_m}\left(1 - \left(1 + \dfrac{m}{2}\right)\left(\dfrac{\varepsilon E_f}{S_c}\right)^{m+1}\right)} \tag{8}$$

Solving Equation (8) gives the creep strain of the composite. However, it cannot be solved analytically; numerical methods must be applied to calculate the creep rupture life and the time to the onset of tertiary creep. Figure 20 compares the creep curves predicted by different models, which include McLean's, Curtin-McLean's, and Du-McMeeking's model. Figure 21 compares the creep rupture time predicted by these three models. Comparisons of measured creep curves of a SiC fiber-reinforced Ti-6Al-4V composite with model predictions are shown in Figure 22 [29].

Song et al. [30] proposed a "fiber-shell model" for predicting the creep response of a continuous fiber-reinforced composite with fiber damage. The fiber-shell model simulates the creep of continuous fiber-reinforced composites by using a cylindrical cell consisting of a single broken fiber and a shell of the fiber material embedded in an elastic power law creeping matrix. This model predicts the effect of broken fibers on the local stress distribution and creep response under both longitudinal and transverse loading. The time-dependent creep behavior of the composite is calculated using finite element. However, the stress distribution requiring a time-dependent stress analysis is

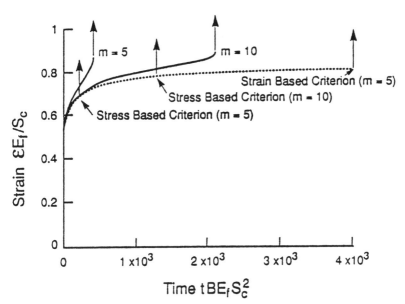

Figure 20. *The creep curves predicted by different models for a material with a ratio of fiber to matrix Young's modulus $E_f/E_m = 3$, a creep exponent of the matrix $n = 3$, two different Weibull moduli of the fiber m, two different ratios of interfacial sliding stress to the characteristic fiber stress τ_0/S_c, and a fiber volume fraction $f = 0.35$ when the composite is subjected to a constant uniaxial loading such that $\sigma/S_c = 0.3$. The arrow indicates the creep rupture.*

not taken into account in this study. Also, the interface between the fiber and matrix is assumed to be perfectly bonded. Figure 23 shows the predicted creep curves in the composite as a function of microstructural and damage parameters [30]. It has been demonstrated that when fibers are broken, the increase in the overall creep strain of the composite and the axial stress in the intact fibers can be significant. Also, the matrix plasticity has very limited effect on the creep behavior of the composite.

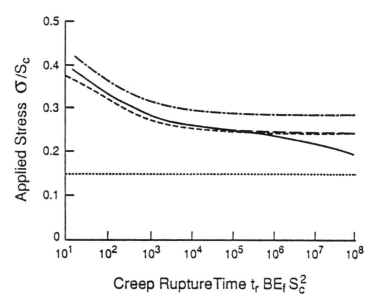

$$n = 3, \quad m = 5, \quad f = 0.35, \quad E_f/E_m = 3, \quad L_s/D = 250$$

——— Fiber Relaxation Model $(\tau_0/S_c = 0.01)$

·——— Curtin - McLean Model

——·—— Strain Based Rupture for McLean's Model

————— Stress Based Rupture for McLean's Model

············ Fiber Bundle Strength

Creep RuptureTime t_r BE_f S_c^2

Figure 21. The applied stress as a function of creep rupture time predicted by different models for a material with a ratio of fiber to matrix Young's modulus $E_f/E_m = 3$, a creep exponent of the matrix $n = 3$, a Weibull modulus of the fiber $m = 5$, a ratio of interfacial sliding stress to the characteristic fiber stress $\tau_0/S_c = 0.01$, and a fiber volume fraction $f = 0.35$.

Recently, Lee et al. [31] proposed a model to predict the creep behavior of fiber-reinforced titanium matrix composite reinforced with noncreeping fibers. This model takes several microstructural damage parameters into account, which include fiber fracture and fiber-matrix interfacial debonding. In this model, the fiber strength also follows a statistical function, which has a Weibull form, and the analysis of stress recovery is also based on Curtin's model [27]. The load transfer

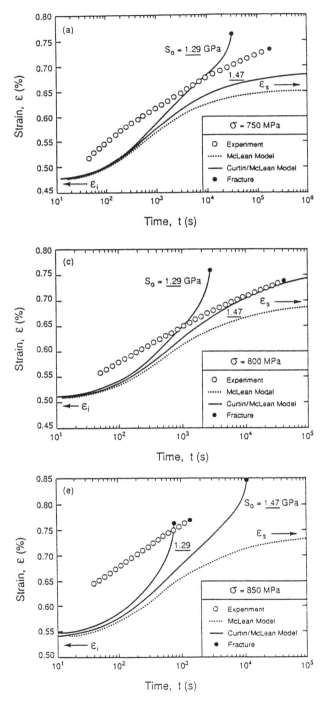

Figure 22. Comparisons of measured creep curves and model predictions for a SiC/Ti-6Al-4V composite.

305

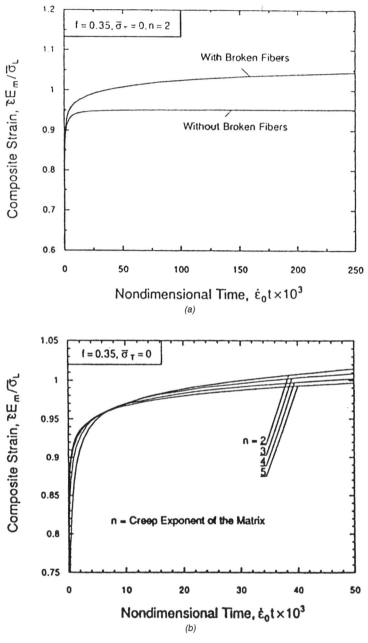

Figure 23. (a) The predicted creep curves of composites under uniaxial tension with and without fiber damage; (b) the effect of creep index n of the matrix on the creep strain; and (c) the effect of crack spacing on creep strain of a composite subjected to longitudinal tensile loading.

306

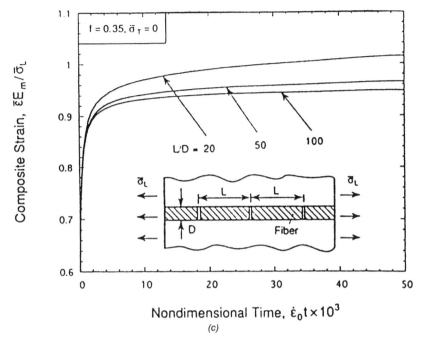

Figure 23. (cont.) *(a) The predicted creep curves of composites under uniaxial tension with and without fiber damage; (b) the effect of creep index n of the matrix on the creep strain; and (c) the effect of crack spacing on creep strain of a composite subjected to longitudinal tensile loading.*

analysis is based on the "global load-sharing" concept. In simulating the creep behavior of a multiple filament composite, the composite is divided into $n_f \cdot n_c$ unit cells; n_f is the number of fibers, and n_c is the number of cells in each fiber. If there are n_c^b fiber breaks occurring in any given row of cells, the stress in the matrix and fibers adjacent to the broken fibers increases due to load transfer. The load-sharing rule in this model is based on the rule of mixture:

$$\sigma_o = \frac{n_f - n_c^b}{n_f}\, \sigma_f V_f + \sigma_m V_m + \Delta\sigma_o \qquad (9)$$

where the additional stress $\Delta\sigma_o$ is simply equal to the stress once carried by the broken fibers, and the term $n_f/(n_f - n_c^b)$ accounts for the fact that stress redistribution in the fiber takes place in only $n_f - n_c^b$

fibers. By considering the bonding conditions, the total stress in one particular fiber is derived to be:

$$\bar{\sigma}_{f(j)} = \frac{\sigma_f}{L}\left[L - n_b^f\delta + \frac{\sigma_f V_f \delta}{\sigma_0} \sum_{k=1}^{n_c - n_b^f} \frac{n_{c(k)}^b}{n_f - n_{c(k)}^b}\right] \qquad (10)$$

for perfect bonding situation. In this relation, n_b^f is the number of broken cells in fiber f, and the summation term adds together the stress transferred from adjacent fiber to each of the k nonbroken cells in the fiber. In the situation where debonding occurs, the average stress is given by:

$$\bar{\sigma}_{f(j)} = \frac{\sigma_f}{L}\left[L - n_b^f(\delta - \delta_b) + \frac{V_f}{\sigma_0}[\sigma_f(\delta - \delta_b) + \sigma_{fb}\delta] \sum_{k=1}^{n_c - n_b^f} \frac{n_{c(k)}^b}{n_f - n_{c(k)}^b}\right] \qquad (11)$$
$$- \frac{\sigma_{fb}\, n_b^f\delta}{L}$$

Once $\bar{\sigma}_f$ is known, the average stress of the matrix, $\bar{\sigma}_m$, can then be obtained through the rule of mixture. Knowing all the values of $\bar{\sigma}_m$ will enable Equation (7) to be solved, which then gives the strain of composites. Figure 24 shows a comparison between the predicted and experimental results for an SCS-6/Ti-6Al-4V composite at 427°C and at applied stresses between 965 and 1241 MPa [32].

SUMMARY REMARKS

The progress that has been made in understanding the creep behavior of fiber-reinforced titanium matrix composites is compiled in this chapter. Experimental investigation has clearly identified that the creep behavior of the TMCs is critically influenced by fiber fracture, interfacial debonding, and environment-assisted matrix cracking. Several theoretical models have been proposed to predict the creep strain and creep rupture life. However, attempts to compare the experimental and predicted creep strain and rupture lives have been largely unsuccessful. This is due to insufficient basic understanding about the creep deformation processes and their dependence on constituent properties of the composites. A systematic investigation to better understand the creep behavior and the accompanying microstructural damage evolution over a range of anticipated operating conditions is needed. Appropriate creep models incorporating these

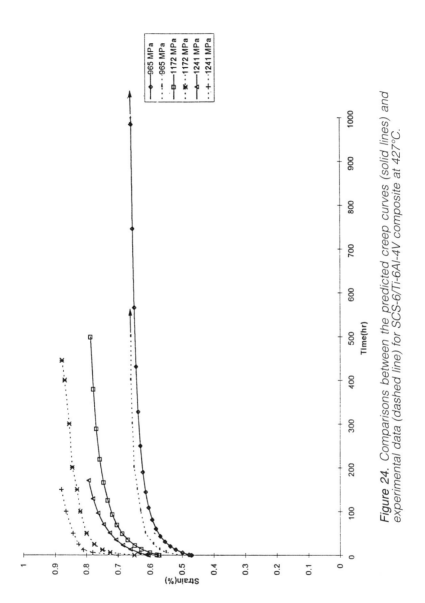

Figure 24. Comparisons between the predicted creep curves (solid lines) and experimental data (dashed line) for SCS-6/Ti-6Al-4V composite at 427°C.

309

time-dependent damage features that occur during creep deformation process should be developed and validated.

ACKNOWLEDGMENTS

This work has been supported by the National Science Foundation (DDM-9057030) and the Air Force Office of Scientific Research (F49620-93-1-0320, Dr. Walter Jones is the program monitor). Thanks are also due to Dr. P. C. Wang, H.-Y. Huang at UCLA, Dr. Z. Z. Du at UCSB and Drs. M. L. Gambone and S. W. Schewenker at AF Wright Laboratory for helpful discussion.

REFERENCES

1. D. G. Backman and J. C. Williams, "Advanced Materials for Aircraft Engine Applications", *Science, 255,* 1992, p. 1082.
2. C. H. Ward and A. S. Culbertson, "Issues in Potential IMC Applications for Aerospace Structures", An Overview of Potential Titanium Aluminide Composites in Aerospace Applications", in *Intermetallic Matrix Composites III,* Materials Research Society, Vol. 350, 1994, p. 3.
3. J. M. Larson, S. M. Russ and J. W. Jones, "Possibilities and Pitfalls in Aerospace Applications of Titanium Matrix Composites", in *NATO AGARD Conference on Characterization of Fiber Reinforced Titanium Metal Matrix Composites,* AGARD, Bordeaux, France, 1993.
4. D. B. Miracle, P. R. Smith and J. A. Graves, "A Review of the Status and Developmental Issues for Continuously Reinforced Ti-Aluminide Composites for Structural Applications", in *Intermetallic Matrix Composites III,* Materials Research Society, Vol. 350, 1994, p. 3.
5. J. M. Larsen, W. C. Revelos and M. L. Gambone, "An Overview of Potential Titanium Aluminide Composites in Aerospace Applications", in *Intermetallic Matrix Composites II,* Materials Research Society, Vol. 350, 1992, p. 3.
6. J. Doychak, "Metal and Intermetallic Matrix Composites for Aerospace Propulsion and Power Systems", *JOM, 44*[6], 1992, p. 46.
7. A. G. Evans and F. Zok, "The Physics and Mechanics of Fiber-Reinforced Brittle Matrix Composites", *Journal of Materials Science, 29,* 1994, p. 3857.
8. J.-M. Yang and S. M. Jeng, "Deformation and Fracture of Ti- and Ti3Al-Matrix Composites", *JOM, 44*[6], 1992, p. 52.
9. S. W. Schwenker, "Longitudinal-Creep Behavior of Fiber-Reinforced Titanium-Matrix Composites", Ph.D. Dissertation, University of Dayton, 1994.
10. M. McLean, "Creep Deformation of Metal Matrix Composites", *Composites Science and Technology, 23,* 1985, p. 37.

11. P. K. Wright, "Creep Behavior and Modeling of SiC/Titanium MMC", in *Titanium Matrix Composites,* edited by P. R. Smith and W. C. Revelos, WL-TR-92-4035, Wright Research and Development Center, Wright-Patterson AFB, OH, 1992, p. 251.

12. M. L. Gambone, "The Fiber Strength Distribution and its Effect on the Creep Behavior of a SiC/Ti-1100 Composite", Ph.D. Dissertation, University of Virginia, 1995.

13. M. Khobaib, in *Titanium Aluminide Composites,* edited by P. R. Smith, S. J. Balsone and T. Nicholas, WL-TR-91-40201991, Wright Research and Development Center, Wright-Patterson AFB, OH, 1991, p. 450.

14. M. Khobaib, "Damage Evolution in Creep of SCS-6/Ti-24Al-11Nb Metal Matrix Composites", *Proceedings of the American Society for Composites—Sixth Technical Conference,* 1991, p. 638.

15. M. L. Gambone, "Fatigue and Fracture of Titanium Aluminide", Vol. II, WRDC-TR-89-4145, Wright Research and Development Center, Wright-Patterson AFB, OH, 1990.

16. R. A. Mackay, P. K. Brindley and F. H. Froes, "Continuous Fiber-Reinforced Titanium Aluminide Composites", *JOM, 43*[5], 1991, p. 23.

17. S. M. Jeng, P.-C. Wang and J.-M. Yang, "Tensile and Flexural Creep Behavior of Fiber-Reinforced Titanium Matrix Composites", *Proceedings of the American Society for Composites—Eighth Technical Conference,* 1993, p. 551.

18. S. M. Jeng and J.-M. Yang, "Creep Behavior and Damage Mechanisms of SiC Fiber-Reinforced Titanium Matrix Composite", *Materials Science and Engineering, A171,* 1993, p. 65.

19. H.-P. Chiu, J.-M. Yang and J. A. Graves, "Effect of Fiber Coating on Creep Behavior of SiC Fiber-Reinforced Titanium Aluminide Matrix Composites", *Journal of Materials Research, 9*[1], 1994, p. 198.

20. S. M. Jeng, J.-M. Yang and J. A. Graves, "Effect of Fiber Coating on the Mechanical Behavior of SiC Fiber-Reinforced Titanium Aluminide Composites", *Journal of Materials Research, 8*[4], 1993, p. 905.

21. J.-M. Yang, S. M. Jeng and C. J. Yang, "Fracture Mechanisms of Fiber-Reinforced Titanium Alloy Matrix Composites. Part I. Interfacial Behavior", *Materials Science and Engineering, A138,* 1991, p. 155.

22. R. M. McMeeking, Models for Creep of Ceramic Matrix Composites", in *High Temperature Mechanical Behavior of Ceramic Composites,* edited by S. V. Nair and K. Jakus, Butterworth-Heinemann, 1995, p. 409.

23. M. McLean, "Modeling of Creep Deformation in Metal Matrix Composites", *Fifth Interfactional Conference on Composite Materials,* edited by W. C. Harrigan, Jr., J. Strife and A. K. Dhingra, TMS, Warrendale, PA, 1985, p. 37.

24. M. McLean, "Mechanisms and Modes of High Temperature Deformation of Composites", in *High Temperature High Performance Composites,* edited by F. D. Lemkey et al., Materials Research Society, V. 120, 1988, p. 67.

25. S. Goto and M. McLean, "Role of Interfaces in Creep of Fiber-Reinforced Metal Matrix Composites. I. Continuous Fibers", *Acta Metallurgica et Materialia, 39* (1991), p. 153.

26. A. Kelly and G. J. Davies, *Metallurgy Reviews, 10*[37], 1965, p. 1.

27. W. A. Curtin, "Theory of Multiple Fiber Fractiure in Ceramic Matrix Composites", *Journal of American Ceramic Society, 74*[11], 1991, p. 2837.

28. Z. Du and R. McMeeking, "Creep Models for Metal Matrix Composites with Long Brittle Fibers", *Journal of Physics and Mechanics of Solids, 43*[5], 1995, p. 701.

29. C. H. Weber, Z.-Z. Du and F. W. Zok, "High Temperature Deformation and Fracture of a Fiber Reinforced Titanium Matrix Composite", *Acta Metallurgica et Materialia, 44*[2], 1996.

30. Y. Song, G. Bao and C. Y. Hui, "On the Creep of Unidirectional Fiber Composite with Fiber Damage", *Acta Metallurgica et Materialia, 43,* 1995, p. 2615.

31. S. Lee, S. M. Jeng and J.-M. Yang, "Modeling and Simulation of the Effect of Fiber Fracture on Creep Behavior of Fiber-Reinforced Metal Matrix Composites", *Mechanics of Materials, 21,* 1995, p. 303.

32. H.-Y. Huang and J.-M. Yang, unpublished results.

8

Fatigue Crack Growth

HAMOUDA GHONEM

Mechanics of Materials Laboratory
Department of Mechanical Engineering
University of Rhode Island
Kingston, RI 02881

INTRODUCTION

BECAUSE TITANIUM METAL matrix composites (MACS) reinforced with
Sic fibers are intended for designs involving load reversals at elevated
temperatures, an important issue in their potential success is the understanding of their fatigue damage tolerance characteristics. In this
type of assessment, the propagation of fatigue cracks is recognized as
a critical damage mode. This is due to the fact that different processing
methods used to fabricate this class of composites inevitably lead to the
formation of microscopic defects, which under cyclic loadings, could
readily nucleate microcracks [1–4]. The presence of these defects, particularly in regions of interfaces within the composite, would lead, during cyclic loadings, to a significant reduction in the crack initiation life
of the composite. The growth behavior of an existing crack, therefore,
becomes a critical factor in determining the residual strength of a composite. Efforts have been made to examine damage processes associated with the growth of a single dominant fatigue crack in composites.
In MACS, loaded in the fibers' direction, the propagation of a crack is
governed by properties of the fiber, matrix, and fiber-matrix interface
[2,5–8]. In composites with strong interfaces and weak fibers, the crack
growth is dominated by fiber and matrix fracture; an example of this is
the B_4C-B/Ti-6Al-4V composite. In contrast, composites with weak interfaces and strong fibers are susceptible to crack branching at the
fiber-matrix interface. Titanium matrix composites reinforced with silicon carbide-type fibers, SCS-6 or SM1240, are characterized by strong
fibers, weak interface, and compressive residual stresses at the fiber-

313

matrix interface. In these composites, the fatigue crack growth process, under mode I loadings, is frequently accompanied by interface debonding, which allows the crack to advance through the matrix leaving unbroken fibers in the wake of the crack, thus bridging the crack surface. In this cracking mode, the crack tip driving force is shielded by the load carried by the bridging fibers. Studies on several unidirectional fiber-reinforced MACS, including SCS-6/Ti-6Al-4V, SCS-6/Ti-15V-3Al-3Cr-3Sn, SCS-6/Ti-24Al-11Nb, SCS-6/Ti-25Al-10Nb-3Cr-1Mo, and SM1240/TIMETAL®21S showed that fiber bridging is an operative damage mechanisms under loading conditions of practical interest [9–16]. The extent of fiber bridging during the crack growth process depends, in addition to the applied stress range, stress ratio, and the fiber strength, on the strength properties of the fiber-matrix interface. This was shown in the work of Gayda et al. [17] on SCS-6/Ti-15-3 composites at elevated temperature and in the work of McMeeking and Evans [7] on SCS-6/Ti-Al composites. Furthermore, the degradation of thermally cycled MACS has been primarily attributed to the mismatch of the coefficients of thermal expansion between fiber and matrix phase and to the formation of reaction products at the fiber-matrix interface, which leads to a stress-strain amplification at the interface [18]. The interfacial failure would occur if the interphase strength is less than the strength of the matrix and the transverse strength of the fiber. This failure is facilitated in titanium matrix composites by the use of Sic-type fibers that are coated with single or multiple carbon-rich layers designed to accommodate progressive debonding and sliding along the interface. For an optimum crack growth resistance in a composite system, the right balance between interfacial strength and fiber bridging must be obtained. Although a weakly bonded interface is desirable for increased resistance to fatigue crack growth, a strong interface is generally required for high transverse strength. If the interface is too strong, the degree of fiber bridging is lessened, and the crack grows fairly quickly. However, if the interface is too weak, there is much more bridging, yet less closure force is achieved. For SCS-6/Ti-6Al-4V composite system, it was found that decreasing the applied stress range increases the interface stress range over which substantial increase in crack growth life can be obtained.

Efforts have been made to examine different aspects of the fatigue crack mechanisms in titanium matrix composites. This chapter, divided into three sections, describes results of these efforts with an objective of providing a detailed physical and mathematical view of the crack growth processes in titanium matrix composites reinforced with silicon carbide-type fibers. The first section focuses on the response of the crack growth rate to different loading conditions, particularly test temperature and frequency of the loading cycle. In this section the characteristics of various parameters associated with the fatigue crack bridging phenomenon, namely, the crack opening displacement, the bridging fiber stress, the fiber-matrix frictional shear stress, and the debonding length along a bridging fiber, are discussed on the basis of available experimental results and specifically those obtained in the author's laboratory. The second section of the chapter is an outline of different mathematical models that were developed to identify the crack tip driving force under the condition of crack bridging. Although an attempt is made to avoid exhaustive mathematical derivations, emphasis is placed on the general structure of these models and their key parameters. The third section defines the conditions at which the crack bridging and acceleration transition occurs. This transition is described in terms of a failure criterion for bridging fibers located in the wake of the crack tip.

PHYSICAL ASPECTS OF FATIGUE CRACK BRIDGING PROCESS

This section presents a qualitative description of the fatigue crack growth process in MMCs with emphasis on the influence of elevated temperatures and loading frequencies on the crack growth behavior. The physical parameters governing the crack growth process under crack bridging condition and their relationships to the observed behavior are described.

Crack Growth Rate

The physical aspects of the fatigue crack growth process in MMCs have been examined by several authors [6,7,9–11,19–23]. The fatigue fracture process in a typical center-notched specimen of an SiC-Ti MMC, loaded parallel to the fiber orientation and for applied load levels that are insufficient to cause immediate fiber fracture at the crack tip, is dominated by matrix cracking and crack bridging. In this mode of cracking, the high-strength bridging fibers carry part of the applied load, thus lowering the crack tip driving force. The growth mechanism under this condition involves a process in which the increase in the matrix crack length is accompanied by an increase in the number of fibers bridging the crack. This process would continue up to a steady-state condition produced by the balance between creating more bridging fibers as the crack length increases and the fracture of these fibers as more stress is transferred to them by the cracked matrix.

The following event of crack growth arrest or acceleration, which is associated with the fracture of one or several bridging fibers, particularly those fibers farthest away from the crack tip, depends on the nature of this balance.

Basic characteristics of bridged cracks in Ti-MMCs have been studied by a number of authors. Davidson [9] measured the crack growth rate in an SCS-6/Ti-6Al-4V composite ($[0°]_4$, V_f = 42%) using center-notched specimens subjected to stress ratio of 0.1. For an applied load range, $\Delta\sigma_a$, varying from 118 to 175 MPa, the crack growth rates ranged from 10^{-10} to 10^{-7} m/cycle. Growth rate acceleration or deceleration with increasing crack length depended on the stress level. At low applied stresses, however, the crack arrested. In addition, the growth rates were relatively insensitive to the number of bridging fibers in the crack wake. Fiber-matrix interface debonding reached up to 1 mm from the crack plane and occurred at several locations, including between fiber and coating, within the coating layers, and between coating and matrix. Cotterill and Bowen [24] have also observed the arrest of bridging cracks in an SCS-6/Ti-15-3 composite tested with stress ratios R = 0.1 and 0.5 at room temperature. The fatigue crack growth rates were slower at R = 0.1, and the crack length at arrest is lower than that at R = 0.5. Although the fatigue crack growth rates were initially similar, fiber breakage occurred more readily at a higher load ratio, resulting in a faster growth rate and shorter life. In the fatigue crack growth study of SCS-6Ti-6Al-4V composite, Bowen et al. [13] observed that increasing $\Delta\sigma_a$ from 100 to 125 MPa, for the same stress ratios (0.1 or 0.5), or increasing the stress ratio for the same $\Delta\sigma_a$, has a minimal effect on the crack growth rate. However, for R = 0.1, the onset of catastrophic failure is promoted at smaller crack lengths, similar to that of SCS-6/Ti-15-3 composite [23,24]. Work on an SM1240/TIMETAL®21S $[0]_6$ composite [23], however, did not show crack arrest (loading conditions: $\Delta\sigma_a$ = 270 MPa, R = 0.1, and frequency = 10 Hz). Larsen et al. [25] in their work on an SCS-6/TIMETAL®21S $[0]_4$ composite ($\Delta\sigma_a$ varied from 270 to 668 MPa and R = 0.1) observed that crack bridging was significant for stress ranges up to 300 MPa, whereas the extent of bridging was limited or nonexistent at the higher stresses.

Effect of Elevated Temperature

Limited work on the fatigue crack growth behavior of MMCs has been carried out at elevated temperatures. The observed crack growth rate from these data, however, appear to differ. Cotterill and Bowen

[24] measured the crack growth rate of an SCS-6/Ti-15-3 composite using SEN specimen in three-point bending tests. Comparison of the 25 and 500°C tests (initial $\Delta K = 16$ MPa$\sqrt{m}$ with $\Delta \sigma_a$ of 13 MPa, $R = 0.5$, and a frequency of 10 Hz) showed that the crack growth rate is higher at higher temperatures. In addition, at 500°C, the initial deceleration of crack growth is interrupted by discrete events of fiber failure, which lead to instantaneous rise in the crack growth rate. The fatigue response of a single edge-notched specimen made of an SiC/Ti-6Al-4V composite exhibits a similar trend of higher crack growth rate at 550°C (0.5 Hz) compared with that obtained at ambient temperature [14]. Barney et al. [26] performed a study on the fatigue performance of the same composite by three-point bending tests. The tests were carried out at a frequency of 10 Hz, $\Delta K_a = 12$ MPa$\sqrt{m}$, $R = 0.5$, at room temperature, and at 500°C. At room temperature, the crack growth rate, da/dN, was found to decrease as the crack length increases until the crack arrested, thus indicating that significant fiber bridging occurred. At 500°C, although the da/dN was higher than that at room temperature, it increased rapidly with the increase in the crack length due to the presence of limited, if any, crack bridging. Measurements of fatigue crack growth of an SM1240/TIMETAL®21S composite using center-notched specimens were carried out in Reference [23]. The tests were performed at room temperature, 500, and 650°C for loading frequencies of 10, 0.1, and 0.02 Hz with $\Delta \sigma_a = 273$ MPa and $R = 0.1$. The common feature of the growth rate curves, shown in Figure 1, is the presence of an initial crack bridging stage followed by repeated events of crack growth acceleration and retardation, see also References [24,27]. Focusing on the initial crack bridging stage, one observes that, opposite to results of Ibbotson et al [14] and Cotterill and Bowen [24], the crack growth rate, for the same loading frequency, decreases as the temperature increases. Furthermore, the slope of the crack growth rate curve, which is a measure of the build-up of the crack tip shielding force, is lowest in the room temperature test and highest in the 650°C test. This suggests that, for the same crack length, the crack tip shielding force increases as temperature increases.

The observed differences in the effect of elevated temperature on the bridging fatigue crack growth rate of several MMC systems discussed above could be explained in terms of the balance between the response of different matrix materials to temperature and the degree of debonding along the fiber-matrix interface. The mechanical response of the titanium alloys Ti-15V-3Al-3Cr-3Sn, Ti-6Al-4V, and TIMETAL®21S at elevated temperatures vary to a great extent in

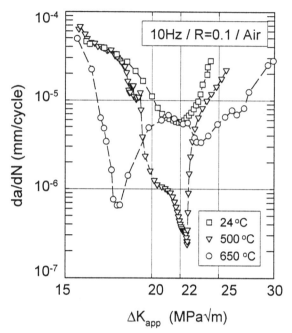

Figure 1. Fatigue crack growth of SM1240/Timetal-21S composite at room temperature, 500 and 650°C for loading frequencies of 10 Hz using center-notch specimens.

modifying the stress and strain states in the composite. The different variations in chemical composition of the interphase zone in each composite results in different interphase characteristics, including interphase shear strength, surface roughness of the debonded interface, and debonding length. Consequently, these differences are expected to produce different crack growth rate responses. The effect of cyclic bending induced by the three-point bend test [24] and the SEN specimen geometry [14] attenuates the fracture of bridging fiber in the crack wake, resulting in higher crack growth rates. Such effect, however, is absent in the fatigue loading of a center crack specimen [16].

The different fracture processes encountered in each of the crack growth stages, namely; bridging, transition, and acceleration, are evident in the appearance of the fracture surface features. The micrograph shown in Figure 2, for the 500°C test of an SM1240/TIMETAL®21S composite, displays two different fracture zones, the first for the crack bridging stage and the second for the repeated crack growth acceleration or retardation stage. The first zone has excessive

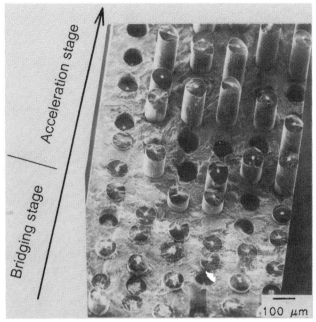

Figure 2. Micrograph of different fracture processes encountered in each of the crack growth stages, namely; bridging, transition, and acceleration.

fiber core damage, which may have resulted from the application of a large number of fatigue cycles and prolonged thermal exposure. The pulled out fibers in this zone are relatively short. In examining the interfacial region of bridging fibers in the composite mentioned above, Marshall et al. [28] and Zheng and Ghonem [16] observed that debonding always occurred within the carbon coating layer or between the carbon and the TiB_2 layers at any test temperature. Figure 3 is an example of a debonded carbon layer at a plane located immediately under the fracture surface of a room temperature test. The second zone in Figure 2 shows no fiber core damage and displays longer fiber pull out length. An example of the unstable fracture zone is Figure 4, which shows a high degree of matrix deformation surrounding the pulled out fibers.

The effect of prior thermal exposure on the fatigue crack growth behavior of an SM1240/TIMETAL®21S and B_4C-B/Ti-6Al-4V composites was examined in References [23,29], respectively. In the former study, the crack growth rates for both as-received and specimen aged for 42 hours at 650°C are measured at room temperature (see Figure 5). Al-

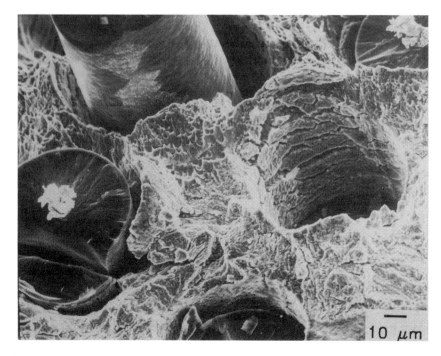

Figure 3. Example of an unstable fracture zone with matrix deformation surrounding the pulled out fibers.

though both samples display identical crack growth rate, the duration
of crack bridging is prolonged for the aged sample. This result indicates a delay in the fracture of bridging fibers, which triggers the
transition from deceleration to acceleration crack growth rate. This
delay is explained as follows. In addition to the expected change of microstructural properties of the matrix or degradation of the interphase due to thermal exposure, thermal aging of a metal matrix composite, particularly at a temperature as high as 650°C, could result in
decreasing the compressive residual stress acting in the fiber. A separate study on SCS-6/TIMETAL®21S [30] showed that the decrease in
the mean stress for a particular applied stress range, although it does
not influence the growth rate in the crack bridging stage, it causes a
delay in reaching a condition of crack growth rate acceleration. The
effect of aging on a B_4C-B/Ti-6Al-4V composite was investigated using
a single edge-notched (SEN) specimen [29]. In the as-received specimen, the crack grows in mode I, whereas in a thermally exposed (7

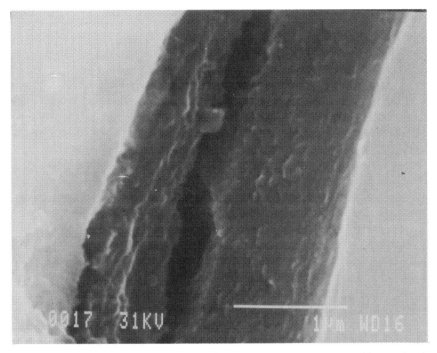

Figure 4. Example of a debonded carbon layer at a plane located immediately under the fracture surface of a room temperature test.

days at 500°C) specimen, crack growth occurred both across the fiber and along fiber-matrix interfaces. Thermal exposure was found to increase the amount of crack growth in the interface. Interface debonding occurred while the main crack is relatively far from the interface (typically 80 to 100 μm at $\Delta K = 32$ MPa$\sqrt{\text{m}}$). In the as-received specimen, on the other hand, a crack as close as 20 μm at $\Delta K = 22$ MPa$\sqrt{\text{m}}$ to the interface did not cause debonding. This reflects a higher interface strength in the latter case. Chan [31] also proposed that the lowering of the interphase shear strength value due to thermal exposure would favor extensive interface cracking and crack bridging mechanisms over fracture of fibers, provided that such thermal exposure does not degrade the fiber strength. Although different shear strength levels of the interphase region could be obtained through different processing methods, aging studies on an SCS-6/TIMETAL®21S composite for duration up to 200 hours at 650°C, however, showed no degradation of interphase strength due to thermal exposure in vacuum condition [32].

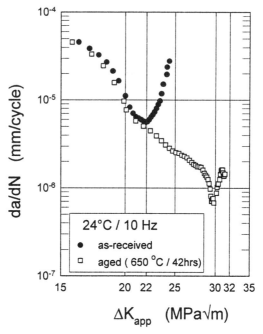

Figure 5. Room temperature crack growth rates for specimens in as-received condition and aged for 42 hours at 650°C.

Effects of Loading Frequency

Time-dependent effects on high-temperature fatigue crack growth behavior in high-strength structural materials are generally described in terms of creep and/or environmental degradation processes. The individual contribution of these two processes depends, in addition to the strengthening characteristics of the alloy and temperature level, on the frequency of the loading cycle. The lower the frequency is, i.e., the longer the cycle duration, the higher the ability of environment and/or creep to contribute to the overall damage process. The situation is no different in the case of SiC-TiMACS, which is further complicated by the differences in the flow characteristics of the main composite constituents. Although the deformation response of the SiC-fibers reinforcement remains elastic for temperatures up to 800°C [33], the titanium alloy matrix readily flows at temperatures as low as 400°C [34]. A number of experimental studies have been carried out to determine the effect of the loading frequency on the crack growth rate of Ti-MACS. Bain and Gambone [35] worked on an SCS-

6/Ti-6Al-4V composite at 318°C subjected to loading frequencies of 20, 0.33, and 0.033 Hz. They showed that, although there was no significant difference in the fatigue crack growth rates for the first two frequencies, the crack growth rate for the 0.033-Hz test increased by a factor of 40. This increase was attributed to environmental damage affecting the matrix material. They did not, however, observe differences between fracture surface features of specimens tested at the three different frequencies. Ibboston et al. [36] studied the crack growth response of the same composite using single edge-notched specimens. Tests were carried out at both room temperature and 550°C for two loading frequencies: 10 and 0.5 Hz. They observed that the crack growth rate increases with an increase in temperature and a decrease in the loading frequency. Cotterill and Bowen [24] measured the crack growth rate in an SCS-6/Ti-15-3 composite at 350 and 500°C for three frequencies: 10, 2, and 0.5 Hz tested at a stress ratio of 0.5. Although no difference was observed in the crack growth rates, da/dN, at 350°C, a rise in da/dN was seen at 500°C between 10 and 2 Hz. Their explanation for this increase in da/dN was the possible aging of the matrix alloy at 500°C, a phenomenon that does not occur at 350°C. During the aging of Ti-15-3 soft α-precipitates form preferentially along the grain boundaries, causing the crack tip to follow an intergranular fracture path.

The work on SM1240/TIMETAL®21S composite in Reference [23] included tests at 500 and 650°C at frequencies of 0.1 and 0.02 Hz. Results in the form of da/dN versus applied stress intensity factor are shown in Figures 6(a) and 6(b). In the first crack bridging stage, these results show no difference between crack growth rates of specimens tested at high frequencies (>10 Hz). At both temperatures, however, similar to results of Reference [24], da/dN increases with the decrease in the loading frequency. To account for the time variations associated with different loading frequencies, results of Zheng and Ghonem [23] for $\Delta K_{app} = 20$ MPa$\sqrt{m}$, were replotted in Figure 7 in the form of crack growth speed, da/dt, versus frequency. This figure illustrates the dependency of the crack growth rate on both temperature and frequency and shows that for the same temperature, the crack speed, for frequencies greater than 10 Hz, decreases as the frequency decreases. This result could be interpreted in terms of the increase in the crack tip shielding or the increase of crack tip blunting due to matrix relaxation at lower frequencies. Zhang and Ghonem [37] expanded the work of Reference [23] by measuring the 500°C crack growth rate for loading frequency of 0.1 Hz with a hold time of 10 sec imposed at the maximum load level. Results of this work com-

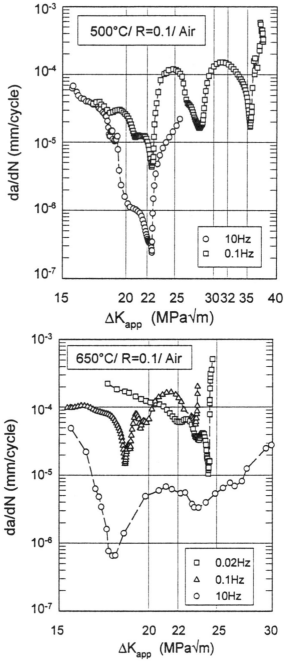

Figure 6. Crack growth rate, da/dN, versus applied stress intensity factor range for SM1240/Timetal-21S composite at (a) 500°C and (b) 650°C for loading frequencies of 10, 0.1 and 0.02 Hz.

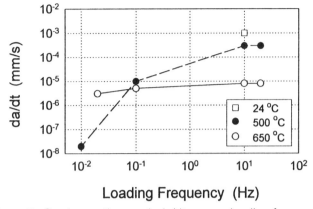

Figure 7. Crack growth speed, da/dt, versus loading frequency.

pared with those of 10 Hz without hold time are plotted in Figure 8. It shows the crack growth rate decreases in the presence of hold time. In interpreting this result, the authors suggested that the hold time during a loading cycle produces creep effects that contrary to environmental effects, see Reference [23], lead to a reduction of the crack tip driving force. To examine this idea, they measured the crack opening displacement (COD) loops for the tests with and without hold time (see Figure 9). The loops corresponding to tests with hold time reveal crack closure at a load level higher than the minimum applied load. The reason for this closure phenomena is explained as follows. During a low-frequency loading cycle, time-dependent flow and associated stress relaxation occur in the matrix phase. The matrix stress relaxation leads to an increase of the matrix-to-fiber load transfer, which results in an increase of the bridging fiber stress and, consequently, the crack tip shielding. The complex aspect of the low-frequency loading occurs, however, in the composite region ahead of the crack tip. In this region, a condition of matrix-fiber isostrain is required to maintain the fiber-matrix bonding. Depending on the degree of the inability of the fiber to accommodate the matrix strain, a compressive residual stress could be developed in the matrix phase of this region. This stress, when relaxed due to the matrix fracture as it is reached by the propagating crack, would result in a matrix surface displacement, producing a crack tip closure. This closure, which occurs at an applied stress level higher than that corresponding to the minimum load of the loading cycle, results in a decrease of the applied stress range and, consequently, the crack tip driving force. This concept was investigated in Reference [37] through an experimental procedure in which

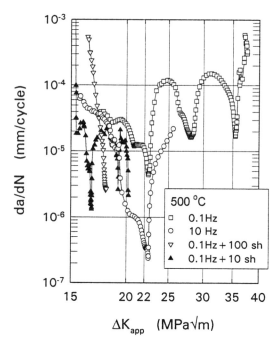

Figure 8. Crack growth rate at 550°C for both 0.1Hz with hold time 10 seconds and 10 Hz without hold time. [22].

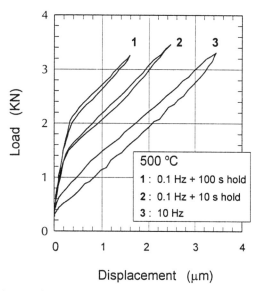

Figure 9. Crack opening displacement loops for specimens tested with and without hold time.

326

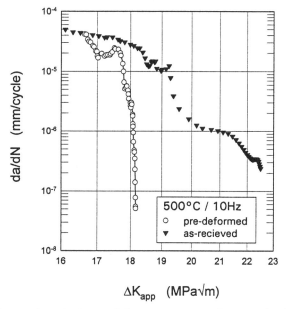

ΔK_{app} (MPa√m)

Figure 10. Comparison between fatigue response of as-received and predeformed specimens: (a) number of fatigue cycles versus crack length, (b) crack growth rate versus applied stress intensity factor in the first crack bridging stage.

the influence of prior creep deformation on the subsequent fatigue crack growth was examined. This was achieved by comparing the 10-Hz crack growth of a specimen that included 0.12% creep deformation prior to fatigue testing to that of a specimen without prior creep deformation. Results in Figure 10(a) show that the number of fatigue cycles required to achieve the same crack length is larger and the crack growth rate [see Figure 10(b)] is lower in the predeformed specimen. These results are supported by the presence of closure in the COD loops of the predeformed specimen. Crack closure was not observed in the loops of the nondeformed specimen.

FACTORS INFLUENCING CRACK BRIDGING PROCESS

Efforts have been made to identify the different factors influencing the crack bridging process with the aim of determining the crack tip driving force. These efforts have largely followed models that employ concepts of linear fracture mechanics coupled with micromechanical interpretation of balance of forces acting on the debonded region of bridging fibers [6,7,9,10,19–22]. Although the majority of these models may not be directly applicable to fatigue crack growth in composites with ductile matrices, they do, however, identify the crack opening displacement range, Δu, the fiber bridging stress range, Δp, the fiber-matrix frictional shear stress, τ_s, and the interface debonding length, l_d, as being critical governing parameters for the fatigue crack bridging process. Attempts have been made to study each of these parameters and their form of correlations with mechanical and material properties of the composite. This section will describe these attempts and their results.

Crack Opening Displacement

The experience gained from studying the fracture processes of high-strength monolithic materials shows that the crack opening displacement (COD), particularly near the crack tip region, is a measure of the crack tip driving force. In addition, COD has the advantage, in spite of its expected small magnitude in bridging cracks, of being a measurable parameter. This provides the possibility of establishing an experimentally based methodology to analyze the crack bridging phenomenon. A number of studies have been carried out in this regard. For example, Davidson [9], in his study of fatigue cracks in SCS-6/Ti-6Al-4V and SCS-6/Ti-14Al-21Nb composites considered that a residual crack opening displacement (CODr) is caused by the relief of axial residual stresses due to the occurrence of sliding between the matrix and the bridging fiber. Using a stereo-imaging technique, the CODr was measured as a function of the crack length and the applied stress, and a conclusion was made that the CODr could be used to compute the slip distance along the corresponding bridging fiber. Crack opening displacements due to the applied load were also measured [9] and were shown to be a function of the square root of the distance behind the crack tip. Telesman et al. [38] used a loading stage mounted inside a scanning electron microscope chamber to measure the Δu profiles of cracks propagating in an SCS-6/Ti-24Al-11Nb composite and used the results to compare with those obtained using different analytical methods. John et al. [39], using a SCS-6/TIMETAL®21S composite at room temperature, measured the COD along a fatigue crack using an IDG system and, similar to the work of Davidson [9], measured the CODr values for the same crack length using a scanning electron microscope. A procedure was then developed to generate the absolute COD profiles along the crack length at different stress levels. Zheng and Ghonem [23] measured, also using an IDG system, the COD at positions along different bridged crack lengths in an SM1240/TIMETAL®21S composite tested in air environment at three different temperatures: 24, 500, and 650°C. Their data, which were not resolved into residual and absolute components, are plotted in Figure 11. They show that for this particular composite system, $\Delta u(x)$ increases with temperature. Because the axial residual stress is inversely proportional with temperature, the difference in the absolute COD values at these temperatures would be higher than that shown in the previous figure. The increase in COD with temperature is possibly due the change in the debonding strength, τ_d, of the fiber-matrix interface with temperature; see Reference [32]. This, in turn, permits, for the same crack length, a larger fiber-matrix relative displacement as temperature increases. Furthermore, in the work of Reference [23], by selecting the value of COD at a unit distance from the crack tip to represent the crack tip opening displacement, Δu_T, the relationship between Δu_T and da/dN, for the different test temperatures mentioned above is plotted in Figure 12. Because Δu_T was observed to increase with the increase in the crack growth rate, the authors concluded that, unlike ΔK_{app}, the crack tip opening displacement is a measure of the crack tip driving force in MMCs. The need of COD measurements to assess the crack growth process produces a set of problems, one of which is the difficulty in measuring a displacement in the range of a few microns. In addition, variations in residual stresses due to nonuniformity of fiber spacing in composites could produce significant errors in COD measurements.

Bridging Fiber Stress

The second important parameter in the crack bridging process is the bridging fiber discrete stress, ΔS. The significance of this parameter comes from the fact that theoretical models developed to calculate the effective crack tip driving force, as will be seen later, base their derivations on the idea that the effect of bridging fibers on the crack tip stress intensity factor can be computed by treating the loads carried by these fibers as a pressure applied along the crack wake. The direction of this pressure is opposite to that of the applied stress. The pressure concept averages the fiber loads on individual fibers, resulting, therefore, in a distribution whose discrete values, Δp, at locations of individual fibers is calculated by multiplying the fiber traction by the fiber volume fraction. The key to use these models is, therefore, knowing the continuous function describing the fiber pressure distribution along the bridged crack length. Davidson [9] assumed, on the

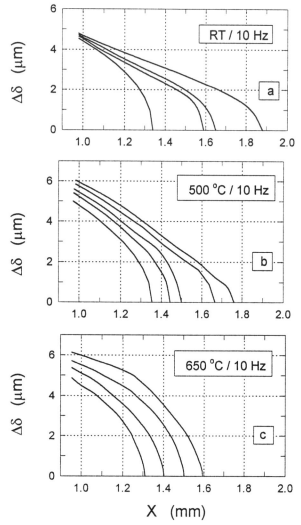

Figure 11. Crack opening displacement (COD) at positions along different bridged crack lengths in a SM1240/Timetal21S composite tested at three different temperatures: 24°C, 500°C, and 650°C.

basis of experimental observations related to COD and slip length along bridging fibers, that the fiber pressure is constant. Using this simplifying assumption, the fiber stresses in the Ti-6Al-4V matrix composite were calculated; their values were found to be in the range of 1.1 to 4 GPa and approximately independent of the distance from

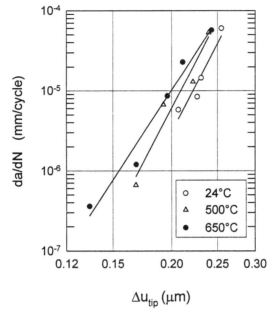

Figure 12. Relationship between crack tip opening displacement, Δu_T, and crack length, a, for the different test temperatures.

the crack tip. John et al. [39], employing the weight function method, calculated the bridging fiber stress distribution, Δp, by correlating predicted and measured crack opening displacements. Their work showed that the bridging fiber stress distribution at the maximum applied fatigue stress is not uniform but increases rapidly from a low value at the crack tip to a near constant away from the crack tip. Chiang et al. [22] applied a modified shear lag model and showed that Δp attains a finite value at the crack tip. Zheng and Ghonem [23], using experimentally measured COD values, employed a linear elastic fracture mechanics-based relationship to calculate the fiber pressure distribution as a function of crack length and temperature. Results for three temperatures (24, 500, and 650°C) are shown in Figure 13. A feature of these distributions is the presence in all test cases of a minimum finite value of Δp at the crack tip. Also, the average value of Δp increases proportionally to the increase in both temperature and the crack length. Furthermore, for the same test temperature, the difference between the maximum and minimum bridging fiber pressure for a particular crack length decreases as the crack length increases. This tendency of Δp to approach a uniform level is also influenced by

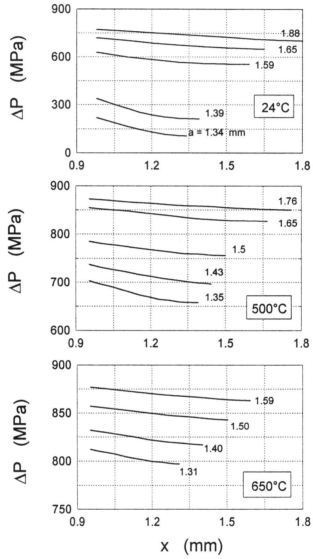

Figure 13. Fiber pressure distribution as function of crack length and temperature 24°C, 500°C, and 650°C.

the temperature as can be seen in Figure 13. The maximum value of ΔS_f is reached at the end of the crack bridging stage and acts on the first bridging fiber located at the crack mouth. These values for the three temperatures are 776 MPa (24°C), 873 MPa (500°C), and 877 MPa (650°C). Figure 14 shows the evolution of ΔS_f calculated at the

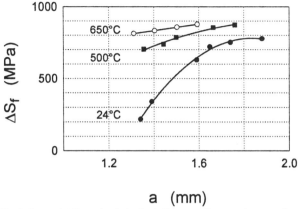

a (mm)

Figure 14. Evolution of ΔS_f calculated at the crack mouth as a function of the crack length and at different temperatures.

crack mouth as a function of the crack length and at different temperatures. ΔS_f seems to converge to an upper limit of about 900 MPa. Because these crack mouth bridging fibers are most likely to be the first to fracture at the end of the crack bridging stage, the 900 MPa would represent the fracture strength of the Sigma fibers. This value is much lower than the monotonic tensile strength of this type of fiber, which is estimated as 3000 MPa level [40]. This discrepancy could be explained by assuming that the fatigue fracture strength of SiC fibers is lower than its monotonic fracture strength. Furthermore, it was speculated that fibers break at distances farthest from the crack tip because of the increased volume of fiber exposed to stress rather than increased stress on fiber [9].

In a recent study by Ghonem [41], the work of Budiansky et al. [42] and Chiang et al. [22] was extended to the cyclic loading condition. In this modified shear lag model, the debonding length of a bridging fiber is divided into an initial debonding length, L_i and a reversed sliding length L_R ($L_i > L_R$). Within this reversed sliding length, L_R, the interface degradation due to the cyclic process, particularly at elevated temperature conditions, would alter the average frictional sliding stress, τ_s. This work, however, assumes that τ_s and L_R are constant during the entire loading and unloading processes. Results of the model produce explicit expressions for both the maximum and minimum bridging fiber stress, S_{max} and S_{min}, respectively. These expressions, in addition to L_i, L_R and τ_s, include parameters representing the postprocessing residual stress field in both fiber and matrix phases and the load ratio, R, of the applied loading cycle. The shielding part of the bridging fiber stress is then calculated as ΔS_{eff}, which is defined in this model as:

$$\Delta S_{eff} = \frac{S_{max} - S_{min}}{S_{max}} \quad , \quad \begin{array}{l} \text{if } S_{min} > 0 \\ \text{if } S_{min} \leq 0 \end{array}$$

Ghonem [30] employed this model in a parametric study to evaluate the sensitivity of ΔS_{eff} in SCS-6/Timetal MMC to different parameters characterizing the applied load cycle. One of his conclusions is that, for the same applied load range, the crack tip shielding increases as the mean stress increases.

Fiber-Matrix Frictional Shear Stress

The frictional shear stress, τ_s, has been the parameter most studied due to the recognition that the mechanics of the fiber-matrix interface region is critical for the understanding of the composite behavior [11,20,31,43]. Different methods of measurement have been applied to determine the magnitude of τ_s. The most common of these is the use of fiber push out and pull out tests, which measure τ_s in terms of the stress required to cause fiber-matrix slippage in thin-slice specimens of the composite under investigation [28,32,44,46,47]. These methods, however, view τ_s as being a single material constant. This view has been utilized in a variety of bridging models to predict the crack tip driving force, the debonding length, and the crack opening displacement. The existence of a gap between the results of these models and corresponding experimental data has focused the attention on examining the characteristics of τ_s and the validity of extending its single value as measured from prefatigued specimens to conditions under which fiber-matrix interfacial damage accumulates [9–11]. Recent work in this area has shown that repeated sliding on a fiber-matrix interface may lead to deterioration of that interface either by fracture of fiber coatings [48,49], wear of asperities, or plastic relaxation of the friction stress on the interface [45]. Indeed, available fiber push out and pull out tests have revealed reduced interface frictional stress in metal, intermetallic, and ceramic composites after a few loading and unloading cycles. However, push out tests performed on fatigued composite specimens have shown either further decease or limited increase in the value of τ_s compared with that obtained from as-received specimens. Furthermore, the variation in debonding mechanisms along the interface as discussed in Reference [11] indicates that the value of τ_s is apparently a function of the location along the debonding length and crack length. These results led several authors [11,23,31] to suggest that τ_s is not a material constant but should rather be viewed as a variable that depends on loading conditions (applied stress level, temperature, and frequency), debonding mechanisms, viscoplastic deformation of surrounding matrix, and environmental attack of the fiber-matrix interface region. Eldridge [50] measured the frictional shear stress in titanium matrix composites at different high temperatures in vacuum-controlled environment. His work showed that the increase in temperature causes a decrease in the value of τ_s. In contrast, pull out tests by Marshall et al. [19] and push out tests by Clyne and Withers [51] both applied to SM1240/TIMETAL®21S, at room temperature, have showed that high-temperature aging prior to testing leads to an increase in the as-received τ_s. These results imply that τ_s is a condition-sensitive parameter. Here, again, similar to the COD measurements, τ_s, although essential to the understanding of the crack bridging mechanism, has a magnitude ranging from 0.9 to 360 MPa [11,32,49,52]. This large variation in the values of τ_s is due to many factors, including differences in test methods: pull out versus push out and test specimen size; test environment, air versus vacuum and composite fabrication processes. Furthermore, measured values of τ_s, using fiber pull out or push out methods, does not account for the degradation of the frictional shear stress due to time or number of loading cycles. In the absence of accurate values of τ_s it would be difficult to close the gap between experimental observations and results of crack growth analytical models.

Fiber-Matrix Interface Crack Length

The crack tip shielding associated with crack bridging phenomena is the outcome of two successive damage events. In the first event, relative displacement between the fiber and the matrix due to the crack opening displacement results in a continuum shear stress distribution, τ, along the interface. This, in turn, initiates interfacial debonding with an initial crack length, which depends on the stresses acting along the interface and the properties of the interphase region. In the second event, the balance of axial forces along this bridging fiber requires that the fiber traction at the plane coinciding with the matrix crack surface exceeds its far-field value. Once the initial debonding length has been established for a bridging fiber, progressive debond-

ing of this interface is governed by a mode II crack propagating between two materials with different elastic constants. An important
parameter in this case is the critical stress intensity factor (or the
critical strain energy release rate) of the interphase [53–55]. The distribution of the debonding length along a bridging crack is such that
the maximum length occurs at the crack mouth and the minimum
near the crack tip. Results of shear lag models show that an increase
in debonding length leads to a higher fiber stress. The occurrence of
fiber-matrix debonding and sliding, therefore, results in bridging traction and a pull out resistance to crack opening [56]. The fiber-matrix
interface debonding accommodates the crack opening displacement,
thus contributing the toughening effect that is provided by the dissi-
pation of energy associated with the debonding process.

Marshall et al. [28] examined the surface of pulled out fibers in an
SM1240/TIMETAL®21S composite and concluded that sliding occurs
between carbon and TiB_2 fiber, coating layers in both as-fabricated
and heat-treated specimens. Similar results were reported in Reference [16]. On the other hand, in an SCS-6/Ti-6-4 composite, cyclic
loading causes debonding to occur between the carbon-rich coating
and the SiC part of the fiber and between the different layers of the
carbon coating [9]. Yang et al. [57] and Eldridge and Ebihara [46], in
separate studies, reached a similar conclusion regarding the locations
of interfacial debonding in an SCS-6/Ti-15-3 composite. They observed
that debonding occurs at locations between the fiber and its multiple
carbon coatings, within the carbon layers, and occasionally between
the coating and the reaction zone. SEM observation of fiber surface of
an SCS-6/Ti-24-11 composite showed, however, that interface debonding always occurs between the carbon coating and the reaction zone of
the composite [46]. From these studies it is apparent that, for the
same temperature level, interfacial debonding in SiC-Ti MMCs could
occur at different locations across the fiber-interphase-matrix region.
This dictates the frictional coefficient of the mating debonded interfaces. In addition, the interphase shear strength is influenced by the
growth kinetics and thermal aging characteristics of the interphase,
as described in the previous section. Although direct measurements of
τ_s are difficult to obtain, compared values of τ_s could be extracted from
results of fiber pull out or push out tests of composite samples.

A recent analysis of fatigue crack growth in an SCS-6/Ti-6Al-4V
composite employing an elastic bridging model [58] concluded that
the interface debonding length, rather than the interface shear stress,
is the defining variable for modeling the bridging fatigue crack

growth in MACS. Modeling of interface debonding could be achieved using classical fracture mechanics concepts [59]. Other approaches are based on the balance between the interfacial shear stress and the shear strength of the interphase [60–63]. The basic requirement to establish this balance is the knowledge of the internal stress states acting along the fiber-matrix interface and the shear strength of the interphase region. Results of a combined experimental-numerical study on the debonding length, employing the balance of interface stresses criterion carried out by Tamin et al. [63] on SCS-6/Timetal composite, show that the extent of the initial debonding length for a particular loading condition is dependent on the internal stress state along the interface, the shear stress required to debond the interphase, and the surface roughness of the debonded interface. Furthermore, the debonding length increases with an increase in temperature and with a decrease in the friction coefficient.

The length of the fiber-matrix debonding interface along the bridged matrix crack has been studied by several authors [9,11,16,59,64]. The slip length, which includes the entire interface damage process zone in an SCS-6/Ti-6-4 composite, measured from the crack plane, was reported to reach up to 1 mm [9]. Bakuckas and Johnson [11] estimated that the debonded length in an SCS-6/Ti-15-3 composite at room temperature ranges from 1500 μm at the first intact fiber closest to the crack mouth to 100 μm at 1.2 mm away from this fiber (crack length, $a = 4.28$ mm). Zheng and Ghonem [16] measured the fiber pulled out length along fracture surfaces of SM1240/TIMETAL®21S composite specimens. This length, which was assumed to be proportional to the debonding length of the corresponding fiber, was found to be temperature dependent, with an average ranging from 120 μm at 24°C to 215 μm at 650°C. These different results concerning the interface debonding length are compared in Figure 15.

Once the fiber-matrix interface of a bridging fiber is debonded, the growth of the resulting crack requires, as mentioned above, the knowledge of mode II driving force of a crack propagating between two materials with different elastic constants. The significance of this knowledge lies in the fact that the increase in the interface debonding length during a particular number of fatigue loading cycles results in an increase in the fiber stress, which, in turn, influences the mode I matrix crack growth rate. Using the oscillatory stress singularities for open tip interface cracks [54,65], Chan and Davidson [53] developed a simple expression for the stress intensity factors ΔK_I and ΔK_{II} of an

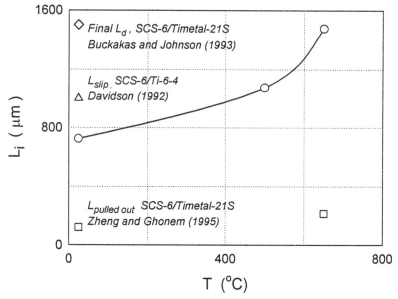

Figure 15. Relationship between debonding length and test temperature.

interface fatigue crack subjected to normal and shear loadings. This is written as:

$$\begin{bmatrix} \Delta K_I \\ \Delta K_{II} \end{bmatrix} = \frac{\sqrt{\pi}}{2\,C\,(\lambda_1^2 - \lambda_2^2)} \begin{bmatrix} \lambda_1 & \lambda_2 \\ \lambda_2 & \lambda_1 \end{bmatrix} \begin{bmatrix} D_2 \\ D_1 \end{bmatrix}$$

where C, λ_1, and λ_2 are functions of the shear moduli and Poisson's ratios for the two elastic media of the interface. D_1 and D_2 are the slopes of the plots representing the cyclic COD in both the normal and shear directions, respectively, versus $\sqrt{r}$, where r is the corresponding distance measured from the crack tip. In this analysis, the derived form of the effective ΔK parameter is identical to that of fatigue cracks in homogeneous materials, i.e.:

$$\Delta K_{eff} = [\Delta K_I^2 + \Delta K_{II}^2]^{1/2}$$

Employing a numerical approach to determine the relative fiber-matrix displacements along an interface crack of a bridging fiber in an SCS-6/Timetal composite, Tamin et al. [66] calculated the COD in the shear direction of this crack, assuming the absence of a normal

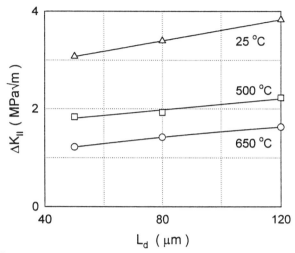

Figure 16. Mode II stress intensity factor, ΔK_{II}, as function of the interface debonding length, l_d.

COD. ΔK_{II} was then determined as a function of the interface crack length and test temperature; results are shown in Figure 16. Aside from this attempt, which has not been validated experimentally, no other work exists in literature dealing with the fatigue propagation characteristics of interface cracks in Ti MMC. This is possibly due to the fact that, although the debonding length is an important parameter in determining the fiber stress, no experimental technique is available yet to provide accurate measurements of its magnitude.

FIBER BRIDGING MODELS

The development of life prediction methodology for metal matrix composites relies on successful modeling of the fatigue crack growth process. Conventional fracture mechanics characterization of mode I fatigue crack growth behavior is accomplished through the relation between the crack growth rate and the stress intensity factor range. The most common relation for the crack growth rate, da/dN, is the power law function of the applied stress intensity factor range, ΔK_a. The fatigue crack growth behavior of a unidirectional MMC under crack bridging condition, however, shows an initial crack stage in which the crack growth rate decreases as the applied stress intensity factor, or the matrix crack length, increases [9–11,16,24]. In this case, unlike the monolithic matrix alloy material, ΔK_a cannot be treated as a crack tip driving force. A more suitable definition for his force is the stress intensity factor range for the matrix material, ΔK_m. One of the early attempts to deal with the crack driving force in composites are those by Aveston et al. [67] and Budiansky et al. [42]. Using an energy balance approach, they derived an expression for K_m of a long crack that is crack length independent, although it includes basic composite parameters. This driving force is expressed as:

$$K_m = \sqrt{\frac{E_m\, G_m}{1 - v_m^2}} \tag{1}$$

where E_m and G_m are the elastic modulus of the matrix phase and the mode I matrix crack extension energy release rate, respectively, and v_m is the matrix Poisson ratio. Other attempts to calculate the effective crack tip driving force base their formulations on the idea that the discrete bridging fiber traction along the crack length acts in opposite direction to the applied stress, thus providing a shielding force for the crack tip. Modeling of this concept is achieved by superimposing the solution of a crack subjected to a far-field applied stress range, $\Delta\sigma$, and a crack subjected to a change in the closure pressure, $\Delta p(x)$, which is related to the discrete fiber traction range, $\Delta S(x)$, at the crack plane through the relationship:

$$\Delta p(x) = V_f \Delta S(x) \tag{2}$$

where V_f is the fiber volume fraction in the composite. For a center crack configuration with the distance, x, measured from the center of the crack of length $2a$, the homogenized composite stress intensity factor range, ΔK_c is given by [68]:

$$\Delta K_c = Y\Delta\sigma \sqrt{\pi a} + 2Y \sqrt{\frac{a}{\pi}} \int_{a_0}^{a} \frac{\Delta p(x)}{\sqrt{a^2 - x^2}}\, dx \tag{3}$$

where a_o is the nonbridged initial crack length and Y is a geometric factor for finite specimen width correction. The continuum half crack opening displacement range, $\Delta u_c(x)$, for the same crack configuration is written as:

$$\Delta u_c(x) =$$

$$\frac{4\,\phi\,\Delta S}{E'} \sqrt{a^2 - x^2} + \frac{4\,\phi}{E'\pi} \int_{a_0}^{a} \Delta p(x') \log \left| \frac{\sqrt{a^2 - x^2} + \sqrt{a^2 - x'^2}}{\sqrt{a^2 - x^2} - \sqrt{a^2 - x'^2}} \right| dx' \tag{4}$$

where ϕ is a correction factor for finite width, E' is the modulus of an orthotropic material containing a crack normal to the loading direction, and x' is an integration variable. The function, $\Delta u_c(x)$, in the above equation is assumed to be equal to the discrete $\Delta u(x)$ in the matrix phase of the composite in some models [7,16,19]. McCartney [6], however, reported that $\Delta u_c(x)$ is the change in displacement within the slip region only and should be related to $\Delta u(x)$ by the relation:

$$\Delta u(x) = \left(1 + \frac{V_f E_f}{(1 - V_f) E_m}\right) \Delta u_c(x) \tag{5}$$

where E_f is the elastic modulus of the reinforcing fiber.

Because in crack bridging conditions the cracking of a composite material is limited to the matrix only, the fatigue crack tip driving force is assumed to be the effective stress intensity factor in the matrix of the composite, ΔK_{eff}. McMeeking and Evans [7] and Zheng and Ghonem [16] suggested that ΔK_{eff} is equal to ΔK_c. Marshall et al. [19], assuming a condition of isostrain in the composite constituents near the crack tip, suggested that the relationship between the discrete ΔK_{eff} and the continuum ΔK_c could be considered to be the driving force for homogenized composites. The matrix stress intensity factor could then be written as:

$$\Delta K_{eff} = \frac{E_m}{E_c} \Delta K_c \tag{6}$$

where E_c is the modulus of the composite in the longitudinal direction. On the other hand, McCartney [6], using an energy balance approach for a bridged crack, assumed that the fracture surface energy of the matrix, γ_m, and of the composite, γ_c, are related through the fiber volume fraction, V_f, such that:

$$\gamma_c = (1 - V_f) \gamma_m \tag{7}$$

Rewriting this equation in terms of the stress intensity factor range yields (note that $\gamma = K^2/E$):

$$\Delta K_{eff} = \sqrt{\frac{E_m}{E_c (1 - V_f)}} \Delta K_c \tag{8}$$

where E_c is the elastic modulus of the composite system. It is not possible to validate any of the above criteria on physical ground, and the differences between them, as was shown in References [11,69] are quite significant.

Turning attention to Equation (3), a requirement for its solution is the knowledge of the bridging fiber stress distribution along the crack length, $\Delta p(x)$. Several approaches have been attempted for this closure pressure, one of which is the shear lag type model. This model is based on the idea of the load transfer from the cracked matrix to the fibers through relative sliding between the fiber and the matrix over a region where the shear stresses acting along the interface exceed the strength of the interphase. In developing the original formulations for this approach, Marshall et al. [19] determined, through the use of a force equilibrium on a concentric cylinder, the closure pressure in terms of a parameter λ and the relative fiber-matrix half crack opening displacement range, $\Delta u(x)$. The closure pressure is written as:

$$\Delta p(x) = \left(\frac{\Delta u(x)}{\lambda}\right)^{1/2} \tag{9}$$

and the parameter λ is given by:

$$\lambda = \frac{R\,(1 - V_f)\,E_m}{4\,\tau_s\,V_f^2\,E_f\,E_c} \tag{10}$$

where R is the radius of the fiber and τ is the fiber-matrix frictional shear stress. In the above equation, a length of the debond or sliding interface is treated as a frictional interface with a constant frictional shear stress, τ_s. This stress, as mentioned before, can be obtained through various methods, such as fiber push out test [46,32], fiber pull out test [70,71], or by calibrating the measured crack opening profiles with the best-fit values of τ [10,11]. It was noted, however, that the use of a single-value τ_s neglects the interfacial wear occurring during fatigue cycles, which, in turn, decreases the value of τ_s [28,72]. In addition, Equation (10) suggests that the fiber traction vanishes at the crack tip.

In contrast to the force-based formulation described above, McCartney [6], performed an energy balance calculation for a bridging fiber and showed that the closure pressure described in Equation (9) should be modified to:

$$\Delta p(x) = \left(\frac{\Delta u(x)\,\zeta}{\lambda}\right)^{1/2} \tag{11}$$

where ζ is given by:

$$\zeta = \frac{E_c}{(1 - V_f)E_m} \tag{12}$$

Another approach to determine the closure pressure function considers the fiber bridged specimen as a structure whose members in the crack wake (i.e., bridging fibers) can carry tensile loads created by normal stress, bending stress, or both. The force balance in this fiber pressure model is derived through the strength of material approach to stress calculation in a beam subjected to bending or tensile forces, or both. The closure pressure is assumed to be equal to the stress carried by the fibers in the bridged region averaged out over the total bridged area. For a single edge-notched specimen geometry, the closure pressure is given by:

$$\Delta p(x) = \Delta \sigma \left(\frac{w}{(w - a_o)} + \frac{6wa_o(0.5(w - a_o) - (x - a_o))}{(w - a_o)^3} \right); \qquad a_o \leq x \leq a \tag{13}$$

where ΔS is the remotely applied stress range, w is the width of the specimen, a_o and a are the initial notch length and the local crack length, respectively, and x is the distance along the crack measured from the free surface. It is clear that the above equation, which originated from an analysis of fast fracture in steels [73], does not require the knowledge of the frictional shear stress, τ, and is independent of the crack opening displacement. The above equation was adopted by Ghosn et al. [20] in the analyses of composites under cyclic loading conditions.

The relationship between the closure pressure function, $\Delta p(x)$ and the crack opening displacement range $\Delta u(x)$ of Equations (11) and (13) can now be combined with Equation (4), and also with Equation (5) if one considers McCartney's energy approach. In the latter case, the resulting nonlinear integral equation is then solved numerically for the closure pressure that is required to determine the composite stress intensity factor range using Equation (4). The Bueckner weight function method and the finite element method could also be employed in conjunction with the closure pressure function to determine both the fatigue crack tip driving force and the crack opening displacement profiles.

A few experimental based attempts were made to calculate the effective stress intensity factor range by combining experimental results and simple fracture mechanics equations. In Reference [16], e.g., the crack opening displacements were measured at positions along different crack lengths at different temperatures. These values were substituted in the LHS of Equation (4), which was solved numerically to obtain the fiber pressure distribution for each crack length; see Figure 13. This distribution was then used in Equations (3) and (6) to calculate the effective stress intensity factor range. Results of their work for 10-Hz loading frequency at temperatures of 24, 500, and 650°C indicate that ΔK_{eff} is inversely proportional to both crack length and temperature. These trends are consistent with the observed decelerating crack growth rate behavior of bridged cracks in titanium matrix composites as seen, e.g., in Figure 1.

In another study, Nguyen and Yang [58] considered that the traction $T(x)$ in a bridging fiber is the sum of two components, namely, an elastic stress component, $T_{el}(x)$, resulting from elongation of the elastic fiber due to crack opening displacement, $u(x)$, and a component, denoted as $T_{fric}(x)$, corresponding to the frictional forces, τ_s, acting along the fiber-matrix debonding interface of length l_d. The fiber traction function, $T(x)$ is thus written as:

$$S(x) = S_{el}(x) + S_{fric}(x)$$

$$= E_f \frac{u(x)}{l_d(x)} + \frac{2\tau_s l_d(x)}{R} \tag{14}$$

The closure pressure distribution, $\Delta p(x)$, is then obtained from Equation (2). In the above equation, the crack opening displacement profile, $u(x)$, and the distribution of fiber-matrix debonding lengths, $l_d(x)$, along a bridged crack length are determined experimentally [9]. Equation (3) is then employed to calculate the effective crack tip driving force.

This review of the different models addressing the driving force of a bridging crack shows that, although the majority of the models agree on the use of classical fracture mechanics concepts as a base for their mathematical derivations, differences exist in the fiber pressure and displacement distributions, which are requirements for the solutions of these models. Simple experimental methods to validate any of their corresponding values do not currently exist. Furthermore, all the models consider the frictional shear stress, τ_s, an important parameter in their derivations and assume it is constant with time. In reality, the degradation of τ_s with time, temperature, and with fatigue reversals appears to be an important consideration that, measured by the wide and different values reported by many authors, as discussed earlier in this chapter, is difficult to model.

BRIDGING-ACCELERATION FATIGUE
CRACK GROWTH TRANSITION

The bridging fatigue crack growth in Ti MMCs is characterized by a decrease in the growth rate with an increase of the applied stress intensity factor range, or crack length, indicating an initial fully crack bridging stage. This bridging fatigue crack growth stage could occupy larger than 30% of the fatigue life of the test specimens [16]. The end of this crack growth retardation stage, corresponding to the attainment of the minimum crack growth rate, is followed either by a crack arrest or a transition to an accelerated crack growth process. In the latter case, this transition point in the crack growth rate curve was attributed to the breakage of bridging fibers, particularly those located near the crack mouth [16,24]. These fibers have experienced the largest number of fatigue cycles and have possibly suffered the severest damage in the form of cyclic wear induced by the fiber-matrix interfacial frictional shear stress [31]. Meanwhile, during cyclic loading of the composite, particularly at elevated temperature, the time-dependent matrix deformation becomes more pronounced as the stresses in the matrix phase relax. This stress relaxation is accompanied, for equilibrium requirements, by the simultaneous buildup of stresses in the adjacent fiber through the load transfer process across the fiber-matrix bonded interface [74]. These two damage events of stress buildup in a bridging fiber and continuous strength degradation of the fiber, which occur simultaneously, interact and eventually result in breakage of bridging fibers at the transition stage of the fatigue crack growth. The failure of any of the bridging fibers would result in redistribution of stresses among unbroken bridging fibers, resulting in a decrease of the crack tip shielding force, thus triggering a condition of crack growth acceleration.

The idea that the fatigue life of a bridging fiber is governed by time- and load cycle-dependent effects taking place during the fatigue crack growth process forms the basis for formulating a fracture criterion for fibers bridging fatigue cracks. This criterion, developed by Tamin and Ghonem [75], is based on the notion, as described above, that the buildup of stresses in the adjacent fiber through the load transfer across the fiber-matrix bonded interface occurs concurrently with a continuous deterioration of the fiber residual strength due to cyclic related effects, including wear and surface abrasion induced by the frictional shear stress acting along the fiber-matrix debonded interface. This criterion is schematically illustrated in Figure 17, where the stress state in ceramic fiber could be represented by the axial component or the maximum principal stress component. As the cyclic load-

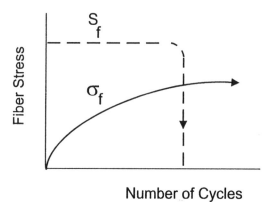

Number of Cycles

Figure 17. Elements for a bridging fiber failure criterion; the build-up stresses in the fiber through the load transfer across the fiber/matrix bonded interface and the concurrent decrease in the fiber residual strength.

ing proceeds, the increased stress level in a bridging fiber could reach the value of its decreased fiber residual strength; causing fiber fracture. The current strength of this particular bridging fiber, therefore, is located at the intersection of the fiber stress and the residual strength curves as illustrated in Figure 19. Once these two curves are established for a given loading condition, prediction of the bridging fiber life during the fatigue crack propagation can be made.

The damage process leading to the fracture event of a bridging fiber at the bridging-acceleration transition in the fatigue crack growth rate of Ti MMC can now be described with the aid of the fracture criterion. The requirement for this criterion is the knowledge of the stress evolution in a bridging fiber and the corresponding residual strength degradation of the fiber throughout the fatigue cycles. The stress experienced by a bridging fiber is a function of applied load, crack length, number of fibers bridging the matrix crack, frictional shear stress acting along the debonding interface; and the load partition between uncracked and cracked regions of the composite. In addition, the evolution characteristics of the bridging fiber stress are also influenced by residual stress state, temperature, loading frequency, and material variables, such as microstructure of the matrix phase. Thermal residual stresses arising from CTE mismatch of the constituents phases influence the load transfer characteristics by altering the fiber-matrix interphase properties [76–80]. The combination of chemical bonding and thermally induced clamping results in high interphase shear strength, especially at a low temperature. At elevated temperature, the transfer of load is further modified by various inelastic processes occurring at or in the vicinity of fiber-matrix interface, such as matrix plasticity, interface debonding, and frictional sliding.

Analytical modeling of crack bridging, such as shear lag analysis and fiber pressure model indicate that the maximum axial stress in the bridging fiber is at the section that coincides with the matrix crack plane. The fracture sites of the bridging fibers, however, revealed through fractographic analysis of Ti MMC, are generally located above and below the matrix crack plane [9,23]. In some studies where the surface damage of the fiber is governed by the frictional wear, which is assumed to be uniform along the slip length, the probable fracture site in the bridging fiber can then be determined statistically [43]. In addition, the calculated shear stress gradient is highest at the cross section coinciding with the tip of fiber-matrix debonding length due to sudden transition from debonded to fully bonded interface. Because the residual stress field at this section is less affected by the debonding interface, the buildup of bridging fiber stress is most pronounced here. On the other hand, the compressive residual stress in the bridging fiber acts in retarding the time-dependent increase of the fiber axial stress. The cross section coinciding with the tip of fiber-matrix debonding length then becomes a critical location for fiber fracture. During fatigue loading at a lower frequency, the composite is subjected to higher stresses for a duration of time longer than that due to a higher frequency loading. In addition, the rate of load transfer to, and the stress buildup in bridging fibers is faster as the temperature level increases. Consequently, a combination of high-temperature and low-frequency loading facilitates the load transfer from the matrix to the bridging fiber through the enhancement of matrix creep deformation. It indicates that the buildup of stress in a bridging fiber is a time-dependent process. A more pronounced time-dependent effect on bridging fatigue crack growth is revealed through hold time tests of SM1240/TIMETAL®21S composite specimens at an elevated temperature of 500°C [37].

Some studies on fatigue behavior of MMCs show that the SiC-reinforcing fibers failed at stress levels much lower than their preprocessing tensile strength [81]. The in situ strength of SiC-type fibers commonly used as reinforcement in titanium-based MMCs ranges from 3000 to 4000 MPa [82]. However, the strength of bridging fibers, calculated on the basis of the shear lag analysis, was about 1000 to 2000 MPa [10,11], whereas that determined from the measured bridging fiber strain ranges between 1000 and 4000 MPa [9]. This difference in the fiber strength can be explained by the knowledge that the brittle SiC fiber is sensitive to surface defects to the extent that a minimum fiber surface abrasion can significantly reduce the fracture toughness of the fiber [83].

To employ the criterion for predicting the life of bridging fibers, the residual strength curve must be established as a function of loading cycles. Tamin [84] measured the residual strength of SCS-6 fibers at two temperatures: 500 and 650°C. Results are shown in Figure 18 in the form of residual strength versus the number of load cycles at various maximum applied fatigue loads. The loading frequency for the tests reported in this figure is 10 Hz with a stress ratio of 0.1. These curves indicate that the fatigue strength is maintained until the fracture event occurs and that the onset of fiber fracture is almost instantaneous. Although the strength of the fiber is shown to be insensitive to test temperature, temperature influences the fracture process of the fiber through a rapid initiation of cracks in the outermost coating layer of the fiber; an example of this is shown in Figure 19. A typical spacing of circular cracks observed for the two tests mentioned above is shown in Figure 20. In this work, the crack density is defined as the inverse of the mean distance between the cracks along

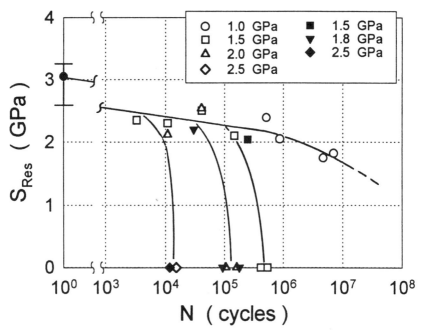

Figure 18. Fiber residual strength versus number of loading cycles.

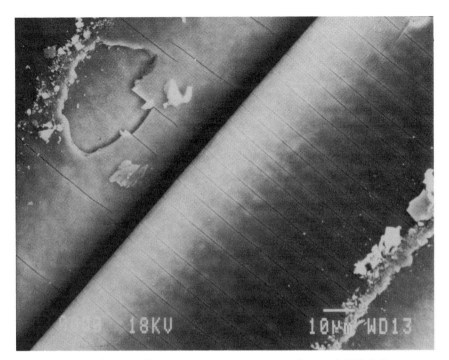

Figure 19. Surface fatigue cracks in the coating layer of SCS-6 fibers.

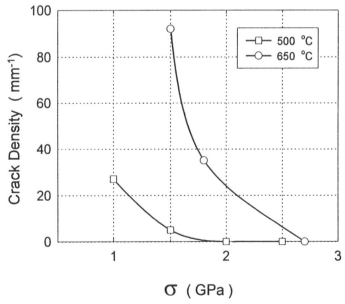

Figure 20. Fiber's fatigue crack density versus temperature.

the fiber. Absent from this study is the effect of loading frequency on the residual strength of fibers.

SUMMARY AND FUTURE WORK

This chapter discusses the physical and mathematical aspects of the bridging fatigue crack growth in titanium metal matrix composites. Effects associated with operating temperature and loading frequency are discussed. For the role of temperature, several authors show experimentally that it has a major influence on the crack growth rate; they differ, however, on the trend of this influence. For the role of frequency, it is accepted that at high test temperatures, where time-dependent mechanisms dominate, the crack growth rate is inversely proportional with temperature. The influence of fatigue-creep interaction was studied by observing the composite response to loading cycles, including hold time durations. This type of cycle is expected to promote matrix creep in composite regions ahead of the crack tip. The limited work carried out in this regard shows that the decrease in the crack growth rate with frequency is associated with crack tip closure brought about by the redistribution of the internal stresses due to the time-dependent matrix flow. The basic characteristics of various parameters governing the crack bridging process are also described and correlated with the crack growth process. These parameters include crack opening displacement, fiber pressure distribution, frictional shear stress, and fiber-matrix interface debonding length along bridging fibers. The significance of each of these parameters with respect to the crack growth behavior is discussed. Several types of fiber bridging models, including fiber pressure model, shear lag analysis and its variations, and combined mathematical-experimental approach are reviewed. These analytical models, which combine fracture mechanics and micromechanical analysis, attempt to model the influence of crack bridging on composites fatigue crack growth behavior. They focus on formulating the closure pressure function to quantify the reduction of the crack tip driving force due to crack bridging. The last part of the chapter describes the crack growth stage at which the deceleration-acceleration transition of the crack growth rate occurs. This was achieved by establishing a failure criterion for bridging fibers based on the balance between the fiber stress and the fiber fatigue residual strength.

The scope of results covered in the chapter, although showing that the subject of crack growth in titanium matrix composites is complex and involves the interplay of several material- and time-dependent processes, points out areas that, in the opinion of the author, require further research. For example, the success in formulating a valid fracture criterion of bridging fibers would permit an accurate prediction of the minimum crack length at which a mode of crack growth rate acceleration starts. This length evaluated for a particular loading condition could be used as a design tool to provide the tolerance limit for the use of the composite under consideration. A basic element in the formulation of such a criterion is the fiber fatigue residual strength as a function of the history of operating temperature, environment and loading characteristics, particularly those of the mean stress. Influence of prior exposure to parameters used during composite consolidation, such as pressure and temperature, on fiber strength should also be investigated.

Another important subject is related to the fiber-matrix interface debonding during crack bridging. The significance of the debonding process is due to the fact that fracture energy made available by the externally applied load on a notched specimen is shared between mode I matrix crack and mode II fiber-matrix interface crack. An interdependent relationship thus governs growth behavior of these two cracks. An important step in quantifying this relationship is the understanding of the damage initiation and propagation in the interphase layer. The initiation of damage in this layer is affected by the localization of stresses as the matrix crack tip approaches the interface. This localization process depends on the interphase shear strength, residual stress state, and the loading frequency. The latter determines the nature of the fatigue-creep interactions occurring during the loading cycle. The initial debonding length, estimated through a debonding criterion, will thus be a function of the fiber stress and the properties of the interphase at the test temperature. Compressive residual force present at the interface, induced by the composite consolidation process, could cause the surfaces of the debond crack to come in contact with one another, introducing a frictional sliding force along the debonded surfaces. Propagation of the debond crack is also a relevant subject and requires the basic understanding of crack growth mechanisms along interfaces. The driving force of the debond crack, which is the shielding component of the matrix crack, is a function of a host of parameters, including the interphase strength, fracture and friction properties, and the stress state at the crack tip and along the crack faces. Quantifying of the initiation and propagation of the interface debond crack must be based on understanding of the basic interface properties and the factors influencing such properties.

The factors of most concern relate to the processing parameters, which include consolidation temperature, pressure, and postprocessing cooling rate. These variables result in the alteration of the interface debonding strength, either chemically or due to variations in the clamping radial residual stresses, and therefore influence the subsequent behavior of the composite during fatigue loadings. Attempts should be made to optimize the interface strength via the control of processing parameters to produce the desired fracture mode at the selective operating loading conditions.

REFERENCES

1. Cox, B. N., Marshall, D. B. and Thouless, M. D., "Influence of Statistical Fiber Strength Distribution on Matrix Cracking in Fiber Composites", *Acta Metall.*, Vol. 37, No. 7, pp. 1933–1943, 1989.
2. Sensmier, M. D. and Wright, P. K., "The Effect of Fiber Bridging on Fatigue Crack Growth in Titanium Matrix Composites", in *Fundamental Relationships Between Microstructures and Mechanical Properties of Metal Matrix Composites,* M. N. Gungor and P. K. Liaw, eds., TMS, Warrendale, PA, pp. 441–457, 1990.
3. Wright, P. K., Nimmer, R., Smith, G., Sensmier, M. and Brun, M., "The Influence of Interface on Mechanical Behavior of Ti-6Al-4V/SCS-6 Composites", in *Interfaces in Metal-Ceramics Composites,* R. Y. Lin, R. J. Arsenault, G. P. Martins and S. G. Fishman, eds., TMS, Warrendale, PA, pp. 559–581, 1989.
4. Naik, R. A. and Johnson, W. S., "Observation of Fatigue Crack Initiation and Damage Growth in Notched Titanium Matrix Composites", *Composite Materials: Fatigue and Fracture,* pp. 753–771, 1989.
5. Davidson, D. L., Chan, K. S., McMinn, A. and Leverant, G. R., "Micromechanics and Fatigue Crack Growth in an Alumina-Fiber-Reinforced Magnesium Alloy Composite", *Metallurgical Transactions,* Vol. 20, pp. 2369–2378, 1989.
6. McCartney, L. N., "Mechanics of Matrix Cracking in Brittle-Matrix Fiber-Reinforced Composites", *Proceedings of the Royal Society London,* Vol. 409A, pp. 329–350, 1987.
7. McMeeking, R. M. and Evans, A. G., "Matrix Fatigue Cracking in Fiber Composites", *Mechanics of Materials,* Vol. 9, pp. 217–227, 1990.
8. Soumelidis, P., Quenisset, J. M. and Naslain, R., "Effect of the Filament Nature on Fatigue Crack Growth in Titanium Based Composites Reinforced with Boron, $B(B_4C)$ and Sic Filaments", *Journal of Material Science,* Vol. 21, pp. 895–903, 1986.
9. Davidson, D. L., "The Micromechanics of Fatigue Crack Growth at 25°C in Ti-6Al-4V Reinforced with SCS-6 Fibers", *Metallurgical Transaction,* Vol. 23A, pp. 865–879, 1992.
10. Telesman, J., Ghosn, L. J. and Kantzos, P., "Methodology for Prediction of

Fiber Bridging Effects in Composites", *Journal of Composites Technology and Research,* Vol. 15, No. 3, pp. 234–241, 1993.

11. Bakuckas, J. G. and Johnson, W. S., "Application of Fiber Bridging Model in Fatigue Crack Growth in Unidirectional Titanium Matrix Composites", *Journal of Composites Technology and Research,* Vol. 15, No. 3, pp. 242–255, 1993.

12. Jeng, S. M., Allassoeur, P. and Yang, J. -M., "Fracture Mechanisms of Fiber Reinforced Titanium Alloy Matrix Composites. V. Fatigue Crack Propagation", *Material Science and Engineering,* A154, pp. 11–19, 1992.

13. Bowen, P., Ibbotson, A. R. and Beevers, C. J., "Characterization of Crack Growth in Continuous Fibre Reinforced Titanium Based Composites under Cyclic Loading", in *Fatigue of Advanced Materials,* R. O. Ritchie, R. H. Dauskardt and B. N. Cox, eds., MCE, Birmingham, AL, pp. 379–394, 1991.

14. Ibbotson, A. R., Bowen, P. and Beevers, C. J., "Cyclic Fatigue Resistance of Fiber Reinforced Titanium Metal Matrix Composites at Ambient and Elevated Temperature", *Proceedings of the 7th Titanium Conference,* San Diego, CA, July 1992.

15. Brindley, P. K., Draper, S. L., Eldridge, J. I., Nathal, M. V. and Arnold, S. M., "The Effect of Temperature on the Deformation and Fracture of SiC/Ti-24Al-11Nb", *Metallurgical Transaction,* Vol. 23A, pp. 2527–2540, 1992.

16. Zheng, D. and Ghonem, H., "High Temperature/High Frequency Fatigue Crack Growth Damage Mechanisms in Titanium Metal Matrix Composites", in *Life Prediction Methodology for Titanium Matrix Composites, ASTM STP 1253,* W. S. Johnson, J. M. Larson and B. N. Cox, eds., American Society for Testing and Materials, Philadelphia, PA, 1995.

17. Gayda, J., Gabb, T. P. and Freed, A. D., "The Isothermal Fatigue Behavior of a Unidirectional SiC/Ti Composite and the Ti Alloy Matrix" in *Fundamental Relationships Between Microstructure and Properties of Metal Matrix Composites,* P. K. Liaw and M. N. Gungor, eds., The Mineral, Metal and Materials Society, Warrendale, PA, pp. 497–513, 1990.

18. Kyno, T., Kuroda, E., Kitamuru, A., Mori, T. and Taya, M., "Effects of Thermal Cycling on Properties of Carbon Fiber/Aluminum Composites", *Journal of Engineering Materials and Technology,* Vol. 110, pp. 89–95, 1989.

19. Marshall, D. B., Cox, B. N. and Evans, A. G., "The Mechanics of Matrix Cracking in Brittle-Matrix Fiber Composites", *Acta Metallurgica,* Vol. 33, pp. 2013–2021, 1985.

20. Ghosn, L. J., Kantzos, P. and Telesman, J., "Modeling of Crack Bridging in a Unidirectional Metal Matrix Composite", *International Journal of Fracture,* Vol. 54, pp. 345–357, 1992.

21. Walls, D. P., Bao, G. and Zok, W., "Mode I Fatigue Cracking in a Fiber Reinforced Metal Matrix Composite", *Acta Metall. Mater.,* Vol. 41, pp. 2061–2071, 1993.

22. Chiang, Y.-C., Wang, A. S. and Chou, T.-W., "On Matrix Cracking in Fiber Reinforced Ceramics", *Journal of Mechanics and Physics of Solids,* Vol. 41, pp. 1137–1154, 1993.

23. Zheng, D. and Ghonem, H., "Fatigue Crack Growth of SM1240/Timetal-21S Metal Matrix Composite at Elevated Temperature", *Metallurgical and Material Transactions A,* Vol. 26A, pp. 2469–2478, 1995.

24. Cotterill, P. J. and Bowen, P., "Fatigue Crack Growth in a Fibre-Reinforced Titanium MMC at Ambient and Elevated Temperatures", *Composites,* Vol. 24, No. 3, pp. 214–221, 1993.

25. Larsen, J. M., Jira, J. R., John, R., and Ashbough, N. E., *Crack Bridging in Notch Fatigue of SCS-6/Timetal 21S Composite Laminates, Life Prediction Methodology for Titanium Matrix Composites, ASTM STP 1253,* W. S. Johnson, J. M. Larsen, and B. N. Cox, eds., American Society for Testing and Materials, Philadelphia, PA, 1995.

26. Barney, C., Beevers, C. J. and Bowen, P., "Fatigue Crack Growth in SiC Continuous Fiber Reinforced Ti-6Al-4V Alloy Metal Matrix Composites", *FATIGUE 93,* Vol. 2, J.-P. Bailon and J. I. Dickson, eds., Engineering Materials Advisory Services Ltd., UK, pp. 1073–1078, 1993.

27. Bowen, P., Ibbotson, A. R. and Beevers, C. J., "Fatigue Crack Growth in Fibre Reinforced Titanium Metal Matrix Composites", in *Mechanical Behavior of Materials—V1, Vol. 3,* M. Jono and T. Inoue, eds., *Proceedings of 6th International Conference on Mechanical Behavior of Materials,* Kyoto, Japan, July 29–Aug. 2, 1991, pp. 107–112, 1991.

28. Marshall, D. B., Shaw, M. C., Morris, W. L. and Graves, J., "Interfacial Properties and Residual Stresses in Titanium and Titanium Aluminide Matrix Composites", in *Workshop Proceedings on Titanium Matrix Components,* P. R. Smith and W. C. Revelos, eds., Wright-Patterson AFB, OH, pp. 329–347, 1991.

29. Chan, K. S. and Davidson, D. L., "Effects of Interfacial Strength on Fatigue Crack Growth in a Fiber-Reinforced Ti-Alloy Composite", *Metallurgical Transaction A,* Vol. 21A, pp. 1603–1612, 1990.

30. H. Ghonem, "Isothermal and Thermomechanical Fatigue Crack Growth in Metal Matrix Composites", Mechanics of Materials Laboratory, University of Rhode Island, Report: MML-95-4, 1996.

31. Chan, K. S., "Effects of Interface Degradation on Fiber Bridging of Composite Fatigue Cracks", *Acta Metall. Mater.,* Vol. 41, pp. 761–768, 1993.

32. Osborne, D. and Ghonem, H., "High Temperature Interphase Behavior of SiC Fiber-Reinforced Titanium Matrix Composites", Mechanics of Materials Laboratory, University of Rhode Island, Report: MML-95-1, March 1995.

33. DiCarlo, J. A., "Creep of Chemically Vapor Deposited SiC Fibers", *Journal of Material Science,* Vol. 21, pp. 217–224, 1986.

34. Wright, P. K., "Creep Behavior and Modeling of SCS-6/Titanium MMC", *Workshop Proceedings on Titanium Matrix Components,* P. R. Smith and W. C. Revelos, eds., WL-TR-92-4035, pp. 251–276, 1992.

35. Bain, K. R. and Gambone, M. L., "Fatigue Crack Growth of SCS-6/Ti-6-4 Metal Matrix Composites", in *Fundamental Relationship Between Microstructure and Mechanical Properties of Metal Matrix Composites,* P. K. Liaw and M. N. Gungor, eds., The Minerals, Metals and Materials Society, Warrendale, PA, pp. 459–469, 1990.

36. Ibbotson, A. R., Beevers, C. J. and Bowen, P., "Damage Assessment and Lifing of Continuous Fiber-Reinforced Metal Matrix Composites", *Composites,* Vol. 24, No. 3, pp. 241–247, 1993.

37. Zhang, T. and Ghonem, H., "Time-Dependent Fatigue Crack Growth in Titanium Metal Matrix Composites", *Fatigue and Fracture of Engineering Materials and Structures,* Vol. 18, No. 11, pp. 1249–1262, 1995.

38. Telesman, J., Kantzos, P. and Brewer, D., "In Situ Fatigue Loading Stage Inside Scanning Electron Microscope", *Lewis Structures Technology— 1988,* NASA CP 3003, Vol. 3, pp. 161–172, May 1988.

39. John, R., Kaldon, S. G. and Ashbaugh, N. E., "Applicability of Fiber Bridging Models to Describe Crack Growth in Unidirectional Titanium Matrix Composites", in *Workshop Proceedings on Titanium Matrix Components,* P. R. Smith and W. C. Revelos, eds., Wright-Patterson AFB, OH, WL-TR-93-4105, pp. 270–290, 1993.

40. Le Petitcorps, Y., Lahaye, M., Pailler, R. and Naslain, R., "Modern Boron and SiC CVD Filaments: A Comparative Study", *Composites Science and Technology,* Vol. 32, pp. 31–55, 1988.

41. Ghonem, H., "Bridging Fiber Stress in Metal Matrix Composites—An Analytical Model", *Durability of Composite Materials,* R. C. Wetherhold, ed., The American Society of Mechanical Engineers, New York, 1995.

42. Budiansky, B., Hutchinson, J. W. and Evans, A. G., "Matrix Fracture in Fiber-Reinforced Ceramics", *J. Mech. Phys. Solids,* Vol. 34, No. 2, pp. 167–189, 1986.

43. Thouless, M. D., Sbaizro, O., Sigl, L. S. and Evans, A. G., "Effect of Interface Mechanical Properties on Pullout in a SiC Fiber-Reinforced Lithium Aluminum Silicate Glass-Ceramic", *J. Am. Ceram. Soc.,* Vol. 72, pp. 525–532, 1989.

44. Marshall, D. B. and Oliver, W. C., "Measurement of Interfacial Mechanical Properties in Fiber Reinforced Ceramic Composites", *J. Am. Ceram Soc.,* Vol. 70, pp. 542–548, 1987.

45. Makin, T. J., Warren, P. D. and Evans, A. G., "Effects of Fiber Roughness on Interface Sliding in Composites", *Acta Metall Mater.,* Vol. 40, pp. 1251–1257, 1992.

46. Eldridge, J. I. and Ebihara, B. T., "Fiber Push-out Testing Apparatus for Elevated Temperatures", *Journal of Materials Research,* Vol. 9, No. 4, pp. 1035–1042, 1994.

47. Watson, M. C. and Clyne, T. W., "The Use of Single Fiber Pushout Testing to Explore Interfacial Mechanics in SiC Monofilament-Reinforced Ti—II. Application of the Tests to Composite Material", *Acta Metall Mater.,* Vol. 40, pp. 141–148, 1992.

48. Kantzos, P., Ghosn, L. and Telesman, J., "The Effect of Degradation of the Interface and Fiber Properties on Crack Bridging", *HITEMP Review,* Vol. 2, Cleveland, OH, pp. 32-1–32-14, 1992.

49. Warren, P. D., Mackin, T. J. and Evans, A. G., "Design, Analysis and Application of an Improved Push-Through Test for the Measurement of Interface Properties in Composites", *Acta Metall Mater.,* Vol. 40, pp. 1243–1249, 1992.

50. Eldridge, J. I., "Fiber Push-Out Testing of Intermetallic Matrix Composites at Elevated Temperatures", in *Proceedings of the Symposium on Intermetallic Matrix Composites II*, D. Miracle, J. Graves and D. Anton, eds., Materials Research Society, Vol. 273, pp. 325–330, 1992.

51. Clyne, T. W. and Withers, P. J., *An Introduction to Metal Matrix Composites*, Cambridge University Press, London, 1993.

52. Yang, J. M., Jeng, S. M. and Yang, C. J., "Interfacial Properties Measurements for SiC Fiber-Reinforced Titanium Alloy Composites", *Scripta Metallurgica*, Vol. 24, pp. 469–474, 1990.

53. Chan, K. S. and Davidson, D. L., "Driving Forces for Composite Interface Fatigue Cracks", *Engineering Fracture Mechanics*, Vol. 33, No. 3, pp. 451–466, 1989.

54. Hutchinson, J. W., Mear, M. E. and Rice, J. R., "Crack Paralleling an Interface Between Dissimilar Materials", *Journal of Applied Mechanics*, Vol. 54, pp. 828–832, 1987.

55. Suo, Z. and Hutchinson, J. W., "Interface Crack Between Two Elastic Layers", *Int. J. Fracture*, Vol. 43, pp. 1–18, 1990.

56. Thouless, M. D. and Evans, A. G., "Effects of Pull-out on Toughness of Reinforced Ceramics", *Acta Metall Mater.*, Vol. 36, p. 517, 1988.

57. Yang, J. M., Jeng, S. M. and Yang, C. J., "Fracture Mechanisms of Fiber-Reinforced Titanium Alloy Matrix Composites. Part I. Interfacial Behavior", *Material Science and Engineering*, Vol. A138, pp. 155–167, 1991.

58. Nguyen, T.-H. B. and Yang, J.-M., "Elastic Bridging for Modeling Fatigue Crack Propagation in a Fiber-Reinforced Titanium Matrix Composite", *Fatigue Fract. Engng. Mater. Struct.*, Vol. 17, No. 2, pp. 119–131, 1994.

59. Hutchinson, J. W. and Jensen, H. M., "Models for Fiber Debonding and Pullout in Brittle Composites with Friction", *Mechanics of Materials*, Vol. 9, pp. 139–163, 1990.

60. Aveston, J. and Kelly, A., "Theory of Multiple Fracture of Fibrous Composites", *Journal of Materials Science*, Vol. 8, pp. 352–362, 1973.

61. Lee, J. W. and Daniel, I. M., "Deformation and Failure of Longitudinal Loaded Brittle-Matrix Composite Materials", *Proceedings of the Tenth Symposium on Composite Materails: Testing and Design*, G. C. Grimes, ed. ASTM, pp. 204–221, 1992.

62. Cho, C., Holmes, J. W., and Barber, J. R., "Distribution of Matrix Cracks in a Uniaxial Ceramic Composite", *Journal of the American Ceramics Society*, Vol. 75, No. 2, pp. 316–324, 1992.

63. Tamin, M. N., Osborne, D. J. and Ghonem, H., "Influence of Interfacial Properties on Fiber Debonding in Titanium Metal Matrix Composites", *Fatigue and Fracture at Elevated Temperature*, A. Nagar and S. Mall, eds., The American Society of Mechanical Engineers, New York, NY, AD-Vol. 50, pp. 121–134, 1995.

64. Cox, B. N. and Marshall, D. B., "Crack Bridging in the Fatigue of Fibrous Composites", *Fatigue Fract. Engng. Mater. Struct.*, Vol. 14, No. 8, pp. 847–861, 1991.

65. Rice, J. R., "Elastic Fracture Mechanics Concepts for Interface Cracks", *J. Appl. Mech.* 55, pp. 98–103, 1988.

66. Tamin, M., Osborne, D. J., and Ghonem, H., "A Criterion for the Fracture of Bridging Fibers in Metal Matrix Composites", *Intl. Conf. on Ceramic and Metal Matrix Composites, CMMC 96,* San Sebastian, Spain, 9–12 September, 1996.

67. Aveston, J., Cooper, G. A. and Kelly, A., "Single and Multiple Fracture", *Conference on the Properties of Fibre Composite,* National Physical Laboratory, IPC Science and Technology Press, pp. 15–26, 1971.

68. Tada, H., Paris, P. C. and Irwin, G. R., *The Stress Analysis of Cracks Handbook,* Del Research Corporation, St. Louis, MO, 1985.

69. Gao, Yu-Chen, Mai, Yiu-Wing, and Cotterll, "Fracture of Fiber-Reinforced Materials," *Journal of Applied Mechanics and Physics,* V. 39, pp. 550–572, 1988.

70. Hsueh, C.-H., "Interfacial Debonding and Fiber Pull-Out Stresses of Fiber-Reinforced Composites. III. With Residual Radial and Axial Stresses", *Material Science and Engineering,* A145, p. 135, 1991.

71. Hsueh, C.-H., "Interfacial Debonding and Fiber Pull-Out Stresses of Fiber-Reinforced Composites. IV. Interpretation of Fiber Pull-out Curves", *Material Science and Engineering,* A149, p. 135, 1990.

72. Kantzos, P., Telesman, J. and Ghosn, L., "Fatigue Crack Growth in a Unidirectional SCS-6/Ti-15-3 Composite", *Third Symposium on Composite Materials: Fatigue and Fracture, ASTM STP 1110,* T. K. O'Brien, ed., American Society for Testing and Materials, Philadelphia, PA pp. 711–731, 1991.

73 Hoagland, R. G., Rosenfield, A. R. and Hahn, G. T., "Mechanisms of Fast Fracture and Arrest in Steels", *Metallurgical Transactions,* Vol. 3, pp. 123–136, 1972.

74. Tamin, M. N. and Ghonem, H., "Evolution of Bridging Fiber Stress in Titanium Metal Matrix Composites at Elevated Temperature", *Advances in Fatigue Lifetime Predictive Techniques,* 3rd Volume, *ASTM STP 1292,* M. R. Mitchell and R. W. Landgraf, eds., American Society for Testing and Materials, Philadelphia, PA, 1995.

75. Tamin, M. N. and Ghonem, H., "Fracture Criterion for Bridging Fibers in Titanium Metal Matrix Composites at Elevated Temperature", *Durability of Composite Materials,* R. C. Wetherhold, ed., The American Society of Mechanical Engineers, New York, NY, MD-Vol. 51, pp. 51–58, 1995.

76. Nakamura, T. and Suresh, S., "Effects of Thermal Residual Stresses and Fiber Packing on Deformation of Metal Matrix Composites", *Acta Metall. Mater.,* Vol. 41, No. 6, pp. 1665–1681, 1993.

77. Jansson, S. and Leckie, A., "Reduction of Thermal Stresses in Continuous Fiber Reinforced Metal Matrix Composites with Interface Layers", *Journal of Composite Materials,* Vol. 26, No. 10, pp. 1474–1486, 1992.

78. Ghonem, H., Wen, Y. Zheng, D., Thompson, M, and Linsey, G., "Effects of Temperature and Frequency on Fatigue Crack Growth in Ti-β21S Monolithic Laminate", *Materials Science and Engineering,* Vol. 161, pp. 45–53, 1993.

79. Tamin, M. N., Zheng, D., and Ghonem, H., "Time-Dependent Behavior of Continuous-Fiber-Reinforced Metal Matrix Composites: Modeling and Ap-

plications", *Journal of Composites Technology and Research,* Vol. 16, No. 4, pp. 314–322, 1994.

80. Rangaswamy, P. and Jayaraman, N., "Residual Stresses in SCS-6/Ti-24Al-11Nb Composite: Part II—Finite Element Modeling", *Journal of Composites Technology and Research,* Vol. 16, No. 1, pp. 54–67, 1994.

81. Wilt, T. E. and Arnold, S. M., "A Computationally-Coupled Deformation and Damage Finite Element Methodology", in *Proceedings of the 6th Annual HITEMP Review,* Vol. 2, Cleveland, OH, pp. 35-1–35-15, 1993.

82. Martineau, P., Lahaye, M., Pailler, R. and Naslain, R., "SiC Filament/Titanium Matrix Composites Regarded as Model Composites. Part I—Filament Microanalysis and Strength Characterization", *Journal of Materials Science,* Vol. 19, pp. 2731–2748, 1984.

83. Bunsill, A. R., 1988, *Fiber Reinforcements for Composite Materials,* Elsevier Science Publications, New York, NY.

84. Tamin, M. N., "Bridging Fatigue Crack Growth Damage Mechanisms in Titanium Metal Composites", Ph.D. Thesis, University of Rhode Island, April 1997.

—————————— 9 ——————————

Notch Strength

W. S. JOHNSON
School of Materials Science and Engineering
Georgia Institute of Technology
Atlanta, GA 30332-0245

C. A. BIGELOW
FAA Technical Center

DAVID M. HARMON
McDonnel Douglas Aerospace

INTRODUCTION

METAL MATRIX COMPOSITES (MMC) have several inherent properties, such as high stiffness-to-weight ratios and high strength-to-weight ratios, which make them attractive for advanced aerospace applications. These composites also have a higher operating temperature range than polymer matrix composites and most conventional metals. Like polymer composites, MMC are notch sensitive. The degree of notch sensitivity depends on notch size and shape, laminate orientation, and material properties. Fiber-matrix interfaces can also play a key role in the mechanical behavior of MMC [1]. The interfaces govern the mode and extent of load transfer between the fiber and matrix. When the interfaces are strong and transmit all loads fully, isolated fiber fractures tend to spread more rapidly to other fibers and hasten failure [2]. To design damage-tolerant structures or to simply understand the effects of fastener holes, the laminate fracture strength must be known for a wide range of laminates, notch geometries, loading conditions, and material properties. To avoid testing all possible parametric combinations, an analytical method for predicting the fracture strength of notched MMC is needed. Over the past two decades, researchers have analyzed a number of notched MMC laminates for a variety of behaviors and have developed an analytical micromechanics-based strength prediction methodology to predict failure of notched MMC. This chapter will review some of the experimental and analytical methodology and present selected results. The micromechanics-based strength prediction methodology combines analytical techniques ranging from a three-dimensional (3D) orthotropic finite element (FE) model of a notched specimen to a micromechanical model of a single fiber. Residual strength predictions will also be made using fiber bridging considerations.

Work with boron/aluminum (B/Al) MMC showed that, in this low yield strength matrix, extensive yielding of the matrix occurred at the notch tips, such that specimens with sharp notches and center holes failed at similar stress levels [3]. In brittle polymeric matrix composites, similar notch-insensitive results have been observed for quasi-isotropic laminates [4]. However, in polymeric composites, the notch insensitivity was caused by extensive matrix cracking and delaminations near the crack tip that significantly reduced the local stress concentration [5]. The fiber-matrix interface plays a particularly significant role in MMC with a matrix having a high yield strength, such as silicon-carbide/titanium (SCS-6/Ti-15-3). Debonding of the fiber-matrix interface was found to be a primary damage mechanism in SCS-6/Ti-15-3 composites [1]. Extensive fiber-matrix debonding occurred such that specimens with sharp notches and center holes failed at similar stress levels [6]. Matrix cracking was found to be a significant factor in predicting the postfatigue residual strength of SCS-6/Ti-15-3 specimens containing center holes [7].

Traditional strength prediction approaches do not lend themselves to MMC when highly nonlinear behavior is observed or when excessive matrix damage (i.e., fiber-matrix debonding, matrix cracking) precedes laminate failure. For example, linear elastic fracture mechanics (LEFM) methods, (see, e.g., Waddoups et al. [8]) would not apply when the material behaves nonlinearly. An LEFM approach could possibly be used in cases of matrix damage without material nonlinearity, but this would require orthotropic stress-intensity factor solutions for extremely complex crack configurations. Two-parameter approaches, such as introduced by Whitney and Nuismer [9,10], use elastic stress solutions for undamaged configurations and attempt to account for damage through the use of some characteristic length. (An extensive review of commonly used fracture models for predicting the notched strength of composite laminates is presented elsewhere [11].) A micromechanics-based strength prediction methodology described herein presents a more direct approach by modeling behavior and damage on a constituent level, thus explicitly accounting for the matrix nonlinearity, fiber-matrix interface debonding, and matrix cracking.

The following sections will describe experimental observations and analytic methodologies for titanium matrix composites. Some work pertaining to boron/aluminum composites will be included. The aluminum matrix work is to illustrate the type of behavior that may be exhibited by a titanium matrix at elevated temperatures where the modulus and yield strength of the titanium are significantly reduced.

EXPERIMENTAL OBSERVATIONS

Mall and his colleagues [12–14] have conducted numerous tests to identify notch strength and the associated damage mechanisms on several different layups, fiber and matrix combinations, and temperatures. Basic observations from this work are described below.

Fiber-Matrix Interfacial Debonding

Early fiber-matrix interface debonding occured in off-axis plies early in the overall loading history to failure. These interfacial failures result in a nonlinear response in the stress-strain curves. Microphotographs of these interfacial failures using an edge replica technique were first shown in Reference [1] and are also shown in Reference [12].

Very little fiber pull out was noticed on the failure surface for room temperature tests. Elevated temperature tests exhibited more fiber pullout [12] as shown in Figure 1. At room temperature, the fiber-matrix interface has significant radial compressive stresses holding the matrix in contact with the fiber. The residual stresses are virtually nonexistent at 650°C; thus, there is less resistance to the fibers pulling out of the matrix. Another contributing factor is that the matrix material has a much higher modulus at room temperature than at 650°C, so that when cracking occurs in the matrix at room temperature, the stress concentration is greater, thus encouraging the crack to propagate straight across the specimen.

Notch Effects

Rattray and Mall [13] conducted notch strength tests on SCS-6/TIMETAL®21S with a quasi-isotropic layup containing various size holes or slitlike notches. Figure 2 shows the stress-strain response of

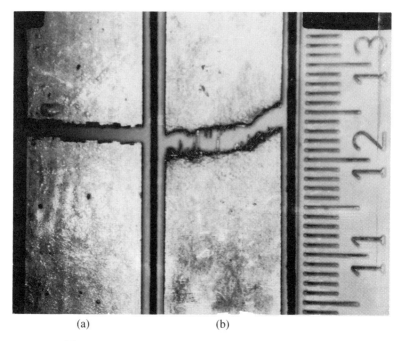

(a)   (b)

Figure 1. Unnotched specimens: (a) RT, (b) 650°C.

several notch geometries at room temperature. The strain was measured locally across the notched region. D is half-notch length, and W
is half-specimen width. Rattray and Mall [13] found that the notched
strength (the remotely applied load divided by the unnotched cross-
sectional area of the specimen) was independent of the notch shape
but dependent on the length of the notch in the width direction of the
specimen. Plots of data taken at room temperature and 650°C are
shown in Figure 3. Notice that all the strength data fall below the net
strength line, indicating that there is indeed a notch effect. One
would expect that the slitlike notch would have a much higher stress
concentration factor than the circular hole and should, therefore, fail
at a lower stress. However, that was not observed. This plot is very
similar to one for B/Al [3]. A rationale for this behavior in titanium
matrix composites (TMCs) will be given late- in the analysis section.

Pin-Filled Holes

Roush and Mall [12] used SCSs-9/TIMETAL®21S laminates with
stacking sequences of $[0/90]_{2s}$ and $[0/\pm45/90]_s$ to assess the effects of

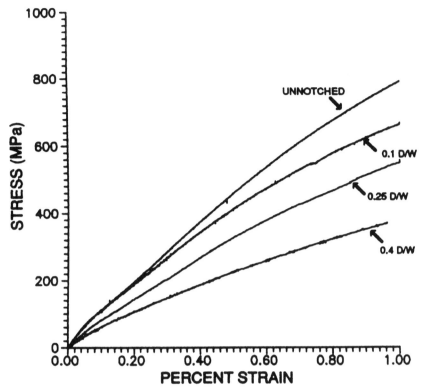

Figure 2. Stress-strain relationship encompassing hole for notched specimen at RT.

pin-filled holes. The specimens had width-to-hole diameter ratios of 6 and were tested at room temperature, 482, and 650°C. The insertion of a pin into the hole had negligible effect on the stiffness and strength, damage initiation, and progression of the laminate at room and elevated temperatures.

Compression

Gooder and Mall [14] conducted a series of tests on SCS-6/TIMETAL®21S to assess the compressive notch strength of quasi-isotropic laminates. They found that the notched strength in compression at 650°C was about half the room temperature notch strength. This was most likely due to the significant decrease in matrix modulus at 650°C, which led to earlier fiber buckling at the edge

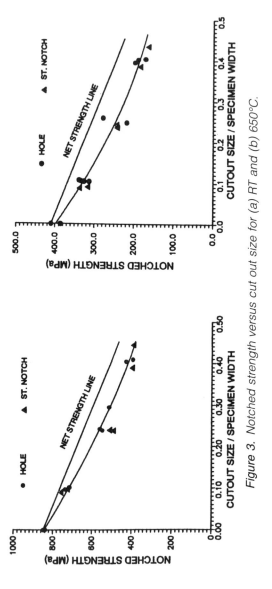

Figure 3. Notched strength versus cut out size for (a) RT and (b) 650°C.

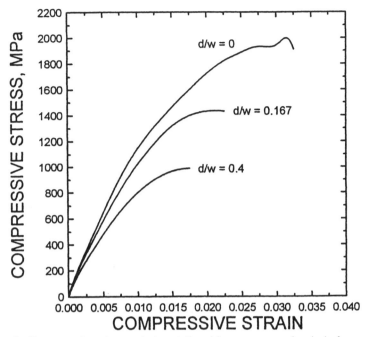

Figure 4. Compression stress-strain relationship encompassing hole for notched specimen at RT.

of the notch. Figure 4 shows the stress-strain response for an unnotched and two notched specimens loaded in compression at room temperature. Comparing these curves to those in Figure 2, you can see that the strain to failure in compression was roughly twice that in tension. The compression failures resulted from fiber fracture due to microbuckling preceded by shear yielding of the matrix. The final failure was in shear at about 45° off the loading axis.

NOTCHED LAMINATE STRENGTH ANALYSES

Notched strength analysis is important because most structural configurations for which metal matrix composites would be considered, will contain some form of hole or notch, resulting in a local stress concentration (i.e., fastener holes, access holes, etc.). The effect this notch has on laminate strength may be important, and there must be methods available to analyze this situation.

The analysis developed was designed to predict the laminate strength with a center notch. The notch may be elliptical or circular in shape, and finite width effects are taken into account. Development of loaded hole analyses and analyses for edge notch effect are subject areas for additional work.

The notched strength analysis combines a ply-by-ply unnotched strength prediction analysis with notched stress analyses and includes two procedures to specifically account for notch effects on failure initiation. A weak matrix lamina is subject to shear matrix yielding and/or cracks that grow parallel to the reinforcing fibers. This yield zone initiates at critical shear locations around the notch perimeter. Typically, the ply will fail once the first fiber breaks at the notch edge. The matrix material is not capable of picking up the load originally carried by the fiber. Conversely, a strong matrix system is not subject to the shear yielding or cracking phenomenon, and the matrix is capable of carrying load carried by a finite number of fibers. Typically, an aluminum or resin matrix composite is considered a weak matrix system and titanium matrix composite is considered a strong matrix system. In addition, it is in the polymer matrix composites that shear cracks occur, whereas only yielding develops in metal matrix composites.

The ply-by-ply unnotched strength analysis is very straightforward. Standard laminated plate theory as described by Jones [15] is used to calculate laminate stiffnesses. The full-stiffness matrix is calculated, including the extension, bending, and coupled terms. This allows for the analysis of nonsymmetric laminates and of laminates subjected to thermal loads. The applied laminate loads, N_x, N_y, and N_{xy}, are multiplied by the stiffness matrix of the laminate to obtain the strains in each ply. From these individual ply strains, the corresponding stresses are obtained and converted to their principal components parallel and perpendicular to the fibers. A failure criterion is then applied to determine which ply reaches its linear elastic limit. Once a ply is predicted to fail, the mode of failure for the ply is determined and the stiffness of the ply is reduced. The mode of failure is determined from the three principal stress component stress-to-strength ratios. The ratio that is maximum is defined as the dominant mode of failure for the ply. This mode of failure dictates how the stiffness reduction is to be performed. If the stress ratio corresponding to the longitudinal strength, X, is maximum, then the fibers are critical and represent the dominant failure mode. In this case, the fiber stiffness for this ply is reduced by a factor of 1000. If, on the other hand, the stress ratio corresponding to the transverse strength, Y, or the shear strength, S, is maximum, then matrix yielding is considered the dominant mode. Therefore, the matrix stiffness would be reduced by a factor of 1000. This factor was chosen to avoid any computational singularities that may occur if a value of 0 was used instead.

The general procedure for predicting notched strength is very similar to that for unnotched strength. The difference in the notched analysis is that the stress state at each point along the notch perimeter must be determined and analyzed. Specifically, given the applied load ratios, N_x, N_y, and N_{xy}, the gross (far-field) ply strains and principal stresses are calculated identically as they were in the unnotched analysis. The local stress state at the notch perimeter is calculated in the form of the tangential stress, σ_θ. The method of determining the tangential stress is discussed below. Mohr's circle is used to calculate the components of stress parallel and perpendicular to the fibers (σ_1, σ_2, and τ_{12}) at each point around the notch (between $\theta = 0°$ and $\theta = 180°$). Failure criterion is then applied to determine the gross applied stress required to failure and the mode of failure. The critical point around the notch perimeter is the point that requires the minimum gross applied stress for failure to occur.

If the failure mode at the critical point is a tension or compression failure of the matrix, then the lamina stiffnesses are reduced just as they were in the unnotched analysis. If the mode of failure is fiber, then before the fiber modulus for this ply is reduced, it must first be determined if the matrix is capable of carrying additional load once the fiber fails. On the other hand, if the mode of failure is matrix shear, then the effect of shear yielding or shear cracking must first be determined before the matrix modulus is reduced.

Once the proper analysis is performed, the stiffnesses for the failed ply are reduced, and the notched strength analysis continues. Final laminate failure is defined to occur once each ply has experienced either a matrix or fiber failure in compression or tension. A ply that fails due to matrix shear is not considered failed and is reanalyzed once the yielding or cracking phenomenon is properly accounted for in the analysis.

A verification of the two strength analyses was achieved by comparing predictions with notched test data for boron/aluminum, (B_4C)B/Ti-15-3, and SCS-6/Ti-15-3 (Figures 5 and 6). These data include unidirectional and cross-plied laminates with different width-to-diameter ratios. These tests were all uniaxial. The biaxial portion of the analysis has not yet been verified.

In these predictions, the only difference in how the analysis proceeds is in how the 0° plies are treated in the aluminum (weak) ver-

Material/ Lay-Up	Width (in.)	Diameter (in.)	Thickness (in.)	Strength	
				Test (ksi)	Analysis (ksi)
(B₄C) B/TI-15-3					
0°	1.513	0.250	0.174	94.53	94.70
	1.510	0.250	0.175	94.89	94.70
SCS-6/TI-15-3					
(0, ±45, 90)s	0.752	0.192	0.062	72.40	65.30
	0.751	0.192	0.062	70.20	65.30
(±60, 90₁₁, ±60)	1.512	0.250	0.117	30.80	32.60
B/A16061					
0°	2.508	0.254	0.171	113.26	101.84
	2.505	0.255	0.172	105.85	101.84
	1.388	0.253	0.172	112.05	95.82
	1.511	0.251	0.172	98.17	95.82
(0₂/±45)s	2.475	0.257	0.172	55.09	56.77
	2.516	0.257	0.173	47.93	56.77
	1.504	0.252	0.173	57.91	53.66
	1.487	0.257	0.172	50.93	53.66
(0₂/90₂)s	2.509	0.254	0.171	46.91	51.81
	2.510	0.253	0.172	43.75	51.81
	1.507	0.254	0.172	50.93	48.82
	1.510	0.261	0.173	41.29	48.82

Figure 5. Uniaxial notched strength predictions are accurate.

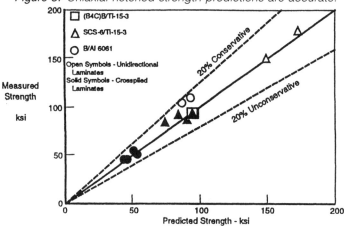

Figure 6. Uniaxial notched strength predictions are conservative.

368

sus titanium (strong) matrix composite. Also, because only the 0° plies are loaded parallel to its fibers, final laminate failure was predicted to occur when the 0° fibers had failed.

Effects of Shear Yielding or Cracking on Composite Notched Strength Analysis

For notched analysis, the local stress state around the hole must be analyzed to determine the failure mode and where the failure will occur. This analysis is done on a ply-by-ply basis. If the matrix material is weak in shear, such as aluminum, a yield zone is likely to form near the hole before fibers at the notch edge fail. The effect of this yield zone on the stress state around the hole must be accounted for to predict the true failure load. In a weak matrix material, laminate failure is predicted to occur when the fibers at the notch edge fail. The matrix material is typically not strong enough to carry the additional load previously carried by the fiber.

A stress analysis around the notch perimeter indicates that the matrix is critical in shear for a weak matrix material. As an example, the stress concentration analysis described below is used to predict the tangential, longitudinal, and shear stress concentration around a finite width boron/aluminum 0° laminate with a center hole subject to uniaxial tension loads (Figure 7). Due to symmetry, results are plot-

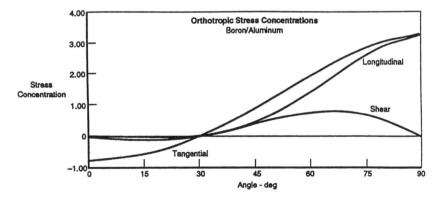

Figure 7. Orthotropic stress concentrations factors around notch perimeter.

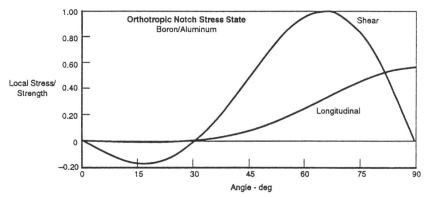

Figure 8. Critical locations around notch are easily determined once notch stress state is known.

ted for only one quadrant. This plot indicates that each of the stress concentrations peak at a different point around the notch. However, if the local stress is normalized with respect to the appropriate strength value, it becomes obvious that the shear stress is most critical. This occurs at an angle of 67° from the load axis (Figure 8).

The shear stress curve in Figure 8 is divided by the matrix shear yield strength. Because the matrix has exceeded its shear yield strength at this point, a yield zone forms, which will continue to grow with additional loading. If the material was a polymer matrix material, the yield zone would actually be cracks that initiate and grow parallel to the fibers.

A notched laminate strength analysis for weak matrix materials is not complete unless it accounts for matrix yielding along the fibers. Accurate notched strength prediction is dependent on modeling the additional load carried in the laminate once matrix yielding starts to occur.

The net effect of matrix yielding is to reduce the stress concentration at the edge of the notch, K_t, and to therefore, allow the laminate to carry more load. This behavior is modeled using a simplified shear lag theory.

Yielding will occur at the point around the hole where the shear stress exceeds the shear strength of the material (Figure 8). As the load in the laminate increases, the matrix material will continue to yield along the fibers. The effect of this yielding is to reduce the amount of load carried by the material directly in line with the notch. Therefore, less load is transferred through the laminate at the notch, and the stress concentration at the edge of the hole is reduced. The matrix will continue to yield until the applied stress multiplied by the

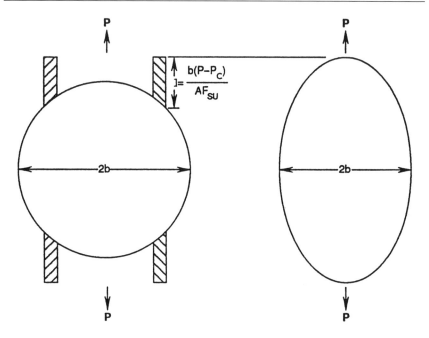

Figure 9. Yield zone is simply modeled.

current stress concentration at the edge of the hole exceeds the material longitudinal tension strength. Theoretically, the matrix could continue to yield all the way to the top of the specimen. If this were to occur, then the failing net section stress should equal the ultimate longitudinal strength of the material.

Matrix yielding along the fibers is modeled by transforming the notch in the laminate to an elliptical notch whose major diameter is parallel to the fibers (Figure 9). The equation for the major diameter was empirically determined to be:

$$R_x^{1.5} = L^{1.5} + R_x^{1.5}$$

By adjusting the geometry of the notch, K_t is decreased at the notch edge. The "L" in the above equation refers to the length of the yield zone and is determined using the following relationship:

$$L = \frac{(\sigma - \sigma_0)2b}{2F_{su}}$$

where σ_0 is the applied stress at which yielding first occurs, σ is the applied stress in the laminate, $2b$ is the notch minor diameter, and F_{su} is the allowable shear stress.

The length of the yield zone from the notch is determined by the amount of load carried around the hole in shear. This load divided by the yield zone length and the specimen thickness is equated to the ultimate matrix shear yield strength.

Although this method of accounting for yielding in 0° plies works very well in a uniaxial analysis, it does not lend itself easily to biaxial analysis. In a biaxial analysis, yielding can occur in any ply regardless of its orientation.

Modeling yielding in every ply is a complicated task. This involves tracking the amount of matrix yielding in every ply separately. Following the modeling scheme described above, each ply would eventually have different notch configurations to account for the matrix yielding in the ply.

Typically, however, the transverse strength is far more critical than the shear strength. Using a uniaxial analysis, it is sufficient to consider matrix yielding effects for 0° plies only because the transverse strength will be critical in the off-axis plies. Similarly, in a biaxially loaded laminate, an off-axis ply is more likely to fail at a point around the notch where the transverse stress component exceeds the transverse strength. Yielding will be most important in 0° plies when longitudinal loading is present and in 90° plies when transverse loading is present. This allows the use of the modified shear lag theory described above.

Effect of a Strong Matrix Material on Composite Notched Strength Analysis

The notched strength method for strong matrix materials, such as titanium, is based on the fact that the fibers at the notch edge will fail before the matrix yields in shear or tension. The objective of this method is to determine how many fibers will fail before the matrix achieves its maximum load-carrying capability. Two assumptions are made in this analysis: (1) the yielding load is the maximum load carried by the matrix and (2) the local stress at the location where the next fiber fails is equal to the lamina longitudinal strength. The load in the ply is obtained by summing the contributions from two areas: (1) in the damaged area where the fibers have failed and (2) where the fibers are still intact:

$$P = 2t(F_y L_d + \int_{L_d+b}^{W/2} \sigma_x(y)dy)$$

In this expression, F_y is the matrix yield strength, "b" is the notch radius perpendicular to the fibers, W is the specimen width, and L_d is the length of the zone where the fibers have failed (Figure 10). The stress distribution through the net section is denoted by $\sigma_x(y)$.

The 0° ply ultimate load is equal to the largest value of P as L_d is incremented through the net section. The exact value of L_d that corresponds to the ultimate load may be determined by differentiating the expression for P with respect to L_d and then solving for L_d.

The expression for the stress distribution is assumed to be an exponential decay of the stress gradient from the notch:

$$\sigma_x(y) = \sigma_g(A + B[1 + \frac{y - b}{\rho}]^{-C})$$

In this expression σ_g is the gross stress and $\rho = a^2/b$ where "b" and "a" are the radius of the notch perpendicular and parallel to the loading, respectively. A, B, and C are constants that are determined through

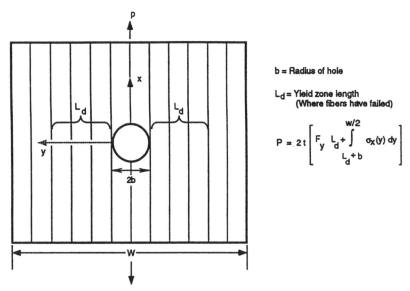

Figure 10. Strong matrix composite notched strength analysis.

the application of three boundary conditions. These conditions state the following:

1. The stress concentration at the edge of the hole must equal K_t:

$$\frac{\sigma}{\sigma_g} @ (y = b) = Kt = A + B$$

2. The stress gradient at the edge of the hole must equal $C1\dfrac{K_t}{p}$ [12]:

$$\frac{d\sigma}{d_y} (y = b) = -C1\frac{Kt}{B} \quad or \quad C = -C1\frac{Kt}{B}$$

3. The load across the net section must equal the applied load:

$$P = wt\sigma_g = 2t \int_b^{W/2} \sigma \, dy$$

Solving for A, B, and C is a nonlinear problem so an iterative procedure is used.

This analysis is extremely simple and has been proven to be very accurate when compared with classical problems. For isotropic materials several studies were made. Comparisons were made to solutions for elliptical slots in infinite width plates [16], Figure 11, for circular

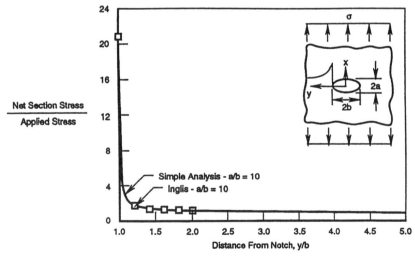

Figure 11. Net section stress prediction in an isotropic infinite width panel.

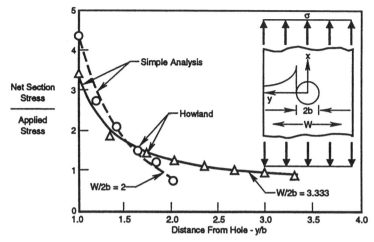

Figure 12. Net section stress prediction in an isotropic finite width panel.

holes in finite width plates [17], Figure 12, and for circular holes in very narrow plates, Figure 13. When the hole diameter is very nearly as wide as the plate, the K_t approaches $2/(1-2b/W)$, as shown by Peterson [18], and the stress is almost linear across the net section (Figure 13).

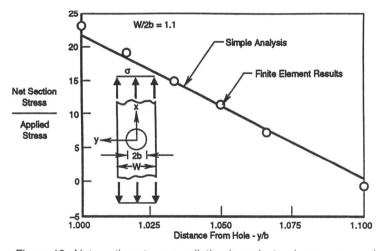

Figure 13. Net section stress prediction in an isotropic narrow panel.

For orthotropic plates with center notches, the analysis is derived to be exact for infinite plates and was verified by comparisons with Lekhnitskii's solutions for highly orthotropic plates. For finite width plates, the analysis was compared with a finite element analysis. In both cases there was excellent agreement [19].

Stresses Around Notch Perimeter

Accurate notched strength prediction is dependent on an accurate representation of the stress state around the notch. In this section, the method used to develop the stress concentrations due to an open notch is discussed. This approach was developed by Finefield, et al. [19].

The notched stress analysis is developed for inplane biaxially loaded laminates with center notches of any elliptical shape. Closed-form equations were derived by Lekhnitskii [20] to describe the local stress at any point on the notch perimeter. This expression was defined for an infinitely wide anisotropic ply:

$$\sigma\theta = \sigma x Fx(\phi,\theta) + \sigma y Fy(\phi,\theta) + \tau xy Fxy(\phi,\theta)$$

The tangential stress, σ_θ, at an angle, θ, around the notch is defined in terms of the gross stresses in the loading directions (σ_x, σ_y, and τ_{xy}), the Lekhnitskii variables (F_x, F_y, and F_{xy}), and the fiber orientation angle, ϕ:

$$Fx(\phi,\theta) = \frac{E\theta}{E1} \{[\cos^2 \phi + (\mu 1\mu 2 - \eta)\sin^2 \phi]\mu 1\mu 2\cos^2\theta +$$

$$[(1 + \eta)\cos^2 \phi + \mu 1\mu 2\sin^2 \phi]\sin^2 \theta - \eta(1 + \eta - \mu 1\mu 2)$$
$$\sin\theta \cos\theta \sin\phi \cos\phi\}$$

$$Fy(\phi,\theta) = \frac{E\theta}{E1} \{[\sin^2 \phi + (\mu 1\mu 2 - \eta)\cos^2\phi]\mu 1\mu 2\cos^2\theta +$$

$$[(1 + \eta)\sin^2 \phi + \mu 1\mu 2\cos^2 \phi]\sin^2 \theta + \eta(1 + \eta - \mu 1\mu 2)$$
$$\sin\theta \cos\theta \sin\phi \cos\phi\}$$

$$Fxy(\phi,\theta) = \frac{E\theta}{2E1} (1 + \eta - \mu 1\mu 2)\{- \eta \cos 2\phi \sin2\theta +$$
$$[(1 - \mu 1\mu 2)\cos 2\theta - \mu 1\mu 2 - 1]\sin 2\phi\}$$

where

$$E\theta = \left[\frac{\sin^4 \theta}{E1} + \sin^2 \theta \cos^2 \theta \left(\frac{1}{G12} - \frac{2v12}{E1} \right) + \frac{\cos^4 \theta}{E2} \right]^{-1}$$

$$\eta = \sqrt{\frac{E1}{G12} - 2v12 + 2\sqrt{\frac{E1}{G12}}}$$

$$\mu1\mu2 = -\sqrt{\frac{E1}{E2}}$$

The angle, θ, is measured from the fiber axis (1 direction). The angle, ϕ, is measured from the x-loading axis. These expressions are greatly simplified when analyzing a $0°$ ply ($\phi = 0$) as follows:

$$Fx\,(0,\theta) = \frac{E\theta}{E1}\, \{\mu1\mu2\cos^2 \theta + (1 + \eta)\sin^2\theta\}$$

$$Fy(0,\theta) = \frac{E\theta}{E1}\, \{(\mu1\mu2 - \eta)\mu1\mu2\cos^2\theta + \mu1\mu2\sin^2\theta\}\}$$

$$Fxy(0,\theta) = \frac{E\theta}{2E1}\,(1 + \eta - \mu1\mu2)\,(-\,\eta \sin2\theta)$$

After rearranging terms, these equations can be written in the following form:

$$Fx(0,\theta) = \frac{E\theta}{2E1} \left[\left(K_{90}^x + K_0^x \frac{E1}{E2} \right) - \left(K_{90}^x - K_0^x \frac{E1}{E2} \right) \cos2\theta \right]$$

$$Fy(0,\theta) = \frac{E\theta}{2E2} \left[\left(K_0^y + K_{90}^y \frac{E2}{E1} \right) + \left(K_0^y - K_{90}^y \frac{E2}{E1} \right) \cos 2\theta \right]$$

$$Fxy(0,\theta) = \frac{-E\theta}{2E1} \left[\left(K_{90}^x - K_0^x \frac{E1}{E2} \right) \eta \sin 2\theta \right]$$

The terms K_{90}^x and K_0^x represent the stress concentrations at the edge ($\theta = 90°$) and at the top ($\theta = 0°$) of the notch in an infinitely wide $0°$ ply subject to loading in the x-direction (which for a $0°$ ply is the same as the 1 direction). Similarly, K_{90}^y and K_0^y are the stress concentration factors in a $0°$ ply subject to loading in the y direction (or 2 direction):

$$K_{90}^x = 1 + \eta \qquad K_0^x = \mu 1 \mu 2 \, \frac{E2}{E1}$$

$$K_0^y = 1 + \eta \sqrt{\frac{E2}{E1}} \qquad K_{90}^y = \mu 1 \mu 2$$

The off-axis ply is rotated and analyzed as a $0°$ ply with applied loads determined through Mohr's circle analysis. In this way, the simplified expressions derived above can still be used.

All of the expressions derived above are valid for only infinitely wide panels. Finite width effects are accounted for by substituting finite width expressions for K_{90}^x, K_0^x, K_{90}^y, and K_0^y. These expressions have been derived for panels with different width-to-diameter ratios and different orthotropic properties [21]. In the following discussion, K_t and $K_{t\,net}$ represent the net section and gross section stress concentration factors at the point $90°$ along the notch from the primary loading axis. Similarly, the stress concentration at the top of the notch, $0°$ from the loading axis, is represented by K_{t0}.

The general expression for a stress concentration at the edge of a notch was obtained by curve fitting data published in Peterson's handbook [18]. These data consisted of graphs of the net section stress concentration factor for an elliptical notch in an isotropic panel (Figure 14). The gross section stress concentration factor is then determined as:

$$Kt = \frac{Kt\ net}{(1 - 2b\,/\,W)}$$

where W is the width of the laminate and "b" and "a" are the notch radii perpendicular and parallel to the loading, respectively. The curve fit for $K_{t\,net}$ was generalized to include orthotropic panels. The expression is written below for a $0°$ ply with longitudinal stiffness, E_1, transverse stiffness, E_2, shear stiffness, G_{12}, and Poisson's ratio, v_{12}:

$$Ktnet = 2 + f_1 f_2^2 + f_1 f_2^4 + 0.634\, f_3\, (1 - f_4^2) + 0.167\, f_3 (1 - f_4^4)$$
$$+ 0.109\, f_3 f_5 (2b\,/\,W)$$

where

$$f_1 = \frac{(1 + 2b\,/\,aeff)}{2} - 1$$

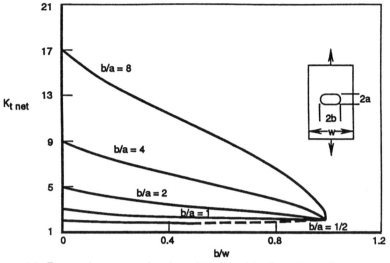

Figure 14. Expressions were developed to calculate the net section stress concentration factor in an isotropic panel with an elliptical notch.

$$f_2 = 1 - 2b / W$$

$$f_3 = \frac{b}{aeff} - 1$$

$$f_4 = 4b / W - 1$$

$$f_5 = 1 - (2b / W)^{100}$$

and

$$aeff = (2a / \eta)[1 - 2b / W]^{0.125(\eta/2-1)}$$

$$\eta = \sqrt{\frac{E1}{G12} - 2v12 + 2\sqrt{\frac{E1}{E2}}}$$

which is the exact solution derived by Lekhnitskii [20] for orthotropic plates of infinite width and degenerates to the well-known K_t of 3 for infinite, isotropic plates.

In composites, the fibers typically have a higher stiffness than the matrix material. The higher stiffness fibers in a 0° lamina will increase the stress concentration at the edge of the hole. Similarly, the stress concentration in an isotropic material can be increased by changing the circular hole into an ellipse with its major axis perpendicular to the loading. To calculate the stress concentrations in notched metal matrix composites, an effective notch dimension, a_{eff}, is defined, which accounts for the material orthotropy.

An extensive check of this curve fit was performed using boundary collocation analysis. The agreement was excellent for a wide range of orthotropic materials [19].

Based on results from finite element and boundary collocation analyses for notched laminates with varying stiffness properties and width-to-diameter ratios, an expression for the stress concentration at the top of the hole, K_{t0}, in a $0°$ ply has also been generated. This expression is:

$$Kt0 = \frac{\sqrt{\frac{E2}{E1}} + 2.8\left(\frac{2b}{W}\right)^2}{\left[1 + \frac{0.00865}{\left(\frac{W}{2b} - 1\right)^{2.76}}\right]}$$

and has been verified using finite element results in Reference 19.

The corresponding finite width stress concentration factors to be used in the Lekhnitskii based, tangential stress equations, described earlier, are given by:

$$K_{90}^x = Kt \qquad K_0^x = Kt0$$

The expressions for K_{90}^y and K_0^y are found by realizing that a $0°$ ply loaded transversely (y-direction) is the same as a $90°$ ply loaded longitudinally (x-direction). Because the angle, θ, around the notch is measured from the fiber axis, K_0^y is calculated by switching the values of E_1 and E_2 and by substituting v_{21} for v_{12} into the expression for K_t:

$$v21 = v12 \frac{E2}{E1}$$

Similarly, K_{90}^y is calculated by making the same switches and substitutions into the expression for K_{t0}.

The expressions derived above provide a completely general expression for the tangential stress at any point around a notch. These ex-

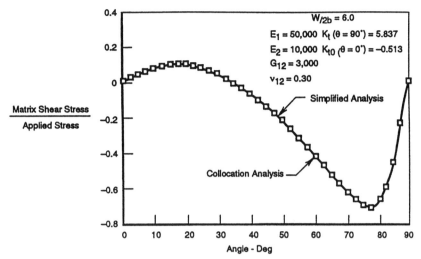

Figure 15. Simplified analysis provides accurate estimates of shear stress around the notch in an orthotropic finite width panel.

pressions account for finite width effects and plate orthotropy. The accuracy of this simplified approach is demonstrated in Figure 15 where the matrix shear stress concentration factor at angles, θ, between $0°$ and $90°$ from the x loading axis is calculated and compared with boundary collocation analysis for a uniaxially loaded $0°$ orthotropic panel. The shear stress concentration factor, Ks, in the loading direction is easily computed from the tangential stress through the application of Mohr's circle:

$$Ks = \frac{\tau 12}{\sigma}$$

Summary

This section addresses an approach using, for the most part, closed-form solutions to determine the notch strength of a composite laminate. Classical lamination plate theory combined with a ply-by-ply strength analysis is used to predict the final ply failure in a laminate subject in three inplane load components. Two methods are included in the analysis procedure to determine the effect of shear yielding or cracking that may occur at the notch perimeter and the effect of a strong matrix material that is able to carry load once a finite number of fibers fail. Stress analyses are described, which predict the stress distribution around the notch perimeter and through the net section. These analyses are based on closed-form solutions developed by Lekhnitskii for infinite width orthotropic plates. Verification of the procedures is provided in the form of strength predictions for center-notched aluminum and titanium matrix composites. The approach is based on classical approximations and is simplistic enough to be incorporated into a design procedure.

MICROMECHANICS-BASED STRENGTH PREDICTION METHODOLOGY

This section presents an approach based on a rather detailed finite element analysis of the notch geometry, loading conditions, and the constituent behavior. This approach can incorporate details of the mechanics, such as matrix nonlinear deformation and fiber-matrix debonding. In general, this approach is more complex than the approach previously described. However, the approach described in this section will give more insight into the failure process and roles of the constituents.

The micromechanics-based strength prediction methodology is based on constituent level behavior and combines three analytical techniques to predict various aspects of the specimen and material behavior of notched MMC. The first analytical technique, a three-dimensional (3D) finite element (FE) analysis (PAFAC [22]) was used to analyze the global behavior of notched specimens with and without damage. The second, a discrete fiber-matrix (DFM) finite element model was used to analyze fiber-matrix interface stresses. The third, a macromicromechanical analysis (MMA) [23], combines the 3D FE analysis with the unit cell DFM model. The MMA was used to analyze notch-tip stress states in notched configurations with no damage. In all cases, the bulk matrix properties were obtained from matrix material that was subjected to the same processing cycles as the composite laminates [1,3].

Three-Dimensional Finite-Element Analysis

The overall specimen behavior was analyzed with a three dimensional finite element program called PAFAC [22], which was developed from a program written by Bahei-El-Din [24,25]. PAFAC (Plastic and Failure Analysis of Composites) uses a constant strain, eight-noded, hexahedral element. Each hexahedral element represents a unidirectional composite material whose fibers can be oriented in the appropriate direction in the structural (Cartesian) coordinate system. The PAFAC program uses the vanishing-fiber-diameter (VFD) material model developed by Bahei-El-Din and Dvorak [26,27] to model the elastic-plastic matrix and elastic fiber based on the constituent properties. Either the Ramberg-Osgood equation [28] or a piecewise linear approximation may be used to model the nonlinear stress-strain curve of the matrix. The fiber is modeled as a linear-elastic material. The PAFAC program predicts fiber and matrix stresses at each element centroid. Using this material model, the analysis calculates the fiber and laminate stresses and predicts when yielding occurs in each element of the finite element mesh. The vanishing fiber diameter material model is described in further detail in Reference [27].

The program PAFAC was used to predict the laminate behavior and first fiber failure. PAFAC predicts first fiber failure to occur when the fiber stresses within an element exceed a given criterion. For the analysis of the B/Al composites [3], a fiber failure criterion based on the axial and shear stress in the fiber was used. Due to the limited computer resources available at the time, the elements at the notch tip were approximately three fibers wide in the FE meshes used to analyze the B/Al laminates. Thus, "first fiber failure" actually indicated failure of all three fibers within an element in the B/Al analyses. In later work [6,7], meshes were refined so that the smallest elements, located next to the notch or hole, were one fiber spacing wide. The appropriate fiber spacing was calculated using the composite fiber volume fraction, the fiber diameter, and the ply thickness. For the analyses of titanium MMC [6,7], a fiber failure criterion based on the axial stress in the fiber was used.

Tests of unnotched SCS-6/Ti-15-3 laminates [1] indicated interfacial debonding in the 90° plies at applied stresses that were as low as 20% of the yield strength of the matrix. To account for the debonding, the PAFAC analysis was modified to include a failure criterion to approximate interfacial debonding in the 90° plies. This interfacial debonding was also found early in the notched tests by Rousch and Mall [12] as previously discussed. When the transverse stress in the elements in the 90° plies reached a specified critical value, the material properties of the 90° plies were modified to represent a ply with a completely debonded interface. For the SCS-6/Ti-15-3 material, the critical transverse stress was chosen to be 155 MPa based on the observed stress-strain behavior of unnotched $[90]_8$ laminates [1].

The PAFAC analysis was also modified to account for interface debonding in the $0°$ plies at the notch tip in a specimen with a sharp notch [6]. The FE meshes were designed to contain one layer of elements per ply where the elements at the notch tip were one fiber spacing wide. The elements next to the notch tip, which were one fiber spacing wide, were modified to include isotropic layers next to the notch and between each ply. To model the fiber-matrix debonding of the $0°$ fiber next to the notch, the elastic modulus of the additional isotropic elements in both $0°$ plies was reduced. The remaining elements in the $0°$ plies were modeled as composite elements with appropriately higher fiber volume fractions. Various debond lengths of the $0°$ fiber were modeled by reducing the modulus of increasing numbers of elements. Further details on modeling the $0°$ fiber debonding are given in Reference [6].

Discrete Fiber-Matrix Model

A discrete fiber-matrix (DFM) model (often referred to as a unit cell model) assuming an infinitely repeating, rectangular array of fibers was used to analyze a single fiber and the surrounding matrix. The MSC/NASTRAN finite element code [29] was used to analyze the DFM model using three-dimensional, eight-noded, hexahedral elements. A piecewise linear approximation of the matrix stress-strain curve was used to model the nonlinear behavior of the SCS-6/Ti-15-3 matrix, and the fiber was modeled as a linear-elastic material. The ply thickness, the fiber volume fraction, and the fiber diameter were used to calculate the dimensions of the DFM model. Compatibility with adjacent unit cells was enforced by constraining the displacements normal to each face to be equal.

Macro-Micromechanical Analysis (MMA)

The third analytical technique, the macro-micromechanical analysis (MMA), was used to analyze notch-tip stress states in notched configurations [23]. The macro-micromechanical analysis combines the 3D homogeneous, orthotropic finite element analysis (PAFAC) of the notched specimen and the discrete fiber-matrix (DFM) micromechanics model of a single fiber. The MMA was used to calculate the stresses in the notch-tip element in the interior $0°$ ply of $[0/90]_{2s}$ laminates assuming a perfectly bonded fiber-matrix interface. The interior $0°$ ply was the location of the highest fiber axial stress predicted by the PAFAC analyses of the specimens. As mentioned, the PAFAC finite element mesh was designed so that the dimensions of the elements next to the notch corresponded to a single fiber spacing. A schematic view of the macro-micro interface for a notched specimen is shown in Figure 16. Displacement boundary conditions from the PAFAC analysis were applied to the DFM mesh to simulate the stress state next to the notch. Further details on the determination of the MMA boundary conditions are presented in Reference [23]. The MMA analysis assumes an undamaged composite.

Boron/Aluminum Predictions

A wide range of laminates containing both circular holes and sharp notches were tested statically until failure [3]. It was found that, due to widespread yielding of the matrix, specimens with the same notch-length-to-specimen-width ratios $(2a/W)$ failed at similar stress levels. This wide scale yielding in the B/Al is very representative of the yielding that may occur in TMCs at temperatures around 650°C and thus this section is included.

PAFAC was used to characterize the elastic-plastic behavior of B/Al laminates [3]. The stress-strain curve of the aluminum was modeled with a Ramberg-Osgood equation [28]. The program predicted the stress-strain response for a variety of laminates containing center holes and sharp notches and predicted laminate failure based on the fiber stress in the element next to the notch. The yield pattern of the aluminum matrix material was predicted by the PAFAC program. Because PAFAC is a 3D analysis, yield patterns were predicted for each ply in a multiple-ply laminate. Figure 17 shows the yield pattern for a B/Al $[0/\pm45]_s$ laminate with a 25.4-mm-diameter center hole, $2a/W = 0.25$, and a fiber volume fraction of 45%. The sketches show the extent of the yielding in the 0 and 45° plies as the applied stress is increased. The yielding in the −45° was very similar to the pattern shown for the 45° layer. The ±45° layers yielded much earlier in the loading history than the 0° layer. Although not shown, the entire specimen (all plies) had yielded at 150 MPa, 56% of the failure stress. As is typical with a low yield strength matrix, widespread yielding occurred before fiber or specimen failure. In unidirectional B/Al laminates containing circular holes or sharp notches, progressive fiber failures were observed and, thus, first fiber failure did not correspond to laminate failure. For unidirectional specimens, PAFAC predictions of first fiber failure corresponded to the experimental observations of first fiber failure. In a B/Al specimen containing cross-plies ($[\pm45]_s$, $[0/\pm45]_s$, $[0_2/\pm45]_s$), the observed first fiber failure corresponded with

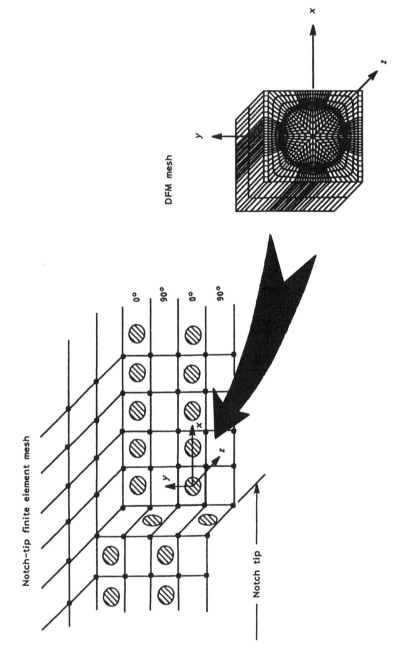

Figure 16. Schematic of MMA interface for DEN specimen for a $[0/90]_{2s}$ notched specimen.

386

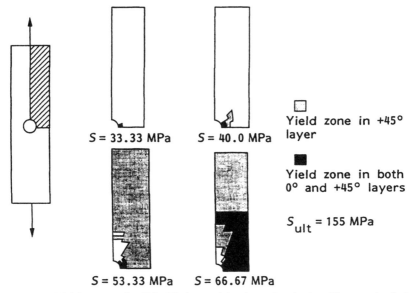

Figure 17. *Yield pattern for [0/±45]ₛ boron/aluminum laminate with a center hole.* 2a/W = 0.25, Vf = 0.45.

laminate failure. Excellent correlation was found between predicted first fiber failure and specimen failure for B/Al laminates containing cross-plies with either circular holes or sharp notches [3]. A typical result is shown in Figure 18, where the applied laminate stress is plotted against the specimen overall strain for a [0/±45]ₛ laminate containing a center hole. The solid line represents the predicted response up to first fiber failure, and the symbols represent the experimental results.

Silicon-Carbide/Titanium Predictions

STATIC STRENGTH

Two specimen configurations of a [0/90]₂ₛ SCS-6/Ti-15-3 material were tested: a center hole (CH) specimen and a double-edge-notched (DEN) specimen. The laminate had a fiber volume fraction v_f of 39%. The two specimen configurations failed at similar stress levels in spite of large differences in their stress concentration factors. The elastic stress concentrations K_T are 3.7 for the CH specimen and 5.7 for the DEN specimen [6]. The static strength of the DEN specimen was 520 MPa; for the CH specimen the static strength was 501 MPa.

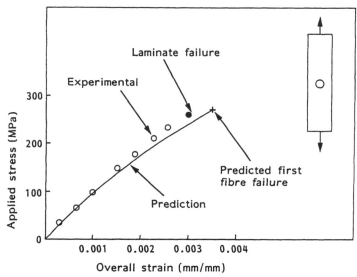

Figure 18. Stress-strain behavior of $[0/\pm45]_s$ boron/aluminum laminate with a center hole. $2a/W = 0.25$, $V_f = 0.45$.

These strengths were unexpectedly close given the difference in the K_T. Fractographs of the failure surfaces next to the notch for both specimen configurations showed a fiber-matrix debond length in the notch-tip 0° fiber of three to four fiber diameters in the DEN specimen, whereas significant fiber-matrix debonding in the 0° ply was not present in the CH configuration [6].

The MMA was used to analyze the stresses in the notch-tip element in the interior 0° ply in both the DEN and CH specimens to determine the fiber-matrix interface stress state. For a unit applied stress ($S = 1$ MPa), the MMA predicted the σ_{rr} stresses shown in Figure 19 for the DEN and CH specimens. The stresses presented are the stresses in the matrix at the fiber-matrix interface calculated at the finite element nodal points. For comparison, the matrix stresses in the interior 0° ply in an unnotched $[0/90]_{2s}$ specimen due to a unit applied stress are also shown. The stresses are presented with respect to the cylindrical coordinate system shown in the figure. Stresses are shown for the plane of symmetry on the xz-plane, i.e. through the center line of the notch or hole. For the two notched configurations, $\theta = 180°$ is the side of the fiber next to the notch.

For interfacial failure, the stress component of primary concern is the radial stress. The peak values of the radial matrix stresses due to

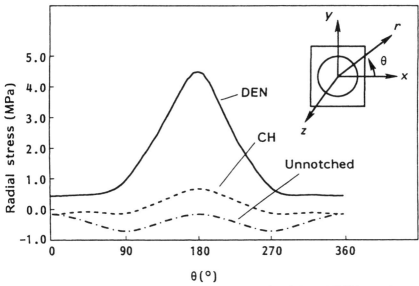

Figure 19. Radial component of interface stress for CH and DEN specimens. SCS-6/Ti-15-3, [0/90]$_{2s}$, V_f = 0.39, 2a/W = 0.33, S = 1 MPa, z = 0.00 plane.

a remote stress of 1.0 MPa are 4.5 and 0.67 MPa for the DEN and CH specimens, respectively. The peak value of the radial stress for the un-notched laminate is –0.17 MPa. The radial stresses for the DEN con-figuration are tensile for all values of θ, whereas for the CH configu-rations, the radial stresses are tensile only from approximately 110 to 250°, and for the unnotched laminate, the radial stresses are com-pressive for all values of θ. Thus, for a given interfacial strength, the interface in the DEN specimen will debond much earlier in the load-ing history than in the CH specimen. For a given load, the 0° fibers next to the notch in the DEN specimen are more likely to have debonded than the 0° fibers next to the hole in the CH specimen.

The PAFAC analysis was used to determine the effect of debonding in the 0° plies on the notch-tip fiber stress concentrations with debonded 90° fiber-matrix interfaces [6]. Figure 20 shows the predic-tions of the 0° fiber stress in the first element next to the notch as a function of applied stress for the DEN and CH specimens. The hori-zontal dashed line indicates the assumed fiber strength of 4200 MPa, and the two vertical dash-dotted lines show the experimental strengths of the two specimens. The fiber strength was calculated from the strain to failure of an unnotched [0/90]$_{2s}$ coupon (e_{ult} = 0.0105

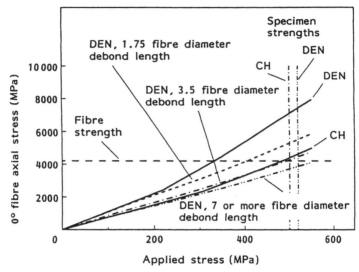

Figure 20. 0° fiber axial stress for CH and DEN specimens with notch tip 0° debonding and debonding of 90° plies. SCS-6/Ti-15-3, [0/90]$_{2s}$, V$_f$ = 0.39, 2a/W = 0.33.

mm/mm) multiplied by the fiber modulus (400 GPa). The solid lines indicate the predicted 0° fiber stress with no debonding in the 0° plies for the DEN and CH specimens. When the axial stress in the first 0° fiber was used as a failure criteria, the analysis predicted the strength of the CH specimen to be 490 MPa, within 2% of the observed strength. However, the strength of the DEN specimen was predicted to 320 MPa, significantly lower than the observed strength of 520 MPa. Debonding at the notch tip in the DEN specimen was simulated as described above. The effect of modeling the 0° debonding in the DEN specimen is also shown in Figure 20. Modeling a 0° debond length of 1.75 fiber diameters (dashed line) reduced the fiber stress in the DEN specimen considerably, increasing the predicted strength to 400 MPa. Modeling a 0° debond length of 3.5 fiber diameters caused the 0° fiber stress to drop nearly to the level of the CH configuration. This debond length is in good agreement with the micrographs of the fracture surface [6] and the predicted strength of 500 MPa is in good agreement with the observed strength of 520 MPa. Modeling a 0° debond length of seven or more fiber diameters, the 0° fiber stress in the notch-tip element in DEN specimen was reduced to a level below that of the CH configuration. These results indicate that, as with the cross-ply B/Al laminates, a first fiber failure criteria can accurately predict notched strengths when fracture is sudden and catastrophic with no progressive fiber failures.

POSTFATIGUE RESIDUAL NOTCH STRENGTH

The development of fatigue damage in SCS-6/Ti-15-3 laminates containing center holes was also investigated [7]. Although the same composite material was used, the laminate used in the fatigue investigations was slightly different than that used for the static loading previously described. For the fatigue loading studies, a $[0/90]_s$ four-ply laminate with a fiber volume fraction of 35.5% was used. Damage progression was monitored at various stages of fatigue loading. In general, fatigue damage consisted of fiber-matrix debonding in the 90° plies with matrix cracks extending from the debonded interfaces. The loading modulus of the composite was reduced by 38% due to the matrix cracking. The matrix cracks observed on the surface were bridged by 0° fibers. No fiber failures were observed due to the fatigue loading.

The postfatigue residual strength was predicted using the PAFAC program. The 0° fiber axial stress next to the hole was predicted for the undamaged virgin and the postfatigue damaged conditions as a function of applied load. As before, the laminate strength was assumed to be the applied load at which the axial stress in the 0° fiber next to the hole reached the fiber strength.

The through-the-thickness matrix cracking was modeled in the PAFAC analysis by reducing the matrix modulus by 69%. This reduction in the matrix modulus corresponded to the measured 38% reduction in the composite longitudinal modulus [7]. The reduced matrix modulus was used only for the postfatigue predictions. As previously mentioned, debonding of the fiber-matrix interfaces in the 90° plies in this composite occurs at relatively low-load levels. The debonding of the 90° plies was included in the predictions for the virgin condition by modifying the material properties of the 90° plies as described above.

The postfatigue residual strength was predicted for a specimen that had been subjected to 200,000 cycles at a load level of 250 MPa. No fiber failures were observed, and the matrix cracking had reached a saturated state. Small regions located directly above and below the center hole had developed no matrix cracks. The matrix modulus of the elements in regions where cracking was seen was reduced. In the regions where no matrix cracks were seen, the material properties were not changed.

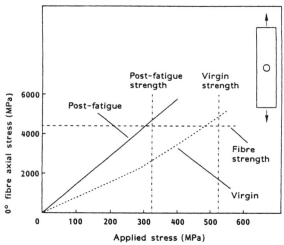

Figure 21. 0° fiber axial stress for post fatigue and virgin conditions. SCS-6/Ti-15-3, [0/90]$_{2s}$, V$_f$ = 0.39, 2a/W = 0.33.

The axial stress in the 0° fiber next to the center hole as a function of the applied stress is shown in Figure 21. The horizontal line indicates the assumed fiber strength of 4400 MPa, and the two vertical lines show the experimental strengths for the two conditions. The fiber strength was calculated from the measured strain to failure of an unnotched [0/90]$_s$ specimen (e_{ult} = 0.0011) multiplied by the fiber modulus (400 GPa). (The fiber strength determined here is slightly different from that determined previously because the laminate used in the fatigue loading was not identical to that used in the static loadings.) As shown in the figure, the applied stress at which the 0° fiber stress equals the fiber strength corresponds closely to the measured strength for both conditions. Using the axial stress in the 0° fiber next to the hole as a failure criterion for laminate failure, both predictions are within 8% of the experimental values. Again, the first fiber failure criterion accurately predicted the specimen strength when failure was sudden and catastrophic and no progressive fiber failures were observed.

Thermal Residual Stresses

The calculations presented above did not explicitly include any thermal residual stresses. Thermal residual stresses were not found to be significant in the B/Al laminates [3]; however, they had a significant effect on the behavior of the SCS-6/Ti-15-3 laminates [1]. The fiber strengths used in the SCS-6/Ti-15-3 predictions were determined from unnotched SCS-6/Ti-15-3 laminates that were processed the same as the notched specimens, and, therefore, the thermal residual stresses were the same in each laminate tested. The thermal residual stresses were implicitly included in the strength predictions because the fiber strengths were determined, including the thermal residual stresses.

CONCLUDING REMARKS

A review of the notch strength of titanium matrix composites has been presented. Experimental studies have shown that the fiber-matrix interface and the matrix modulus (and thus temperature) play a key role in the determination of notch strength. The effects of compressive loading was assessed, revealing that the TMCs reinforced with the SCS-6 fibers had superior strain to failures in compression compared with tensile tests. In general the TMCs were observed to be notch sensitive in terms of notch width to specimen width. However, the TMC notch strength seemed somewhat insensitive to the notch radii.

Two analytical approaches to notch strength predictions were reviewed in some detail. One approach was based more on classical closed-form stress analysis, whereas the second relied more on detailed finite element analysis. The predictive results for either approach was quite good. The approach one would use depends on the level of detail required and the amount of information available.

All in all, as with most composite systems, failure is quite complex, but the reviewed approaches do at least an adequate job of estimating the notch strength for simple notch geometries and loading conditions. Traditional strength prediction approaches do not lend themselves to MMC when highly nonlinear behavior is observed or when fiber-matrix debonding or matrix cracking proceeds laminate failure. The micromechanics-based strength prediction methodology presented here offers a more direct approach by modeling behavior and damage on a constituent level, thus explicitly modeling matrix nonlinearity, fiber-matrix interface debonding, and matrix cracking. Including matrix plasticity in any analysis is not trivial. Predicting interfacial stresses and modeling interfacial failures necessitates great analytical efforts. However, because matrix plasticity, matrix cracking, and fiber-matrix debonding can play an important role in the behavior of TMC, this analytical effort is necessary to advance the understanding and prediction of damage initiation and development in TMC.

REFERENCES

1. Johnson, W. S.; Lubowinski, S. J.; and Highsmith, A. L.: Mechanical Characterization of Unnotched SCS-6/Ti-15-3 Metal Matrix Composites at Room Temperature. *Thermal and Mechanical Behavior of Metal Matrix and Ceramic Matrix Composites, ASTM STP 1080,* J. M. Kennedy, H. H. Moeller, and W. S. Johnson, Eds., American Society for Testing and Materials, Philadelphia, PA, 1990, pp. 193–218.

2. Naik, R. A.; Pollock, W. D.; and Johnson, W. S.: Effect of a High Temperature Cycle on the Mechanical Properties of Silicon Carbide/Titanium Metal Matrix Composites. *Journal of Materials Science,* Vol. 26, 1991, pp. 2913–2920.

3. Johnson, W. S.; Bigelow, C. A.; and Bahei-El-Din, Y. A.: *Experimental and Analytical Investigation of the Fracture Processes of Boron/Aluminum Laminates Containing Notches.* NASA TP-2187, National Aeronautics and Space Administration, Washington, DC, 1983.

4. Daniel, I. M.: Failure Mechanisms and Fracture of Composite Laminates with Stress Concentrations. *Proceedings of the VIIth International Conference on Experimental Stress Analysis,* Haifa, Israel, August 23–27, 1982, pp. 1–20.

5. Harris, C. E.; and Morris, D. H.: A Fractographic Investigation of the Influence of Stacking Sequence on the Strength of Notched Laminated Composites. *Fractography of Modern Engineering Materials: Composites and Metals, ASTM STP 948,* J. E. Masters and J. J. Au, Eds., American Society for Testing and Materials, Philadelphia, PA, 1987, pp. 131–153.

6. Bigelow, C. A.; and Johnson, W. S.: Effect of Fiber-Matrix Debonding on Notched Strength of Titanium Metal Matrix Composites. *23rd National Symposium on Fracture Mechanics, ASTM STP 1189,* R. Chona, Ed., American Society for Testing and Materials, Philadelphia, PA, 1993 (also NASA TM-104131, August 1991).

7. Bakuckas, J. G., Jr.; Johnson, W. S.; and Bigelow, C. A.: Fatigue Damage in Cross-Ply Titanium Metal Matrix Composites Containing Center Holes. Submitted to *Journal of Engineering and Materials Technology,* Nov. 1991 (also NASA TM-104197, March 1992).

8. Waddoups, M. E.; Eisenmann, J. R.; and Kaminski, B. E.: *Macroscopic Fracture Mechanics of Advanced Composite Materials,* 1971, pp. 446–454.

9. Whitney, J. M.; and Nuismer, R. J.: Stress Fracture Criteria for Laminated Composites Containing Stress Concentrations. *Journal of Composite Materials,* Vol. 8, 1974, pp. 253–265.

10. Nuismer, R. J.; and Whitney, J. M.: Uniaxial Failure of Composite Laminates Containing Stress Concentrations. *Fracture Mechanics of Compos-*

ites, ASTM STP 593, American Society of Testing and Materials, Philadelphia, PA, 1975, pp. 117–142.

11. Awerbuch, J.; and Madhukar, M. S.: Notched Strength of Composite Laminates: Predictions and Experiments—A Review. *Journal of Reinforced Plastics,* Vol. 4, 1985, pp. 3–159.

12. Roush, J. T.; and Mall, S., Fracture Behavior of a Fiber-Reinforced Titanium Matrix Composites with Open and Filled Holes at Room and Elevated Temperatures. *Journal of Composites Technology and Research,* Vol. 16, No. 3, July 1994, pp. 201–213.

13. Rattray, J.; and Mall, S.: Tensile Fracture Behavior of Notched Fiber Reinforced Titanium Metal Matrix Composite, *Composite Structure,* Vol. 28, 1994, pp. 471–479.

14. Gooder, J. L.; and Mall, S.: Compressive Behavior of a Titanium Metal Matrix Composite with a Hole. *Composite Structures,* Vol. 31, pp. 315–324, 1995.

15. Jones, Robert M.: *Mechanics of Composite Materials,* McGraw Hill Book Company, New York, 1975.

16. Inglis, C. E.: Stresses in a Plate due to the Presence of Cracks and Sharpe Corners. *Transactions of the Institute of Naval Architects,* Vol. 55, Part 1, 1913.

17. Howland, R. C. J.: "On the Stresses in the Neighborhood of a Circular Hole in a Strip under Tension," *Philosophical Transactions of the Royal Society (London),* Vol. 229A, p. 49, 1930.

18. Peterson, R. E.: *Stress Concentration Factors in Design,* John Wiley & Sons, Inc., New York, 1953.

19. Finefield, M. A.; Harmon, D. M.; Saff, C. R.; and Harter, J. A.: "Thermomechanical Load History Effects in Metal Matrix Composites," U.S. Air Force Technical Report WL-TR-94-3015, January 1994.

20. Lekhnitskii, S. G.: *Anisotropic Plates,* Gordon and Breach Science Publishers, 1968.

21. Harmon, D. M.; Saff, C. R.; and Sun, C. T.: "Durability of Continuous Fiber Reinforced Metal Matrix Composites," U.S. Air Force Technical Report AFWAL-TR-87-3060, McDonnell Douglas, 1987.

22. Bigelow, C. A.; and Bahei-El-Din, Y. A.: Plastic and Failure Analysis of Composites (PAFAC). LAR-13183, COSMIC, University of Georgia, 1983.

23. Bigelow, C. A.; and Naik, R. A.: A Macro-Micromechanical Analysis of a Notched Metal Matrix Composite. *Composite Materials, Testing and Design, ASTM STP 1120,* G. Grimes, Ed., American Society for Testing and Materials, Philadelphia, PA, 1992, pp. 222–233.

24. Bahei-El-Din, Y. A.: Plastic Analysis of Metal-Matrix Composite Laminates. Ph.D. Dissertation, Duke University, 1979.

25. Bahei-El-Din, Y. A.; Dvorak, G. J.; and Utku, S.: Finite Element Analysis of Elastic-Plastic Fibrous Composite Structures. *Computers and Structures,* Vol. 13, No. 1–3, June 1981, pp. 321–330.

26. Bahei-El-Din, Y. A.; and Dvorak, G. J.: A Review of Plasticity Theory of Fibrous Composite Materials. *Metal Matrix: Composites: Testing, Analy-*

sis, and Failure Modes, ASTM STP 1032, W. S. Johnson, Ed., American Society for Testing and Materials, Philadelphia, PA, 1989, pp. 103–129.

27. Dvorak, G. J.; and Bahei-El-Din, Y. A.: Plasticity Analysis of Fibrous Composites. *Journal of Applied Mechanics,* Vol. 49, 1982, pp. 327–335.

28. Ramberg, W.; and Osgood, W. R.: Description of Stress-Strain Curves by Three Parameters. NACA TN 902, 1943.

29. MSC/NASTRAN, Version 6.6, MacNeal-Schwendler Corporation, 1991.

30. Johnson, W. S.: "Fatigue of Continuous Fiber Reinforced Titanium Metal Matrix Composites," *Mechanical Fatigue of Advanced Materials,* Ritchie, Cox, and Dauskardt, eds., MCEP Publishers, 1991.

10

Micromechanical Analysis and Modeling

D. ROBERTSON AND S. MALL

Department of Aeronautics and Astronautes
Air Force Institute of Technology
Wright-Patterson Air Force Base, OH 45433

INTRODUCTION—UNIQUE FEATURES OF TMC MODELING

THE DIFFICULTIES ENCOUNTERED in attempting to model titanium matrix composites (TMCs) is directly related to the application environment. As a high-temperature material for aerospace applications, TMCs are expected to experience extreme thermal and mechanical loads, such as is found in propulsion systems and hypersonic airframe components. Also, the loading in these extreme environments is, in general, highly cyclic both mechanically and thermally. Therefore, fatigue failure is of primary importance. Further complicating the problem are the properties and behavior of the constituents. For instance, the mismatch of the coefficients of thermal expansion of the fiber and matrix, combined with high processing and usage temperatures, can produce significant thermally induced stresses. Temperature-dependent characteristics of the constituents also become important due to the large temperature range, and time-dependent viscoplastic deformation of the matrix material is likely to occur in the combined high-temperature, high-stress environment. Additionally, the fiber-matrix interface for such materials can be characterized as relatively weak, so interfacial failure and subsequent separation and slip between the constituents occur at low applied stresses.

All of these characteristics are difficult to model under monotonically applied loads, but exacerbating the problem further is the fact

397

that any model must be able to perform a thermomechanical fatigue analysis through several hundred or even thousands of cycles. The reason this is necessary is because many of the nonlinear mechanisms will not stabilize until numerous load cycles have been accomplished. The constituent microstresses throughout this process may change significantly as the load is redistributed between the various plies and then subsequently between the fiber and matrix, but to predict the material's strength and/or fatigue life, good approximations of the constituent microstresses are required. Thus, it is important that these nonlinear mechanisms be properly accounted for in a micromechanics model.

Due to the highly nonlinear nature of the problem, even if an appropriate model is developed, convergence on a solution may be difficult to achieve. In fact, selection of the nonlinear material model for each constituent may have more impact on the solution than the choice of constitutive relations for controlling the interaction between the fiber and matrix (usually referred to as micromechanics models). For nonlinear problems, small variations in the input parameters can produce large variations in the solution. An example might be the permanent strain in an elastic-plastic material during monotonic loading where a small percent increase in the maximum stress will produce a large percent increase in the permanent strain. Therefore, care must be taken in choosing, developing, and solving the nonlinear material relations.

DISCRETE MICROMECHANICAL MODELS

In light of the many nonlinearities that occur for TMCs, as discussed above, obtaining a continuum-based micromechanical model that can be applied for general use (i.e., various layups, loading environments, etc.) is very difficult. Therefore, most of the analyses in the literature are discrete models that assume a given field variable has a known distribution over a specified region (usually assumed constant). This enables inelastic material models to be incorporated into the micromechanics in a straightforward manner. As expected, in most cases, these approaches can be related to the finite element method through an appropriate choice of elements, grid, and constraints. Also, many of these have been embedded into computer programs that are readily available. The main advantages of these computer programs over a traditional finite element program are the speed, ease-of-use, and not having to be concerned about gridding considerations.

Although many different discrete analyses have been used, this chapter will primarily limit the discussion to those that have been embedded into readily available computer programs and the finite element method. The micromechanics models that are available in computer programs consist of the multicell model, the vanishing fiber diameter model, the concentric cylinder approach, and the method of cells [1–4]. Brief overviews and comparisons of these micromechanic models were presented by Bigelow et al. [5] and Kroupa et al. [6]. Much analytical work has been performed with these techniques, and they have been found to be very useful in predicting the material behavior and examining the constituent microstresses. The following sections summarize the models by discussing their constitutive relations, nonlinear capabilities, results, and how they have been used to analyze metal matrix composites.

Multicell Model

The multicell model (MCM) proposed by Hopkins and Chamis [1,7] employs a representative volume element (RVE) that consists of a single fiber, interphase, and surrounding matrix. The response of the RVE is determined by partitioning it into various regions as shown in Figure 1. Equivalent rectangular areas for the fiber, matrix, and interphase are then used in a one-dimensional analysis to independently examine the composite behavior in each direction. For instance, solving for the transverse normal response of the composite may be visualized by replacing each rectangular area by an equivalent spring stiffness. Therefore, this approach is able to predict three-dimensional composite behavior through a one-dimensional analysis. These micromechanics equations form the basis of the METCAN (*MET*al matrix *C*omposite *AN*alyzer) analysis code [8].

As an example, the transverse normal modulus can be determined by partitioning the RVE into regions A, B, and C. Equivalent linear dimensions are first obtained for the analysis from the following equations:

$$a_f = D\sqrt{\frac{\pi}{4}} \qquad a_d = D_0\sqrt{\frac{\pi}{4}} \qquad a_l = D_0\sqrt{\frac{\pi}{4V_f}} \tag{1}$$

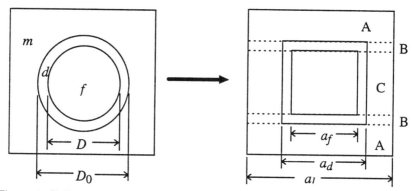

Figure 1. RVE for the multi-cell (METCAN) micromechanics model.

Then from equilibrium with the lamina stress in the direction perpendicular to the fiber:

$$\sigma_l a_l = \sigma_A(a_l - a_d) + \sigma_B(a_d - a_f) + \sigma_C a_f \qquad (2)$$

and upon differentiating with respect to the strain, which from continuity of displacement must be the same for the three regions produces:

$$E_l a_l = E_A (a_l - a_d) + E_B (a_d - a_f) + E_C a_f \qquad (3)$$

The moduli for the three regions, E_A, E_B, and E_C, are obtained by considering each region as a series of its respective components. For instance, E_A is simply the matrix modulus:

$$E_A = E_m \qquad (4)$$

The modulus of region B contains both matrix and interphase and from a simple elements in series analysis:

$$\frac{a_l}{E_B} = \frac{a_l - a_d}{E_m} + \frac{a_d}{E_d} \qquad (5)$$

Similarly, region C contains fiber, interphase, and matrix, so the modulus for region C is obtained from the equation:

$$\frac{a_l}{E_C} = \frac{a_l - a_d}{E_m} + \frac{a_d - a_f}{E_d} + \frac{a_f}{E_f} \tag{6}$$

This same procedure is followed for each property in each direction to obtain closed-form equations that relate the composite lamina properties to those of the constituents. Also, by adopting an iterative procedure, the lamina properties may be related to a composite laminate through the classical laminated plate theory.

The METCAN code employs the model as described above. However, further improvements to this model have examined the effects of increasing the subdivisions of the RVE [9]. Hence, the cylindrical shape of the fiber is more closely approximated by employing several slices through the fiber. Also, this improvement allows for a refined examination of the stress field variation within the RVE.

Unless the material is within the linear elastic regimen, the non-linear material models employed to account for plastic-viscoplastic deformation, interfacial debonding, and damage within the constituents play an equally important role in predicting the material behavior as the RVE micromechanics equations which relate the stress and strain of the constituents to that of the composite. The nonlinear material model within METCAN describes the nonlinear behavior of each constituent property by relating it to a product of terms with unknown exponents [7]. This technique assumes that a given property, P_M, is related to its reference value, P_{M_0}, at a given temperature, stress, and stress rate by:

$$\frac{P_M}{P_{M_0}} = \left[\frac{T_M - T}{T_M - T_0}\right]^n \left[\frac{S_F - \sigma}{S_F - \sigma_0}\right]^m \left[\frac{\dot{S}_F - \dot{\sigma}}{\dot{S}_F - \dot{\sigma}_0}\right]^l \tag{7}$$

where T_0, σ_0, and $\dot{\sigma}_0$ are the temperature, stress, and stress rate at which the reference value was measured. Also, T_M is the melting temperature, S_F is the fracture stress at T_0, and $\dot{S}_F$ is an appropriately chosen high-stress rate, such as the stress rate for penetration to occur during impact. The exponents in the above terms must be determined from appropriate experiments or estimated.

Additional modifications to the interaction relation of Equation (7) have been accomplished and incorporated into METCAN. These include appending further ratios on the right-hand side to account for effects, such as mechanical and thermal cyclic loading, creep, and temperature rate effects [10].

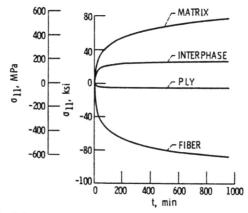

Figure 2. Residual thermal stresses after cool-down from fabrication in a [±45]ₛ
W-1.5ThO₂ fiber reinforced Fe-25Cr-4Al-1Y composite (normal stress in direction
of fiber) [11]

Some sample results of published METCAN solutions and comparisons with experiment are presented in Figures 2–6 [10–11]. A benefit of the one-dimensional simplistic approach within METCAN is the ability to calculate properties, such as the composite strength (Figure 4), which are normally difficult to obtain in a general three-dimensional analysis.

Also, the METCAN code is designed for use with structural analysis codes, such as NASTRAN, and will output information in formatted laminate material property cards that can be used directly with NASTRAN. In addition, METCAN calculates the thermal material characteristics for heat capacity and conductivity. Therefore, the general thermal and mechanical behavior of the material is characterized and available for use in a large-scale structural analysis code if desired.

Vanishing Fiber Diameter Model

The vanishing fiber diameter model represents a simplified but accurate approach for modeling the inelastic behavior of metal matrix composite layups. With this approach, the effects of multiaxial stress-strain relations and inelastic flow rules of the constituents can be examined on the composite behavior. The main assumption in the vanishing fiber diameter (VFD) model is that the fibers possess a vanishingly small diameter even though they occupy a finite volume

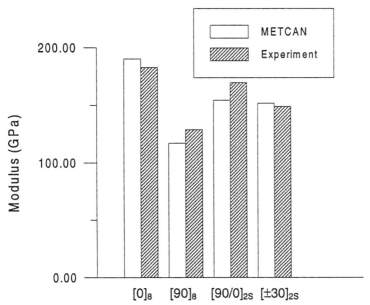

Figure 3. *Room temperature elastic moduli for SCS6/Ti-15-3 [10]*

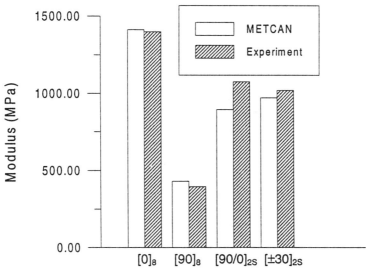

Figure 4. *Ultimate strength at room temperature for SCS6/Ti-15-3 [10]*

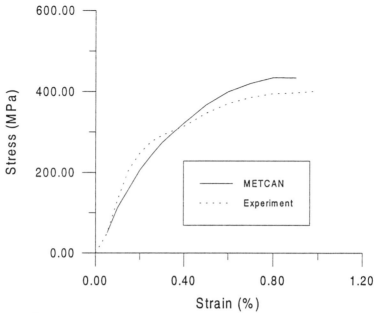

Figure 5. Stress-strain response at room temperature for a $[90]_8$ SCS6/Ti-15-3 composite [10]

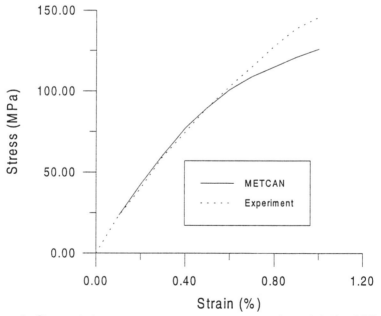

Figure 6. Stress-strain response at room temperature for a $[0/90]_{2s}$ SCS6/Ti-15-3 composite [10]

fraction of the composite [2]. Thus, the fibers do not interfere with matrix deformation in the transverse plane. The RVE for a unidirectional ply is pictorially represented in Figure 7 (axis 1 is aligned with the fiber). Therefore, the fiber constraint on the matrix is in the longitudinal (fiber axis) direction only and hence, the rule-of-mixtures constitutive relations may be applied:

$$\bar{\sigma}_{11} = V_f \sigma_{11_f} + V_m \sigma_{11_m} \tag{8}$$

$$\bar{\sigma}_{ij} = \sigma_{ij_f} = \sigma_{ij_m} \quad \text{for } ij \neq 11 \tag{9}$$

$$\bar{\varepsilon}_{11} = \varepsilon_{11_f} = \varepsilon_{11_m} \tag{10}$$

$$\bar{\varepsilon}_{ij} = V_f \varepsilon_{ij_f} + V_m \varepsilon_{ij_m} \quad \text{for } ij \neq 11 \tag{11}$$

This model was introduced to approximate the inelastic behavior of metal matrix composites by Dvorak and co-workers [2,12]. The fiber material is assumed to be linear elastic, although the properties may be a function of temperature. Therefore, all inelastic strains in a given ply are controlled by the matrix material inelastic deformation. Also, due to the fact that there is no local constraint in the transverse direction (perpendicular to the fiber), constituent microstresses in the transverse direction will not develop unless the ply is loaded trans-

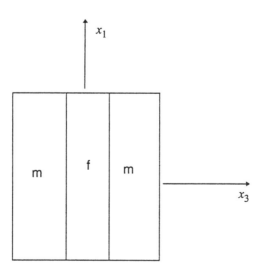

Figure 7. Representative volume element (RVE) for the vanishing fiber diameter model (note: the same RVE is used for the 1-2 plane as well) [5]

versely as indicated by Equation (9). This makes modeling interface conditions other than a perfect bond difficult. However, the constitutive relations of Equation (8–11) are amenable for employing a three-dimensional inelasticity model for the matrix material. Both elastic-plastic and elastic-viscoplastic models have been used in conjunction with the vanishing fiber diameter model.

A Von Mises yield criterion with Ziegler's modification of Prager's hardening rule has been employed for elastic-plastic analysis within the VFD micromechanics equations [2]. The yield surface for the matrix may be represented by:

$$f(\overline{\sigma}_m - \overline{\alpha}_m) = (\overline{\sigma}_m - \overline{\alpha}_m)^T C(\overline{\sigma}_m - \overline{\alpha}_m) - Y^2 = 0 \tag{12}$$

where $\overline{\sigma}$ is the engineering stress, $\overline{\alpha}$ is the yield surface origin, and C is a 6×6 symmetric matrix where $C_{11} = C_{22} = C_{33} = 1$, $C_{44} = C_{55} = C_{66} = 3$, $C_{12} = C_{13} = C_{23} = -1/2$, and all other $C_{ij} = 0$. The symbol, Y, represents the yield stress.

The hardening rule is:

$$d\overline{\alpha}_m = d\mu_m (\overline{\sigma}_m - \overline{\alpha}_m) \tag{13}$$

where $d\mu_m$ is an incremental scalar multiplier. From the condition $df = 0$, the incremental change in the scalar multiplier is found to be:

$$d\mu_m = \frac{1}{Y^2} (\overline{\sigma}_m - \overline{\alpha}_m)^T C d\overline{\sigma}_m \tag{14}$$

In addition, invoking the normality condition for the inelastic strain increment, the flow rule is:

$$d\overline{\varepsilon}^p = d\lambda \frac{\partial f}{\partial \overline{\sigma}} \tag{15}$$

where $d\lambda$ is an incremental plastic multiplier.

By combining the above yield, hardening conditions, and flow rule for the matrix with the constitutive relations for the fiber, matrix, and composite [Equations (8–11)], the composite yield surface and hardening rule are related directly to the plasticity equations for the matrix. These equations have been embedded into a two-dimensional laminate computer program, AGLPLY [2,13,14].

Equations (12–15) are independent of time. Therefore, to account for creep and rate effects, a viscoplastic material model for the matrix was adopted for use in the vanishing fiber diameter micromechanical model. To this effect, a thermoviscoplastic model similar to that of Eisenberg and Yen was employed [12,15]. In this approach, the total strain is partitioned into its elastic, thermal, and inelastic components where inelastic deformation develops only when the state of stress lies outside an equilibrium yield surface. A Mises-type yield surface is assumed, so inelastic deformation occurs when:

$$f(s_{ij} - a_{ij}) = \frac{3}{2}(s_{ij} - a_{ij})(s_{ij} - a_{ij}) - (Y + Q)^2 > 0 \qquad (16)$$

where s_{ij} is the deviatoric stress, a_{ij} is the origin of the yield surface, Y is the yield stress, and Q is an isotropic hardening function. Also, in this case, the more general tensorial representation is employed for the multiaxial viscoplastic theory.

An effective overstress, R, is required in this viscoplastic theory to calculate the extent of inelastic deformation. Overstress is defined as the distance in stress space between the actual stress point, s_{ij}, and the equilibrium (quasi-static) stress point, s_{ij}^*, which lies on the current equilibrium yield surface. Thus:

$$R = \begin{cases} \sqrt{\frac{3}{2}\left(s_{ij} - s_{ij}^*\right)\left(s_{ij} - s_{ij}^*\right)} & \text{for} \quad f(s_{ij} - a_{ij}) > 0 \\ 0 & \text{for} \quad f(s_{ij} - a_{ij}) \le 0 \end{cases} \qquad (17)$$

where the equilibrium stress point is determined from:

$$s_{ij}^* = \frac{\sqrt{2}(Y + Q)(s_{ij} - a_{ij})}{\sqrt{3(s_{kl} - a_{kl})(s_{kl} - a_{kl})}} + a_{ij} \qquad (18)$$

The inelastic strain rate, $\dot{\varepsilon}_{ij}^I$, is determined by assuming it is directed normal to the equilibrium yield surface with a magnitude that is controlled by a power law of the overstress:

$$\dot{\varepsilon}_{ij}^I = \sqrt{\frac{3}{2}} kR^p n_{ij} \qquad (19)$$

where k and p are temperature-dependent material parameters and n_{ij} is the unit normal from the yield surface:

$$n_{ij} = \sqrt{\frac{3}{2}} \frac{(s_{ij}^* - a_{ij})}{(Y + Q)} \qquad (20)$$

The evolution equations for the hardening variables Q and a_{ij} control the isotropic and kinematic hardening characteristics of the material, respectively. The evolution equation for Q (i.e., isotropic hardening) is [12]:

$$\dot{Q} = q(Q_a - Q)\sqrt{\frac{2}{3}\dot{\varepsilon}_{kl}^I \dot{\varepsilon}_{kl}^I} - b_r|Q - Q_r|^{(n_r - 1)}(Q - Q_r) \qquad (21)$$

where q, Q_a, b_r, Q_r, and n_r are temperature-dependent material parameters.

The kinematic hardening behavior is controlled by the following evolution equation for a_{ij}:

$$\dot{a}_{ij} = \dot{\mu}v_{ij} - c_r(a_{kl}a_{kl})^{\frac{1}{2}(m_r - 1)}a_{ij} \qquad (22)$$

where c_r and m_r are temperature-dependent material parameters and v_{ij} is a unit tensor defined as:

$$v_{ij} = \begin{cases} \dot{s}_{ij} / \sqrt{\dot{s}_{kl}\dot{s}_{kl}} & \text{for } \dot{s}_{ij} \neq 0 \\ n_{ij} & \text{for } \dot{s}_{ij} = 0 \end{cases} \qquad (23)$$

The multiplier $\dot{\mu}$ can be found to be:

$$\dot{\mu} = \sqrt{\frac{2}{3}}\frac{kR^p}{n_{kl}v_{kl}}[H - q(Q_a - Q)] \qquad (24)$$

where H is the tangent modulus of the stress-inelastic strain equilibrium curve at the present value of the effective inelastic strain. The evolution of this parameter is given by:

$$H = H_0 + h\left(\frac{\delta}{\delta_0 - \delta}\right) \qquad (25)$$

where

$$\delta = \sqrt{\frac{3}{2}\left(\bar{s}_{ij} - s_{ij}^*\right)\left(\bar{s}_{ij} - s_{ij}^*\right)} \qquad (26)$$

and $\bar{s}_{ij}$ is the bounding surface stress whose unit normal corresponds to the unit normal of the equilibrium yield surface stress. Thus, δ represents the perpendicular distance between the two surfaces. The bounding surface translates and expands in a similar fashion as the equilibrium yield surface, so the equations controlling the bounding surface are similar to those listed above for the equilibrium yield surface except with new material constants. The value, δ_0, is the distance between the yield surface and the bounding surface at the onset of inelastic deformation.

The above viscoplastic relations provide the inelastic portion of strain, and the thermoelastic stress-strain relations provide the thermal and elastic portions of the strain. Thus:

$$\varepsilon_{ij}^{E} = S_{ijkl}\sigma_{kl} \tag{27}$$

$$\varepsilon_{ij}^{th} = \alpha_{ij}\Delta T \tag{28}$$

where the superscripts, E and *th*, represent the elastic and thermal components, respectively, S_{ijkl} is the compliance, α_{ij} is the coefficient of thermal expansion, and ΔT is the change in temperature.

Equations (16–28), in conjunction with the constitutive relations for the VFD micromechanics model and laminate analysis, are combined in the VISCOPLY computer code. As can be seen in the previous equations, although the representative volume element and, therefore, the micromechanics constitutive relations are simpler in the VFD model than in the multicell model, the VFD model provides a more complex analysis due to the addition of the multiaxial nonlinear material relations as employed in the AGLPLY and VISCOPLY programs.

Figures 8 and 9 present comparisons between AGLPLY predictions and experiment as reported by Pollock and Johnson [16–17]. For the layups and elevated temperature considered, the perfect interfacial bond assumption of the VFD model does not adversely affect the result. Therefore, even though TMCs, in general, exhibit other than perfect bond behavior, if care is taken in the application, the VFD model can realistically imitate the composite behavior in most instances.

Furthermore, if the thermoviscoplastic material behavior is modeled with the VISCOPLY program, then the time-dependent nature of the matrix material may also be included in the composite analysis. Some examples of this are demonstrated in Figures 10 and 11 [18–20]. These figures present a much more complex analysis as both a thermomechanical fatigue (Figure 10) and a selected mission temperature load profile (Figure 11) are examined. The results are again very favorable and demonstrate the capability of VISCOPLY to model realistic load profiles.

To analyze general composite structures on a larger scale, the vanishing fiber diameter model has been further employed in a finite ele-

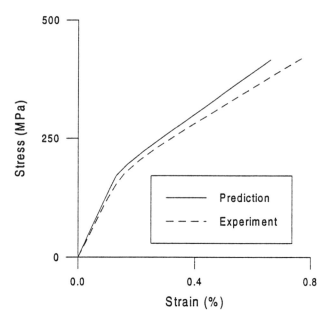

Figure 8. Stress-strain response at 650°C for a [0/±45/90]ₛ layup of the SCS6/Ti-15-3 system [17]

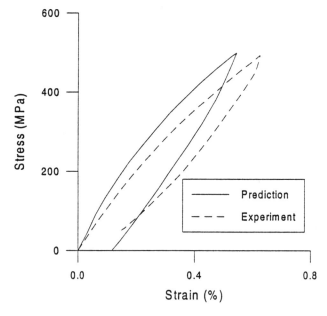

Figure 9. Load-unload response for a [0/90]₂ₛ layup at 650°C of the SCS6/ Ti-15-3 system [17]

410

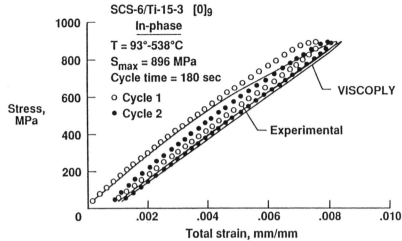

Figure 10. *VISCOPLY prediction of stress-response under in-phrase TMF for unidirectional SCS6/Ti-15-3 [19]*

ment format. To this end, eight-noded solid hexahedral elements are used in which each element represents a unidirectional composite material with arbitrarily oriented fibers. Therefore, any structural layup may be examined by stacking elements of appropriately oriented fibers together in the thickness direction and then creating a 2-D

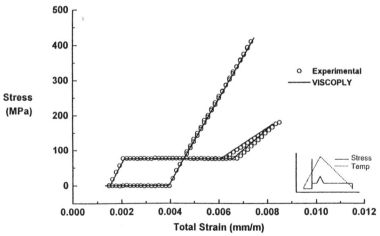

Figure 11. *VISCOPLY predictions versus experiment for cross-ply SCS6/ TIMETAL®21S undergoing a simplified mission profile [20]*

structural grid in the plane of the composite. PAFAC is the elastic-plastic version of this micromechanics and finite element method, whereas VISCOPAC is the thermoviscoplastic version [21,22]. PAFAC has been used to perform postfatigue analysis of cross-ply specimens with center holes [21]. Damage effects (i.e., microcracking) were accounted for by an equivalent reduction in the matrix modulus. The analysis examined the 0° fiber stress concentration next to the center hole, and when this stress exceeded the fiber strength, failure was assumed to occur.

Concentric Cylinder Model

The concentric cylinder approach has been used extensively for micromechanical modeling of composites for many years and is included in most engineering textbooks on composite materials [23]. Some of the main benefits of this approach are obtaining continuous variation of the stress in the radial direction and coupled three-dimensional stress fields while employing only a two-dimensional analysis. On the other hand, it is difficult to model applied transverse or shear loads with this technique. As a result, the concentric cylinder model (CCM) cannot perform general composite laminate analysis. However, the CCM approach can provide important information concerning the stress field in both the fiber and matrix and predict unidirectional behavior. Most other models assume uniform stress fields over relatively large regions, which results in the local stress information being lost. Therefore, a computer code, FIDEP (Finite Difference code for Elastic-Plastic analysis), has been developed, which employs the CCM approach and inelastic material models to perform inelastic analysis of composites [24]. The program supports both unidirectional and cross-ply layups undergoing uniaxial thermomechanical loading. To account for the cross-ply effects with the CCM technique, the 90° ply is modeled as a uniform material with a bilinear elastic-plastic stress-strain response that must be obtained from either another analysis method or experiments [25,26].

In the FIDEP program, the unidirectional composite and the 0° ply of the cross-ply composite are represented as concentric cylinders as shown in Figure 12 [24,27]. In addition, both the fiber and matrix could individually be partitioned into any number of concentric cylinders, thus making the incorporation of an interphase with unique properties from either the fiber or matrix relatively straightforward. Also, some assumptions are necessary for the analysis. A perfect bond is assumed to exist between the various constituents, and a state of

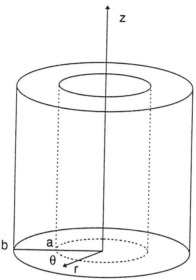

Figure 12. Representative volume for the concentric cylinder model [24,27]

generalized plane strain exists in the axial or fiber direction. Only axisymmetric loading and displacements are allowed on the model, and the constituents are isotropic.

The axisymmetric equilibrium and compatibility equations for the model are:

$$\frac{d\sigma_{rr}}{dr} + \frac{\sigma_{rr} - \sigma_{\theta\theta}}{r} = 0 \tag{29}$$

$$\frac{d\varepsilon_{\theta\theta}}{dr} + \frac{\varepsilon_{\theta\theta} - \varepsilon_{rr}}{r} = 0 \tag{30}$$

If the stress-strain relations are incorporated into Equation (30), then the compatibility relation becomes:

$$\frac{d}{dr}\left[\frac{\sigma_{\theta\theta}}{E} - \frac{\nu}{E}\left(\sigma_{rr} + \nu(\sigma_{\theta\theta} + \sigma_{rr}) - \alpha\,ET + E\left(\varepsilon_{zz} - \varepsilon_{zz}^I\right)\right) + \alpha T + \varepsilon_{\theta\theta}^I\right]$$

$$+ \frac{(1+\nu)}{Er}(\sigma_{\theta\theta} - \sigma_{rr}) + \frac{\varepsilon_{\theta\theta}^I - \varepsilon_{rr}^I}{r} = 0 \tag{31}$$

Equations (29) and (31) are two ordinary differential equations in σ_{rr} and $\sigma_{\theta\theta}$ (ε_{zz} is constant due to the generalized plane strain condition in this direction). These relations form the basis of the CCM model in the FIDEP computer code. The program uses a finite difference formulation to solve the two ordinary differential equations coupled with the thermoelastic stress-strain relations [Equations (27) and (28)] and the chosen inelastic material model.

FIDEP currently supports thermoelastic-plastic materials with strain hardening through a Von Mises yield criteria and Mendelson's modification of the Prandtl-Reuss flow rules [24]. It also supports thermoviscoplastic materials through the Bodner-Partom unified viscoplastic theory [26].

The strain hardening scheme for the thermoelastic-plastic material model in FIDEP requires the change in the inelastic strain for a single-load step increment to be controlled by:

$$\Delta\varepsilon_{ij}^{I} = \frac{1 - \dfrac{(\sigma_{eff})_{i-1}}{3G\varepsilon_{et}}}{1 + \dfrac{1}{3G}\left(\dfrac{d\sigma_{eff}}{d\varepsilon_{eff}^{I}}\right)_{i-1}} [\varepsilon_{ij} - (\varepsilon_{ij}^{I})_{i-1}] \qquad (32)$$

where G is the shear modulus, and the subscript $i-1$ denotes the value at the previous step. The effective stress, σ_{eff}, is given by:

$$\sigma_{eff} = \sqrt{\frac{2}{3}s_{ij}s_{ij}} \qquad (33)$$

where s_{ij} is the deviatoric stress. The effective elastic-thermal component of strain, ε_{et}, is similarly:

$$\varepsilon_{et} = \sqrt{\frac{2}{3}\left(\varepsilon_{ij} - (\varepsilon_{ij}^{I})_{i-1}\right)\left(\varepsilon_{ij} - (\varepsilon_{ij}^{I})_{i-1}\right)} \qquad (34)$$

Thus, if the slope of the effective stress-inelastic strain curve is known, then the three-dimensional hardening characteristics of the material are calculated by Equations (32–34). These calculations occur at each chosen integration point within the finite difference scheme of the FIDEP program. The above relations [Eqs (32–34)] assume isotropic hardening and no time dependence; as such, they are of limited usefulness in examining cyclic load effects. Therefore, the

Bodner-Partom viscoplastic theory has also been incorporated into the FIDEP program.

The Bodner-Partom theory [28] has been modified numerous times since its inception. The specific formulation used in the FIDEP program and one that has been highly successful at modeling titanium alloy neat matrix material is the Bodner-Partom model with directional hardening [29,30]. The inelastic strain rate with this model is given by:

$$\dot{\varepsilon}_{ij}^{I} = D_0 \exp\left[-\frac{1}{2}\left(\frac{Z^{I} + Z^{D}}{3J_2} \right)^{n} \right] \frac{s_{ij}}{\sqrt{J_2}} \tag{35}$$

where D_0 and n are material parameters. The state variables, Z^{I} and Z^{D}, sometimes referred to as the drag stress, control the isotropic and directional hardening characteristics, respectively.

The evolution equation for the isotropic hardening state variable is given by:

$$\dot{Z}^{I} = m_1 \dot{W}_p(Z_1 - Z^{I}) - A_1 Z_1 \left(\frac{Z^{I} - Z_2}{Z_1} \right)^{r_1}$$
$$+ \dot{T}\left[\left(\frac{Z^{I} - Z_2}{Z_1 - Z_2} \right) \frac{\partial Z_1}{\partial T} + (\frac{Z_1 - Z^{I}}{Z_1 - Z_2}) \frac{\partial Z_2}{\partial T} \right] \tag{36}$$

and the evolution of the directional hardening state variable is controlled by:

$$Z^{D} = \beta_{ij}\, \mu_{ij} \tag{37}$$

where the variables β_{ij} and u_{ij} are determined by the equations:

$$\dot{\beta}_{ij} = m_2 \dot{W}_p(Z_3\mu_{ij} - \beta_{ij}) - A_2 Z_1 \frac{\beta_{ij}}{\sqrt{\beta_{kl}\beta_{kl}}} \left(\frac{\sqrt{\beta_{kl}\beta_{kl}}}{Z_1} \right)^{r_2} + \dot{T}\frac{\beta_{ij}}{Z_3}\frac{\partial Z_3}{\partial T} \tag{38}$$

$$\mu_{ij} = \frac{\sigma_{ij}}{\sqrt{\sigma_{kl}\sigma_{kl}}} \tag{39}$$

The constants m_1, m_2, Z_1, Z_2, Z_3, A_1, A_2, r_1, and r_2 in Equations (36) and (38) are material-dependent parameters and control the time-dependent effects and the hardening characteristics. The value, $\dot{W}_p$, is the plastic rate of work given by:

$$\dot{W}_p = \sigma_{ij}\dot{\varepsilon}^I_{ij} \qquad (40)$$

Some examples of the elastic-plastic solutions that have been accomplished with FIDEP are shown in Figures 13 and 14 [24,27]. In these figures, the constituent microstresses in a unidirectional layup of SCS6/Ti-24Al-11Nb composite during cooldown from an assumed zero stress processing temperature of 1010°C are plotted. Figure 13 displays the matrix stress at the fiber-matrix interface as calculated by the FIDEP code compared with results from a 3-D finite element solution. Figure 14 shows the stress as a function of the distance from the center of the fiber at a temperature of 150°C and 700 MPa. This figure represents the results after cooldown and out-of-phase thermomechanical fatigue (TMF) cycling between 150 and 650°C and 70 and 700 MPa. Both figures demonstrate the strong capabilities of the CCM approach for obtaining accurate stress field information (as long as the applied load is axisymmetric).

An example of the thermoviscoplastic capability of FIDEP is demonstrated in Figure 15. In this figure, the axial matrix stress and mechanical strain history at the interface is plotted for the first nine cycles of an in-phase thermomechanical fatigue (TMF) load cycled between 93 and 593°C and 100 and 1000 MPa for two frequencies [26,31]. The frequency effect is very apparent in this plot because the

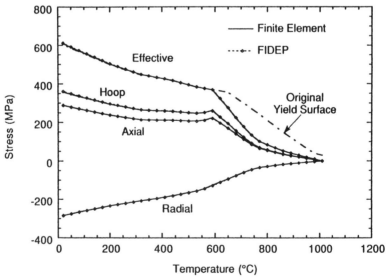

Figure 13. FIDEP predictions of thermally induced stresses in the matrix of a SCS6/Ti-24-11 composite at the interface [24,27]

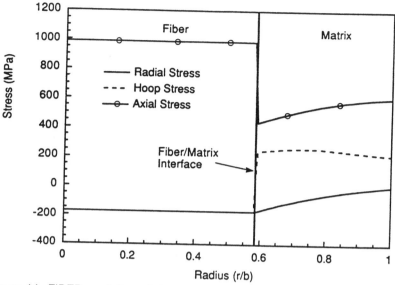

Figure 14. FIDEP prediction of stress as a function of radius for SCS6/Ti-24-11 at 150°C and 80 MPa applied load during an out-of-phase TMF cycle [24,27]

lower frequency (0.0005 Hz) demonstrates extensive inelastic deformation in the first cycle.

Method of Cells

The method of cells micromechanics approach provides both 3-D traction and displacement interaction between the constituents and the capability for modeling a weak fiber-matrix interfacial bond. Also, this can be accomplished within the context of a general applied thermomechanical load on a composite layup. Thus, if the constituent properties and interfacial characteristics are known, the composite behavior for a variety of load environments can be accurately predicted with the method of cells (MOC). Similar to the multicell and VFD model, the MOC model relies on a representative volume element that can be partitioned into regions of specified displacement fields. A regular rectangular array of fibers is assumed in developing the RVE, which is represented by a square fiber and three matrix regions as shown in Figure 16 [32]. This RVE and its associated constitutive relations were first introduced by J. Aboudi in studies [33–35] on inelastic micromechanical relations. They have been further extended to examine general composite layups and many types of con-

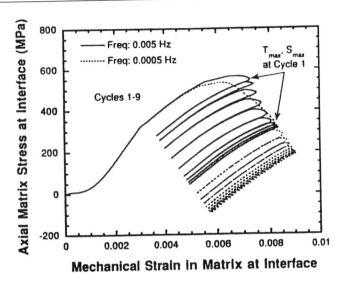

Figure 15. Axial matrix stress and strain at the interface for the in-phase TMF loads as calculated by FIDEP [26.31]

stituent nonlinearities [36–41]. A distinct advantage of the MOC is its ability to examine fiber-matrix interfacial effects that cannot be directly modeled with either the VFD or CCM approach. However, this increased capability results in greater computional intensity.

In the formulation (see Reference [34] for complete derivation), the four subcells are each analyzed in terms of their local coordinate systems whose origins are located at the center of each subcell. A linear displacement field is assumed in the two to three plane, and generalized plane strain is assumed to exist in the fiber (1-axis) direction. Thus, in terms of the local coordinate system, the displacement field for any given subcell ($\beta = 1,2$ and $\gamma = 1,2$) is [34]:

$$\mu_i^{(\beta\gamma)} = w_i^{(\beta\gamma)} + \bar{x}_2^{(\beta)} \, \phi_i^{(\beta\gamma)} + \bar{x}_3^{(\gamma)} \, \psi_i^{(\beta\gamma)} \qquad i = 1,2,3 \qquad (41)$$

where the subscript, i, identifies the given displacement direction, ϕ and ψ are constants, and w represents the displacement of the center of the subcell. Also, terms with repeated β or γ are not summed.

Assuming small strain theory, the strain tensor is:

$$\varepsilon_{ij}^{(\beta\gamma)} = \frac{1}{2}\left[\frac{\partial \mu_j^{(\beta\gamma)}}{\partial x_i} + \frac{\partial \mu_i^{(\beta\gamma)}}{\partial x_j}\right] \qquad i,j = 1,2,3 \qquad (42)$$

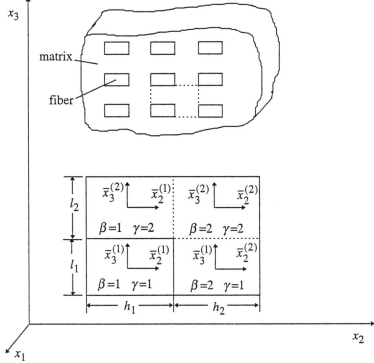

Figure 16. Representative volume element for the method-of-cells microme-chanics model [32]

In addition, from the requirement for displacement continuity between the subcells, the following relations result:

$$w_i^{(\beta\gamma)} = w_i \tag{43}$$

$$h_1\phi_i^{(1\gamma)} + h_2\phi_i^{(2\gamma)} = (h_1 + h_2)\frac{\partial w_i}{\partial x_2} \tag{44}$$

$$l_1\psi_i^{(\beta 1)} + l_2\psi_i^{(\beta 2)} = (l_1 + l_2)\frac{\partial w_i}{\partial x_3} \tag{45}$$

The volumetric average strain of the RVE is equivalent to the composite strain. Therefore:

$$\bar{\varepsilon}_{ij} = \frac{1}{V}\sum_{\beta,\gamma=1}^{2} v_{\beta\gamma}\varepsilon_{ij}^{(\beta\gamma)} \tag{46}$$

where the overbar signifies the composite strain, V denotes the total RVE volume, and $v_{\beta\gamma}$ is the volume of the specified subcell.

From Equations (42–46):

$$\overline{\varepsilon}_{ij} = \frac{1}{2}\left(\frac{\partial w_i}{\partial x_j} + \frac{\partial w_j}{\partial x_i}\right) \tag{47}$$

Continuity of surface tractions with adjacent subcells requires the stresses in the subcells to obey the relations:

$$\sigma_{2i}^{(1\gamma)} = \sigma_{2i}^{(2\gamma)} \tag{48}$$

$$\sigma_{3i}^{(\beta 1)} = \sigma_{3i}^{(\beta 2)} \tag{49}$$

Similar to the strain, the composite stress may also be represented by the volumetric average of the RVE. Thus:

$$\overline{\sigma}_{ij} = \frac{1}{V}\sum_{\beta,\gamma=1}^{2} v_{\beta\gamma}\sigma_{ij}^{(\beta\gamma)} \tag{50}$$

The above equations, along with any appropriate material constitutive relation, provide the basis for the MOC model. In fact, several inelastic and nonlinear elastic material models have been employed with the MOC. For instance, viscoelasticity has been modeled through a power law creep compliance with the Boltzmann integral representation [41]. In this method, the stress may be related to the strain through the Fourier transform, F:

$$F[\sigma_{ij}(t)] = C_{ijkl}(0)\{1 - F[\dot{\zeta}_{klpq}(t)]\}F[\varepsilon_{pq}(t)] \tag{51}$$

where $C_{ijkl}(t)$ are the relaxation functions and are related to $\zeta_{ijkl}(t)$ by:

$$C_{ijkl}(t) = C_{ijkl}(0)[1 - \zeta_{ijkl}(t)] \tag{52}$$

Equation (52) above may be converted to the time domain using the fast-Fourier transform. Also, nonlinear elastic modeling of resin matrix composites has been accomplished through the Ramberg-Osgood material representation [41], which is simply a power law stress-strain relation.

In addition, elastic-plastic deformation with either isotropic, kinematic, or combined hardening has been examined through the use of the Prandtl-Reuss flow rules and a mixed linear hardening rule

[41–43]. In this approach, the Von Mises yield criteria are modified by defining a back stress, Ω_{ij}, so yield occurs when:

$$\sqrt{\frac{3}{2}\left(s_{ij} - \Omega_{ij}\right)\left(s_{ij} - \Omega_{ij}\right)} > \sigma_{ys} \tag{53}$$

where the subscript, *ys*, denotes the yield stress. The flow rule for the inelastic strain is:

$$d\varepsilon_{ij}^{I} = (s_{ij} - \Omega_{ij})d\lambda \tag{54}$$

where $d\lambda$ is an incremental plastic multiplier.

The hardening is controlled by the instantaneous slope of the stress-inelastic strain curve (strain-hardening parameter), H', and the Bauschinger effect ratio, B. The evolution of the yield stress and back stress may be represented by:

$$\frac{d\sigma_{ys}}{d\varepsilon_{eff}^{I}} = (1 - B)H' \tag{55}$$

$$\frac{d\Omega_{ij}}{d\varepsilon_{ij}^{I}} = \frac{2}{3}BH' \tag{56}$$

In the above equations, $B = 0$ corresponds to complete isotropic hardening, whereas $B = 1$ corresponds to complete kinematic hardening.

Viscoplastic deformation of the constituents has been examined through the Bodner-Partom (B-P) unified viscoplastic theory [28–30,35,40] in which both isotropic hardening and directional hardening B-P material models have been used. Because the B-P material model equations were already presented in the previous section on the CCM model, they will not be repeated here. However, the B-P viscoplastic equations, coupled with the three-dimensional interaction capability of the MOC approach, has been a highly successful combination for approximating inelastic composite material response as will be demonstrated in subsequent figures.

The MOC approach has been refined further to examine any number of subcells within the RVE [44]. This more general approach (the generalized method of cells or GMC) allows for more complete stress field information within the RVE as well as a better approximation of the fiber geometry. The RVE using the GMC approach is depicted in Figure 17 [45]. The governing micromechanics equations are a direct

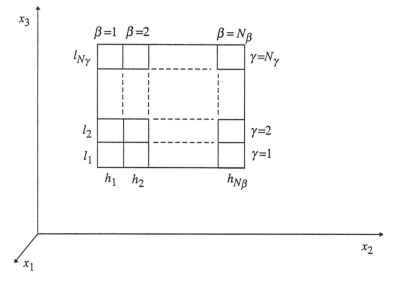

Figure 17. Representative volume element for the generalized method-of-cells micromechanics model [45]

extension of Equations (41–50); only now the subcell identifiers, β and γ, iterate from 1 to the their associated edge grid sizes, N_β and N_γ. Therefore, for the GMC model, Equations (41–43) remain unchanged, and Equations (44–46) are replaced with the following [44]:

$$\sum_{\beta=1}^{N_\beta} h_\beta \phi_i^{(\beta\gamma)} = \frac{\partial w_i}{\partial x_2} \sum_{\beta=1}^{N_\beta} h_\beta \tag{57}$$

$$\sum_{\gamma=1}^{N_\gamma} l_\gamma \psi_i^{(\beta\gamma)} = \frac{\partial w_i}{\partial x_3} \sum_{\gamma=1}^{N_\gamma} l_\gamma \tag{58}$$

$$\overline{\varepsilon}_{ij} = \frac{1}{V} \sum_{\beta=1}^{N_\beta} \sum_{\gamma=1}^{N_\gamma} v_{\beta\gamma} \varepsilon_{ij}^{(\beta\gamma)} \tag{59}$$

Equation (47) also remains unchanged and Equations (48–50) become

$$\sigma_{2i}^{(1\gamma)} = \sigma_{2i}^{(2\gamma)} = \sigma_{2i}^{(3\gamma)} = \ldots = \sigma_{2i}^{(N_\beta\gamma)} \tag{60}$$

$$\sigma_{3i}^{(\beta1)} = \sigma_{3i}^{(\beta2)} = \sigma_{3i}^{(\beta3)} = \ldots = \sigma_{3i}^{(\beta N_\gamma)} \tag{61}$$

$$\overline{\sigma}_{ij} = \frac{1}{V} \sum_{\beta=1}^{N_\beta} \sum_{\gamma=1}^{N_\gamma} v_{\beta\gamma} \, \sigma_{ij}^{(\beta\gamma)} \tag{62}$$

Thus, an RVE with any number of subcells can be analyzed. Of course, increasing the number of subcells also increases the level of computational intensity, particularly, if the material model is very complex, such as a thermoviscoplastic model with many state variables. This requires all the state variables for each subcell to be calculated and saved for each load increment. Also, if detailed stress field information is desired within the RVE, it may be more advantageous to use a traditional finite element approach that correctly ensures complete compatibility between elements as opposed to GMC, which ensures compatibility between subcells only in an average sense. However, if calculations of the composite response with an improved RVE are desired, then GMC offers a straightforward method for such.

The method of cells partitions an RVE into a three-dimensional grid of regions with assumed displacements. For this reason it would be expected for MOC to be similar to the finite element method (FEM). Some comparisons between the calculations from the two approaches using identical gridding have been accomplished [46]. The two approaches do not produce identical results. This is as expected, however, because the assumed displacement field within each element for the MOC approach is not the same in FEM. For instance, MOC assumes a linear displacement field [see Equation (41)] for the brick element, whereas most displacement-based FEM techniques would assume a polynomial containing not only the linear terms but also three second-order cross-terms and a single third-order cross-term. Thus, in terms of degrees of freedom (d.o.f.), each element within the MOC approach contains 12 d.o.f. (4 in each displacement direction). Conversely, most FEM solutions generally assume 24 d.o.f. (8 in each displacement direction) for a brick element. In addition, MOC ensures stress equilibrium is completely satisfied between subcells (or elements), whereas displacement continuity is satisfied only in an average sense. In contrast, displacement-based FEM ensures complete displacement continuity while satisfying nodal equilibrium. Thus, MOC is equivalent to a stress-based FEM solution in which each element possesses a constant stress field. However, the displacement-based FEM approach could be configured to mirror MOC by applying coincident interior nodes and employing appropriate multipoint constraints to ensure a linear displacement field within each element.

In addition, because each of the subcells or regions assumes a constant stress, it would seem reasonable for the MOC equations to also be related to strength of materials techniques. Hence, instead of modeling the problem through assumed displacement fields over specified regions, the equations could be obtained by modeling the RVE as a set of blocks of constant stress. For instance, considering Figure 16 with each subcell possessing a constant strain, the relation for the transverse normal strain, ε_{22}, is readily obtained by inspection to be:

$$h_1\varepsilon_{22}^{(1\gamma)} + h_2\varepsilon_{22}^{(2\gamma)} = (h_1 + h_2)\overline{\varepsilon}_{22} \tag{63}$$

In fact, by performing a similar analysis for all the strain components, the following equations result:

$$\varepsilon_{11}^{(\beta\gamma)} = \overline{\varepsilon}_{11} \tag{64}$$

$$h_1\varepsilon_{i2}^{(1\gamma)} + h_2\varepsilon_{i2}^{(2\gamma)} = (h_1 + h_2)\overline{\varepsilon}_{i2} \tag{65}$$

$$l_1\varepsilon_{i3}^{(\beta 1)} + l_2\varepsilon_{i3}^{(\beta 2)} = (l_1 = l_2)\overline{\varepsilon}_{i3} \tag{66}$$

The above three equations are identical to the MOC equations [Equations (39–41)] that were derived from elasticity. The equations relating the subcell stresses to the composite stresses are also identical to those previously listed [Equations (44–46)].

Due to completely satisfying the equilibrium requirement in the MOC equations, the equations for continuity of displacement [i.e., Equations (39–41) or Equations (64–66)] satisfy displacement continuity only in an average sense. Hence, there will be slip at the interfaces between subcells. To prevent unwanted slip, a modified version of the MOC equations has also been used [43,47]. The modified equations are identical except it is assumed Equations (44–45) apply for the subscript i equal to 2 and 3 only. Instead, for the subscript $i = 1$ the strain is substituted in the equations, so:

$$\varepsilon_{21}^{(1\gamma)} = \varepsilon_{21}^{(2\gamma)} \tag{67}$$

$$\varepsilon_{31}^{(\beta 1)} = \varepsilon_{31}^{(\beta 2)} \tag{68}$$

This ensures complete displacement continuity in the longitudinal direction at the interfaces between subcells. However, this modification necessitates the equilibrium for the longitudinal shear stresses (i.e., σ_{12} and σ_{13}) be satisfied only in an average sense through the RVE. Thus, either the compatibility conditions (i.e., displacement continuity) must be relaxed and satisfied only in an average sense as in the original MOC formulation, or equilibrium must be relaxed and satisfied only in an average sense as in the modified equations. However, either set has been shown to provide a realistic approximation of the composite response [35,48]. Initially, the modified set of equations was referred to as the free transverse shear approach and was derived in a strength of materials fashion [43].

The MOC micromechanics equations, as presented above, apply for a single-ply or unidirectional layup. However, it is important for a micromechanics-based analysis to address multidirectional layups. Therefore, the MOC has been extended to multidirectional layups through various techniques. Some have employed equivalent plastic forces and moments, coupled with the calculated laminate elastic properties [39]. The inelastic strains from the previous increment are integrated through the thickness to obtain the equivalent plastic loads. Other researchers have performed a direct assembly of the inelastic micromechanics equations up through the laminate level [48].

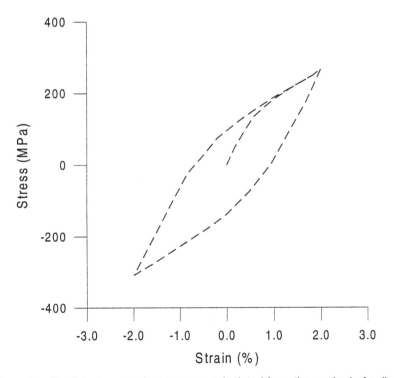

Figure 18. Predicted composite response calculated from the method-of-cells for a tensile-compressive cycle of a $[0/\pm 45]_{2s}$ layup of SCS6/Ti-15-3 [49]

Figure 18 presents sample results of the MOC formulation for a titanium-based MMC laminate undergoing a single-cycle tensile-compressive load [49]. The fiber and matrix are assumed to be stress free at $\Delta T = 0$ and zero applied load. The composite model was then subjected to a temperature change of $\Delta T = -1093°C$ before applying the mechanical load. As observed in the plot, significant plastic deformation occurs in the matrix, producing hysteresis in the stress-strain response. As a further example, Figure 19 presents temperature-dependent calculations for an angle-ply layup where the temperature and strain were sequentially changed as indicated in the figure [50]. The thermomechanical capability is further demonstrated in Figure 20 where an in-phase TMF cyclic load of a cross-ply is analyzed and compared with experimental results [48]. The increasing strain with

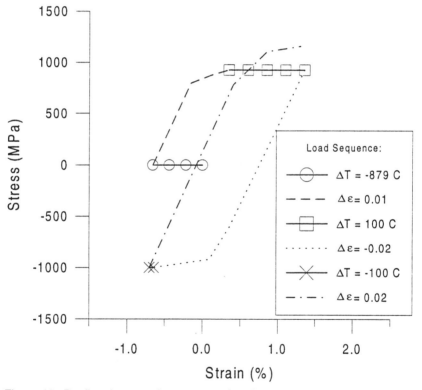

Figure 19. Predicted composite response for alternate thermal and mechanical loading of a [±45]ₛ SCS6/Ti-6-4 composite layup [50]

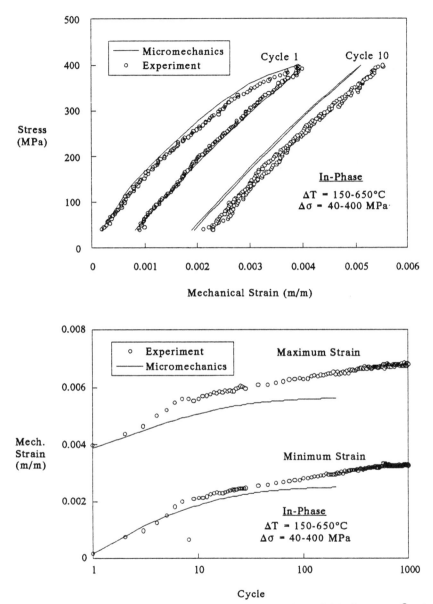

Figure 20. Thermomechanical fatigue response of [0/90]s SCS6/TIMETAL®21S to in-phase loading-anaytical/experimetal comparisons [48]

cycles (strain ratchetting) due to the viscoplastic deformation of the matrix material is readily apparent and matches well with its experimental counterpart up to well over 100 cycles at which point other damage mechanisms, such as fiber and matrix cracking, is likely to dominate the continued cyclic strain increase.

It should be noted that the calculations presented in Figure 20 were performed with the modified MOC equations. However, the modified equations only affect the longitudinal or in-plane shear of the ply, and because these figures involve a cross-ply layup undergoing uniaxial loading, the two sets of equations should produce identical results. The contrast between the two sets of equations can best be demonstrated by analyzing a longitudinal shear load. Figure 21 presents a case in which the in-plane shear response of a $[0/\pm15]_s$ layup of SCS-6/Ti-15-3 composite is calculated using the two approaches [48]. The matrix was assumed to behave according to the original Bodner-Partom viscoplastic material model, which possessed isotropic hardening only [28], and perfect interfacial bonding was also assumed for each analysis. As demonstrated, the figure shows the modified equations predict a stiffer response. This is as expected. For instance, it is well known that a displacement-based finite element solution ap-

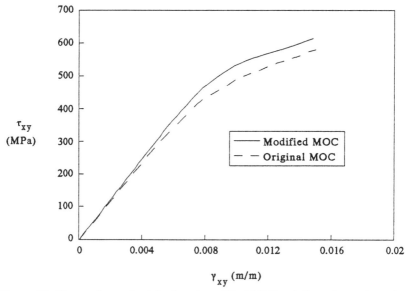

Figure 21. Comparison of original and modified MOC relations for a shear-load in the plane of the laminate for $[0/\pm15]_s$ SCS6/Ti-15-3 [48]

proaches the actual solution from above, producing a stiffer response. On the other hand, a stress-based finite element solution approaches the actual solution from below, producing a weaker response. Thus, when it is considered that the modified equations ensure displacement continuity (displacement-based) while the original equations ensure stress equilibrium (stress-based), then the trend shown in Figure 21 is anticipated.

Moreover, as demonstrated in Figure 20 and in more detailed figures later in this chapter, the MOC approach offers many benefits for composite material analysis. The ability to analyze 3-D stresses within a 3-D RVE provides the researcher with a robust method for including nonlinear material models and unique interfacial characteristics in a composite laminate analysis.

INTERFACE MODELING

As mentioned earlier in this text, titanium-based metal matrix composites (TMCs) possess relatively low fiber matrix bond strengths [51]. This greatly affects the transverse behavior of these materials. In addition, because the stress transfer characteristics across the interface are instrumental in the strength and toughness of the composite, understanding the interfacial damage effects on the composite is extremely important. As such, due to its importance, various techniques for modeling the interface have been employed in the micromechanics models of the previous sections. This section will provide a brief overview of these techniques and present some examples of their use for TMCs.

As early as 1967, modeling of interfacial regions within bonded composite materials was accomplished by assuming the stress at the interface was proportional to the relative displacements of the two surfaces [52]. Thus, two linear elastic parameters were used to characterize the interface by relating the normal and tangential (shear) stresses and displacements across the interface. In addition, most commercially available finite element codes support contact elements for analyzing separation and slip, which occur in many structural applications. Such contact-type elements and their associated finite element codes have been used extensively in TMC micromechanics modeling [53–57]. Also, this same approach has been employed in many analytical studies in which techniques other than FEM were utilized [36,40,58,59]. The contact element relations are generally represented as:

$$\sigma_I = \begin{cases} k_{tens}\, u_I & \text{if } u_I \geq 0 \\ k_{comp}\, u_I & \text{if } u_I < 0 \end{cases} \qquad (69)$$

and

$$\tau_I = k_{shear}\, v_I \qquad (70)$$

where σ_I and τ_I are the interfacial normal and tangential stresses, u_I and v_I are the relative interfacial normal and tangential displacements, and the elastic constants, k, define the tensile, compressive, and shear characteristics of the interface. For most contact problems the compressive elastic constant, k_{comp}, is chosen to be infinite or if that is not possible, extremely large. Also, for an unbonded contact, the tensile elastic constant, k_{tens}, is chosen as zero or given an extremely low value. In addition, the equation for the shear behavior at the interface may alternatively be assumed to follow a Coulomb friction representation to more correctly characterize an unbonded interface.

The interface model described by Equations (69) and (70) has proven very useful in predicting composite material behavior. In fact, the results presented earlier in Figure 20 were obtained by using this approach to characterize the interfacial separation occurring in the 90° ply of the cross-ply layup. If it were not for the contact-type interfacial model that correctly characterizes the interface as extremely weak, the results would not mirror the experiment quite so well. A comparison of assuming a weak (unbonded) interface versus a perfectly bonded interface is depicted in Figure 22 in which a finite element model was employed to examine the transverse normal behavior of a unidirectional TMC [57]. As expected, the addition of a weak bond greatly influences the solution. Although the initial modulus remains essentially the same as for the strong bond case, once the residual thermal compressive stress at the interface is overcome, separation occurs and the composite response becomes substantially weaker. Thus, the transverse load a TMC can withstand at room temperature (assuming a weak interface) is approximately one-third what would otherwise be the case if the interface were strong (perfectly bonded). Several finite element studies have been conducted to analyze the fiber-matrix separation under transverse loading [53–57,60,61]. In addition, some of the simplified micromechanics models have also been used to analyze fiber-matrix debonding by utilizing a contact-type relation [40,45,53,59]. These studies have revealed a strong dependence of the composite response and interfacial separation on the

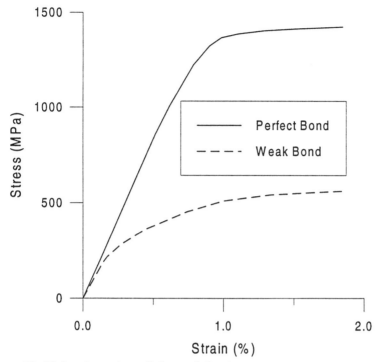

Figure 22. *Finite element predictions of the transverse room temperature response of SCS6/Timetal®21S [57]*

loading temperature. The thermal residual stresses are key to determining interfacial separation, and the viscoplastic behavior of the matrix material is also greatly affected by the loading temperature. Therefore, it is no surprise for the thermal effects to have such a pronounced influence.

Modeling the interface as an unbonded fiber-matrix contact does not allow for a finite interfacial strength. Hence, fiber-matrix separation is always calculated to occur when the residual thermal stresses are overcome. However, the initial fiber-matrix bond is not completely weak because the point of fiber-matrix separation calculated by analytical solutions occur at a lower applied load than observed experimentally as is demonstrated in Figure 23 [53]. Thus, if accurate modeling of the TMC interfacial effects is desired, then more sophisticated interfacial models beyond the simple contact-type element must be sought.

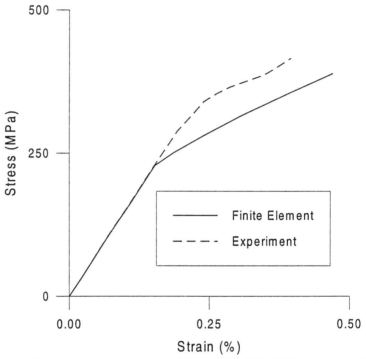

Figure 23. *Room temperature transverse response of SCS6/Ti-6-4 unidirectional composite [53]*

An initial effort to analyze the effects of a finite interfacial bond strength on TMCs through a numerical model was conducted by Gayda and Gabb [62]. In this study, the random nature of fiber-matrix failure was numerically approximated by employing a 128-element grid of one-dimensional elements. Each element alternately contained either all matrix material or a fiber-matrix RVE whose individual fiber content was obtained by alternatively adding and then subtracting a random number between 0 and 0.1 through 50 iterations. Each one-dimensional fiber-matrix element was initially modeled with a perfect interfacial bond. Once a given element exceeded its bond strength, the tensile stiffness for that element was dropped to zero. Thus, with this technique, the characteristics of random fiber packing and the statistical strength distribution of the interfaces could be modeled and their associated effect on the composite response examined. Figure 24 presents the monotonic transverse response as calculated by this technique for three constant interfacial failure strengths

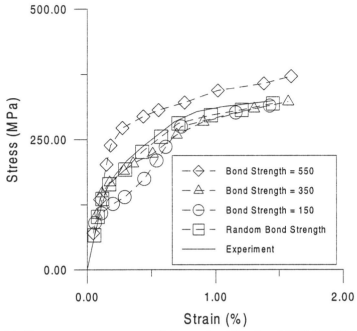

Strain (%)

Figure 24. Transverse response predictions of unidirectional SCS6/Ti-15-3 composite using finite bond strength analytical solutions of a set of random fiber content one-dimensional elements [62]

[62]. A random distribution of bond strengths with values ranging from 150 to 550 MPa was, likewise, considered, and its corresponding response is also presented in Figure 24. This simplified one-dimensional technique for characterizing the effects of the interface on the transverse behavior was shown to be highly successful for both monotonic and fatigue loading. Figure 25 depicts the predicted and experimental stress-strain response at selected cycles of a strain-controlled fatigue test [62]. A bilinear elastic-plastic approximation of the Ti-15-3 material response was assumed for the calculations.

The above approach provides for a finite bond strength followed by progressive interfacial failure but cannot be extended to more general micromechanics approaches. It is not a purely interfacial model. An interfacial model that accounts for progressive interfacial failure and is also more conducive to an RVE micromechanics approach has been proposed for TMCs [63,64]. This model provides a displacement-traction relationship at the interface, which allows for progressive interfacial failure. This relationship was derived based on a statistical

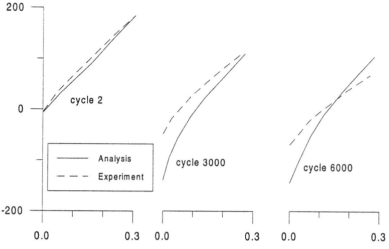

Figure 25. Prediction of strain-controlled transverse fatigue response of SCS6/Ti-15-3 [62]

representation of the average behavior of a conglomerate of individual fiber-matrix interfacial failures. In the formulation, it was assumed that the random nature of the fiber spacing and bond strength could be appropriately represented by a single statistical distribution of the interfacial stresses throughout the composite. The fraction of failed interfaces are then calculated from the portion of the distribution possessing stresses greater than the interfacial failure strength. These failed interfaces will separate when subjected to a normal tensile stress and slip when subjected to a shear stress. Under a monotonically increasing load, the percentage of failed interfaces will increase with increasing load until all interfaces have failed. If the composite is unloaded at any point, then the percentage of failed interfaces will remain constant until the stress once again reaches its previous maximum value. Thus, the percentage of failed interfaces represents the interfacial damage level to be monitored through the statistical representation. By further relating the displacement of the failed interfaces through a linear relation with the stress at the interface if it had remained intact, a general set of expressions for the average interfacial stress and average interfacial displacement for all interfaces throughout the composite may be developed.

A sample plot of the average interfacial stress-displacement behavior normal to the interface during a load-unload-reload sequence is presented in Figure 26 [63]. This interfacial behavior is almost iden-

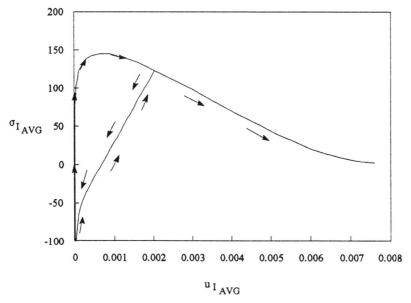

Figure 26. Stress-displacement behavior of the interface assuming a normal statistical distribution for the interfacial stresses and a finite bond strength [63]

tical to what has been previously proposed for general inclusion
debonding [65]. However, the statistically based model provides a
physical justification for such. Also, the behavior depicted in Figure
26 can be approximated with a linear representation as shown in Figure 27. In equation form, the linear approximation of the interface
model normal to the interface is given by:

$$u_I = \begin{cases} S^u(\sigma_I - \sigma_{I_C}) & \text{for } u_I < u_I^* \\ u_{I_f}\left(1 - \dfrac{\sigma_I}{\sigma_{I_f}}\right) & \text{for } u_I \geq u_I^* \end{cases} \tag{71}$$

where

$$S^u = \begin{cases} 0 & \text{for } \sigma_I < \sigma_{I_C} \\ S_{sep}^u = \dfrac{u_I^*}{(\sigma_I^* - \sigma_{I_C})} & \text{for } \sigma_I \geq \sigma_{I_C} \end{cases} \tag{72}$$

$$\sigma_I^* = \sigma_{I_f}\left(1 - \frac{u_I^*}{u_{I_f}}\right) \tag{73}$$

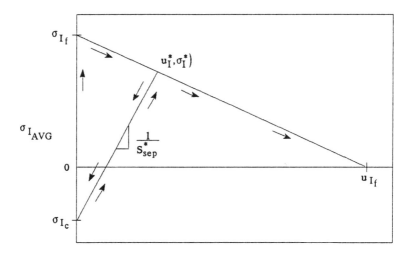

Figure 27. Linear approximation of statistically-based interfacial model for evaluating the stress-displacement behavior at the interface [63]

and

$$u_I^* = \max_{0 \le t \le t_0} [\sigma_I(t)] \tag{74}$$

The quantities σ_I and u_I are the interfacial normal stress and relative displacement, respectively, and the parameters u_{I_f}, σ_{I_f}, and σ_{I_C} are material-dependent properties of the interface.

Equations for the shear behavior are similar and can be represented by:

$$\upsilon_I = \begin{cases} S^\upsilon \tau_I & \text{for } \upsilon_I \le |\upsilon_I^*| \\ \upsilon_{I_f}\left(\dfrac{\tau_I}{|\tau_I|} - \dfrac{\tau_I}{\tau_{I_f}}\right) & \text{for } \upsilon_I > |\upsilon_I^*| \end{cases} \tag{75}$$

where

$$S^\upsilon = \frac{\upsilon_I^*}{\tau_{If}\left(1 - \dfrac{\upsilon_I^*}{\upsilon_{If}}\right)} \tag{76}$$

and

$$v_I^* = \max_{0 \leq t \leq t_0} |v_I(t)| \tag{77}$$

Similar to the previous equations, τ_I and v_I are the interfacial shear stress and displacement, respectively, and the parameters v_{I_f} and τ_{I_f} are material-dependent properties of the interface. Only two properties are necessary to characterize the interfacial shear of the material as opposed to three for the normal behavior. This is due to the reclosure of separated interfaces in the normal direction occurring at a negative interfacial normal stress. Thus, this value is specified as a material property of the interface, σ_{I_C}. It is unnecessary for shear and is an artifact of the linear approximation. If the equations from the statistical formulation were used directly, then only two constants would be necessary because the statistical distribution would correctly account for the portion of interfaces that have not yet reclosed even though the average interfacial stress is zero.

It is reasonable to assume that, once an interface has failed in shear, it will be unable to support a tensile load and vice versa. Therefore, the above interface model can be applied to mirror this by maintaining the same percentage of failed interfaces for both normal shear. Thus, the greater of the two ratios is taken as the appropriate failure percentage and it is then assumed that:

$$\frac{u_I^*}{u_{I_f}} = \frac{v_I^*}{v_{I_f}} \tag{78}$$

The above interface model has been employed with the method of cells micromechanics approach to model TMCs. Example stress-strain responses to load-unload sequences for a 90° SCS6/Ti-15-3 layup are presented in Figure 28 [63]. A strong feature of the interfacial model is its ability to capture the unloading characteristics as the return slope closely matches the experimental data. A further comparison with various angle-ply layups is given in Figure 29 [64]. Of course, such an interface model can only be applied to a micromechanics approach that possesses an appropriate interface region in the RVE, such as the finite element method, the method of cells, the multicell model, and others. It could not be effectively incorporated into a concentric cylinder approach or the vanishing fiber diameter model.

For interfacial debonding in the vanishing fiber diameter model in which the interface cannot be accounted for directly, an alternate ap-

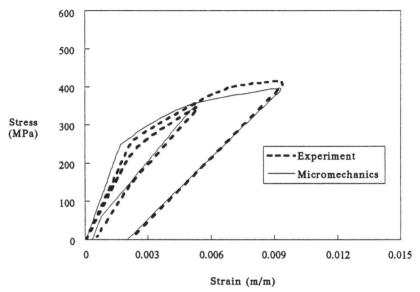

Figure 28. Transverse stress-strain response of unidirectional SCS6/Ti-15-3 as compared to analytical results of a progressive interfacial failure model [63]

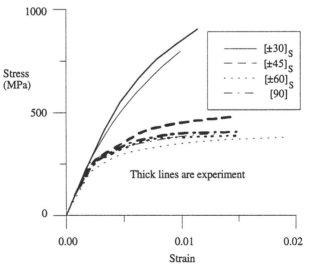

Figure 29. Predictions of the stress-strain response for various SCS6/Ti-15-3 angle-ply layups using a progressive failure model at the interface [64]

438

proach is employed to account for debonding indirectly by reducing the transverse fiber modulus at a specified transverse fiber stress [66]. Alternate criteria consisting of summing the squares of the transverse normal and longitudinal shear stress of the fiber has also been proposed as a method to determine the appropriate point for reducing the transverse fiber modulus in the vanishing fiber diameter model [67]. For concentric cylinder analysis methods, such as in the computer program FIDEP, the concentric cylinder model is applied to the 0° plies only, and the remaining plies are accounted for by inputting an equivalent ply response from a separate analysis [25]. Thus, interfacial failure and debonding effects may be pragmatically accounted for.

Interfacial characteristics near fiber breaks are critical in determining a composite's strength. If the interface is effective at transferring the load to the surrounding matrix and fibers, then the overall toughness of the material will be greatly enhanced. If the interface is too strong, then the crack front will not be diverted at the interface and a rogue crack will develop. On the other hand, if it is too weak, then insufficient stress transfer will occur, also degrading the strength. An excellent discussion of these interfacial mechanics, along with the testing techniques and associated interfacial properties of various TMCs, is presented by Clyne and Watson [68]. Such localized phenomena are difficult to account for in RVE-based micromechanics approaches, which assume all fiber-matrix cells in the composite behave identically. The only method, to date, for including such effects in these micromechanics methods relies on a simplistic shear lag model for fiber breakage and relates the fiber cracks and interfacial shear characteristics near the breaks to an equivalent elastic fiber modulus [69]. Fibers are assumed to possess a failure distribution of the Weibull type. When coupled with a simplistic one-dimensional shear lag model, the equivalent elastic fiber modulus becomes:

$$E_f^* = E_f e^{-\frac{R\alpha}{2\tau}(E_f\varepsilon)^{\beta+1}} \qquad (79)$$

where R is the fiber radius, τ is the interfacial frictional shear stress near a fiber crack, ε is the longitudinal mechanical strain of the fiber, and the quantities α and β are the Weibull scale and shape parameters for the distribution of fiber failures. The above equivalent modulus then replaces the longitudinal fiber modulus for all calculations within the chosen micromechanics method.

Moreover, accounting for the relatively weak fiber-matrix bond strength of TMCs is crucial for properly modeling their behavior. Off-axis response is greatly affected by the interfacial characteristics. Therefore, if the behavior of any layup possessing off-axis plies is to be modeled, it is important for the chosen micromechanics approach to be capable of modeling interfacial characteristics other than perfect bonding.

FINITE ELEMENT MODELING

Due to its wide availability and many other advantages, the finite element method has been used extensively for micromechanical modeling of composites, including TMCs. A tailored RVE possessing most any level of complexity can be analyzed, and the resulting stress field throughout the RVE examined. The grid and choice of element type varies significantly, depending on the type of problem to be examined.

For instance, studies involving unidirectional layups undergoing axisymmetric loading have generally been performed using axisym-

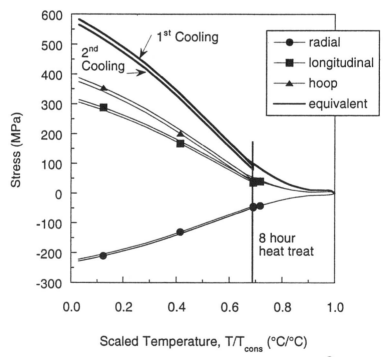

Scaled Temperature, T/T$_{cons}$ (°C/°C)

Figure 30. Cooldown stresses in unidirectional SCS6/TIMETAL®21S as calculated by an axisymmetric finite element model for both with and without heat treatment (T$_{cons}$ = consolidation temperature) [27]

metric elements in a concentric cylinder RVE [70,71]. This allows for a computationally efficient means of accounting for three-dimensional effects. An example of the thermal residual stresses during cooldown calculated using such an approach is presented in Figure 30 [27,71]. This plot presents the matrix stress at the fiber-matrix interface. A further comparison of the axial response to an in-phase TMF is presented in Figure 31 [27,71]. In these figures the fiber was assumed thermoelastic, and the matrix material was modeled as thermoviscoplastic using the Bodner-Partom model with directional hardening. Thus, with only a modest grid and straightforward boundary conditions, the overall composite response and the stress variation in the radial direction can be accurately modeled. However, off-axis loading cannot be readily examined with an axisymmetric finite element grid, and the stress variation associated with angular position around the fiber cannot be examined either. Finite element simulations of such effects necessarily require a more complex RVE and associated element grid.

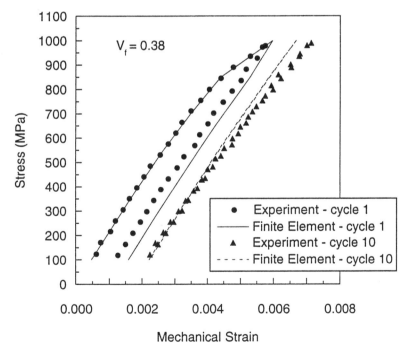

Figure 31. TMF in-phase response (150C–650°C) of undirectional SCS6/ TIMETAL®21S as calculated by an axisymmetric finite element model [27]

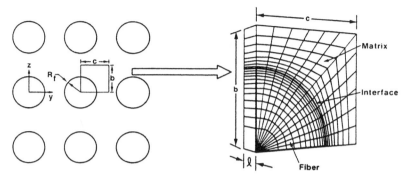

Figure 32. Typical RVE and grid of a finite element micromechanics model for examining transverse effects (Copyright ASTM. Reprinted with permission) [53]

To analyze the behavior of 90° plies, an RVE as pictured in Figure 32 is normally used [53]. A sample grid composed of solid elements is also shown. This type of RVE assumes a regular rectangular array of fibers and has been used by several researchers for TMC modeling [53–55]. Other types of regular arrays, such as hexagonal close-packed, have also been used for composite analysis [12], but the rectangular array has been found to be very successful for TMCs. An example RVE stress field under an applied transverse normal load of 250 MPa is depicted in Figure 33 for an SCS6/Ti-6-4 unidirectional

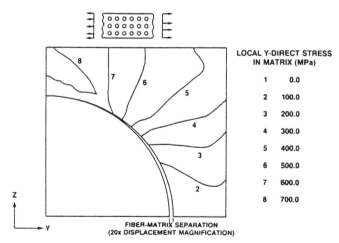

Figure 33. Normal y-direction stress in RVE for a 250 MPa applied transverse load at room temperature assuming a SCS6/Ti-6-4 composite with a weak interface (Copyright ASTM. Reprinted with permission) [53]

composite [53]. Also depicted is the fiber-matrix separation due to the weak fiber-matrix contact. Most finite element micromechanics solutions of TMCs have assumed a weak fiber-matrix contact. Nonetheless, to allow for progressive failure of the interface in a finite element model, Sherwood and Quimby [57] proposed a method of combining rigid beam and 2-D solid elements with contact elements. The thermoplastic properties of the solid elements were used to model the failure characteristics of the interface.

Critical to the finite element solution is the specified boundary conditions. For the calculation shown in Figure 33 and what is typically assumed for transverse normal loading is a state of generalized plane strain in the fiber or x-direction such that the mechanical load in this direction is zero, the left hand and lower faces are not allowed to displace normal to their faces, and all nodes on the right-hand and upper faces are restricted to move equally normal to their faces, respectively (see Figure 33). In certain instances, these boundary conditions have been modified to examine unique characteristics such as stress-free edge effects [61]. A plane stress condition was used near the edge, and a generalized plane strain condition was applied in the center of the specimen. A concentric cylinder model was further employed to determine the residual thermal stress variation along the length of the fiber near the edge. An example of the variation of residual stress with angular position around the fiber at the fiber-matrix interface is

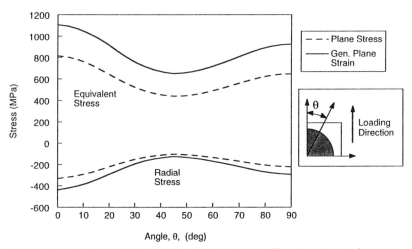

Figure 34. Residual thermal stress in matrix at the fiber/matrix interface versus angular position for unidirectional SCS6/TIMETAL®21S [31]

presented in Figure 34 [31,61]. The difference between the plane stress and generalized plane strain condition is significant and underscores the importance of applying the proper boundary conditions for the problem at hand.

Most micromechanics modeling using the FEM approach is restricted to analyzing unidirectional composites. Developing a finite element grid for a layup of plies with different orientations is difficult and time consuming. However, the next simplest layup for FEM analysis is the cross-ply, and a few studies have been accomplished in this area using fiber-matrix RVEs connected as shown in Figure 35 [57]. Others have used slightly more elaborate grids containing whole fibers [60]. An example calculation of the cross-ply composite stress-strain response for monotonic loading at room temperature is presented in Figure 36 [60]. Gap or contact-type elements were used to model a weak fiber-matrix bond in the 90° ply. The prediction agrees quite well with the experimental result. On the other hand, the prediction is consistently weaker than the experiment that again addresses the contact element's inability to account for a finite interfacial bond strength [53].

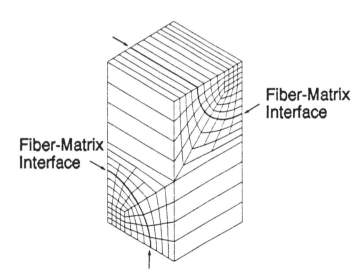

*Figure 35. Sample finite element grid used for analysis of cross-ply laminates (Reprinted from Computers & Structures, **56**, Sherwood, J. A., and Quimby, H.M., "Micromechanical Modeling of Damage Growth in Titanium Based Metal-Matrix Composites," pg 508, Copyright 1995, with permission from Elsevier Science Ltd, The Boulevard, Langford Lane, Kidlington OX5 1GB, UK)*

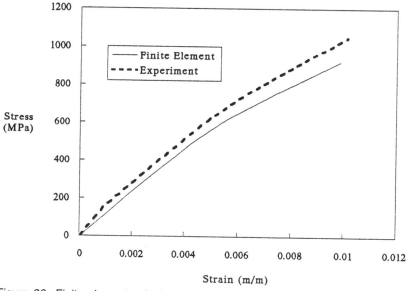

Figure 36. Finite element calculation of the monotonic stress-strain response of cross-ply SCS6/Ti-15-3 at room temperature [60]

OTHER DISCRETE MICROMECHANICS MODELS

The methods presented in this chapter for analyzing the thermo-mechanical-inelastic behavior of TMCs is by no means considered an exhaustive list. Many researchers have developed their own tailored methods to examine various problems. Three such additional methods that have been used in TMC modeling will be discussed in this section.

A simplistic one-dimensional approach for analyzing fiber-matrix separation under transverse normal loading has been used by Nimmer [72] to evaluate thermal residual stress effects on interfacial separation. This method is very similar to the multicell model used in METCAN, but it contains a contact element at the fiber-matrix interface. Thus, the bilinear elastic nature of fiber-matrix separation was examined with a method simple enough for a single-line closed-form solution.

A more complex model capable of examining off-axis effects was proposed by Sun et al [73] and is pictorially represented in Figure 37. This model is similar to the method of cells in that it assumes a constant stress field within each region. However, only three regions are

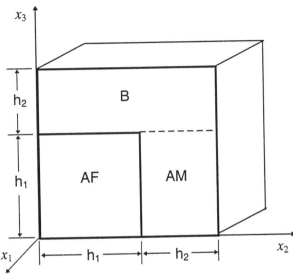

Figure 37. Representative volume element for the micromechanics model used by Sun, et al. [73]

considered instead of four, and only a two-dimensional stress field in the plane of the composite is examined. An elastic-plastic Von Mises-type material behavior was assumed for the matrix. The fiber and matrix normal strains in the direction of the fiber (i.e., ε_{11}) are assumed to be equal, so a state of generalized plane strain exists in this direction. The remaining stresses and strains may be related through simple mechanics of materials, and the resulting equations are very similar to those previously listed for the method of cells. The model was further enhanced by the incorporation of a continuous damage law for reducing the transverse fiber modulus, thus enabling an effective method for approximating fiber-matrix separation.

Figure 38 presents several off-axis predictions of this model along with the experimental comparisons [73]. The excellent correlation demonstrates the effectiveness of such a model.

Another simplistic approach for laminate analysis is to assume the composite behavior is dominated by the 0° plies. Thus, all off-axis plies are simply treated as matrix material [74]. This technique allows for quick calculations and easily extends any micromechanics model designed for only unidirectional behavior to examine composite layups by simply varying the fiber volume fraction.

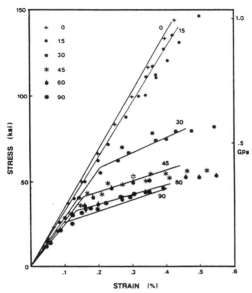

Figure 38. Room temperature off-axis stress-strain response prediction of SCS6/Ti-6-4 with experimental (symbols) comparison (reprinted with permission from "Mechanical Characterization of SCS6/Ti-6-4 Metal Matrix Composite," Sun, C.T., et al., Journal of Composite Materials, 24 (1990). Technomic Publishing Company, Inc., Lancaster, PA, pg 1054)

MODEL COMPARISONS

Because of the numerous techniques available for modeling TMCs, some researchers have performed equivalent calculations with different models and compared their results to identify the various models' associated strengths and weaknesses [5,6]. Such information is useful in choosing the appropriate model for the problem at hand. Although not all the various methods were compared in these studies, a sufficient cross section of different types possessing contrasting levels of complexity and capability were examined to provide a good indication of the model complexity required for various tasks.

Bigelow et al. [5] compared the elastic properties of SCS6/Ti-15-3 as calculated by the vanishing fiber diameter (VFD) model, the multicell model (MCM), the method of cells (MOC) model, and a finite element method (FEM) micromechanics grid. Table 1 lists the calculated composite properties using these methods as reported by Bigelow et al. [5] for a unidirectional and cross-ply layup. Cross-ply results for the MOC and FEM models were not reported because their associated computer programs used for the study did not support layups of mixed orientation. However, the results demonstrate that for elastic properties of unidirectional composites, all models are equally good at approximating the composite. On the other hand, for cross-ply layups, the VFD and MCM begin to deviate slightly from the experimental behavior. This is expected due to fiber-matrix separation of the 90° plies. For additional comparison, the method of cells model for a cross-ply SCS6/Ti-15-3 with the same fiber volume content as in the Bigelow et al. study (32.5%) will produce a normal modulus, $E_L = E_1 = 121$ GPa when a weak interface is assumed and the layup is loaded to 200 MPa after cooldown from an assumed 815°C processing temperature. This compares favorably with the experimental results listed in Table 1 and demonstrates that the physical mechanisms in the composite must be kept in mind when choosing an appropriate model. The VFD and MCM approaches must also include their associated methods for an effective weak fiber-matrix bond if the cross-ply behavior is to be modeled.

An extensive comparison of several micromechanics techniques in predicting the inelastic response of a cross-ply TMC was conducted by Kroupa et al. [6] for thermomechanical fatigue (TMF) loads. Results from the models were also compared with corresponding experimental data. Six computational methods were examined for their accuracy and speed. These consisted of the following computer programs: (1) VISCOPLY, a laminate code based on the vanishing fiber diameter

Table 1. Room temperature laminate properties for SCS6/Ti-15-3 [5].

	[0]$_8$		
	E_L(GPa)	E_T(GPa)	ν
VFD	191	130	0.325
MCM	192	134	0.324
MOC	192	139	0.322
FEM	192	148	0.321
Experiment[Johnson]	186–208	119–123	0.299–0.329
	[0/90]$_{2S}$		
	E_L(GPa)	E_T(GPa)	ν
VFD	161	161	0.263
MCM	163	163	0.266
Experiment[Johnson]	117–148	117–148	0.252–0.253

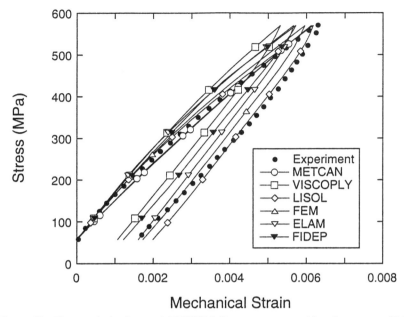

Mechanical Strain

Figure 39. First cycle isothermal (650°C) fatigue response at low frequency (0.01 Hz) for a [0/90]ₛ layup of SCS6/TIMETAL®21S [31]

model; (2) METCAN, the multicell model computer code; (3) FIDEP, the concentric cylinder model with a 90° ply element parallel to the concentric cylinder model; (4) LISOL, a method of cells laminate analysis using the modified equations as described earlier; (5) ELAM, an elementary analysis method described in the previous section where the [90] ply is assumed to possess matrix properties; and (6) FEM, a representative volume element (RVE) finite element grid. These models were subjected to four cyclic load cases consisting of two isothermal (650°C) cases at different frequencies (1.0 and 0.01 Hz), an in-phase TMF case, and an out-of-phase TMF [6].

Figure 39 presents the low-frequency (0.01 Hz) isothermal fatigue response of the first cycle for the six methods listed in the preceding paragraph along with the experimental response [6,31]. A triangular waveform was applied for this and all other fatigue cycles. A maximum stress of 570 MPa was used for the isothermal fatigue cycle at 650°C with a minimum to maximum stress ratio of 0.1. Residual thermal stresses were accounted for in the models through an appropriate cooldown. A significant amount of plasticity occurs during the first cycle for this load case as is observed both experimentally and calcu-

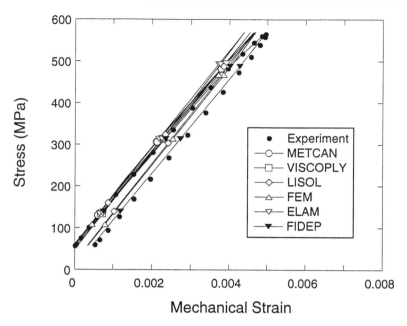

Figure 40. First cycle isothermal (650 °C) fatigue response at high frequency (1 Hz) for a [0/90]ₛ layup of SCS6/TIMETAL®21S [31]

lated by most of the models. Furthermore, Figure 40 depicts the first cycle response for the higher frequency case [6,31]. This plot shows a significantly reduced amount of plasticity for both the models and experiment. The predictive capabilities of the models are, in general, very good for these cases. Only METCAN was unable to account for plasticity effects. This suggests that if time-dependent plasticity effects are to be examined, then the material models for the individual constituents is the critical component rather than the complexity of the micromechanics formulation. Although the micromechanics approaches are each unique, only METCAN used a material model containing no time-dependent plasticity effects. The remaining micromechanics techniques all used more complex viscoplastic material models for the matrix.

Figure 41 presents the in-phase TMF stress-strain response at cycle 10 [6,31]. For the TMF cases, the stress and temperature were cycled simultaneously between 50 and 500 MPa and 150 to 650°C with a cycle period of 180 sec using a triangular waveform. The resulting in-phase response at cycle 10 depicts a significant amount of permanent strain. However, only FEM and LISOL predict a permanent strain

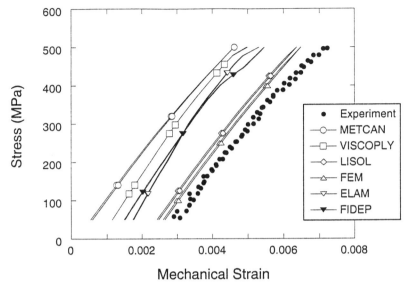

Mechanical Strain

Figure 41. In-phase TMF response at cycle 10 for a cross-ply SCS6/ TIMETAL®21S [31]

comparable with the experiment. These represent the more sophisticated models in terms of fiber-matrix separation of the [90] plies as well as 3-D fiber-matrix constitutive relations coupled with a thermoviscoplastic material analysis. The interaction of these effects over 10 cycles where significant viscoplastic deformation occurs greatly affects the result. If a good approximation of the composite behavior is desired for such loading, then the more sophisticated models are necessary.

On the other hand, Figure 42 presents the out-of-phase TMF response of the 10th cycle [6,31]. Little permanent strain occurs for the out-of-phase case, and, hence, all the models give a reasonable approximation of the composite response.

Comparisons, such as in Figures 39 through 42, provide an assessment of each model's ability to match experimental data. However, more detailed information besides just the composite response is desired in many instances. For example, knowledge of the constituents' microstresses are valuable for failure and life prediction. All of the models provide constituent stress information in the 0° plies, and most provide it for off-axis plies as well. If detailed information of the stress field within the RVE is desired, then the FEM approach is con-

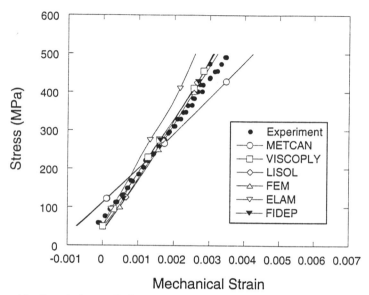

Mechanical Strain

Figure 42. Out-of-phase TMF at cycle 10 for a cross-ply SCS6/TIMETAL®21S [31]

sidered to be the most accurate, but in some cases the researcher only requires the average stress within the fiber and matrix. Therefore, an additional comparison of the six models for calculating the average stress in the matrix of the 0° plies is presented in Figure 43 [6,31]. These calculations are for the first cycle of the in-phase TMF load and represent the stress in the direction of load. The models indicate a significant amount of stress relaxation at maximum load and temperature. If the finite element solution is assumed to be the most accurate, then all models provide a reasonable approximation of the average stress at the end of the cycle.

Although the finite element method is considered the most accurate, its use may be precluded due to the computational intensity. All of the micromechanics techniques described in this chapter, except for the finite element method, require modest computational resources. In fact, many of them were originally configured for use on personal computers, so it is likely any laboratory facility will have the required computer resources.

Furthermore, each technique requires an investment in time before the researcher knows the various strengths and limitations of the particular model. Also, each computer code requires its own support-

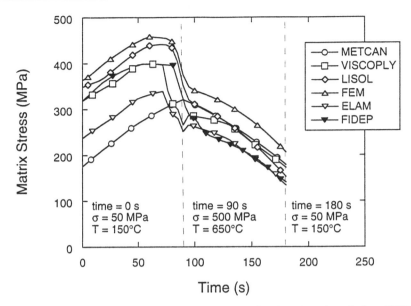

Figure 43. First cycle average matrix stress in the 0° ply of the SCS6/TIMETAL®21S cross-ply layup during in-phase TMF load [31]

ing array of input and output files. Therefore, rather than maintaining many micromechanics models and their computer codes, it may be more advantageous to settle on one or two and become confident in their use. This chapter has sought to introduce the reader to some of the discrete micromechanics models and associated constitutive relations various researchers have used for general analysis of TMC laminates. Before using any of the models, or if more detailed information is desired, the reader should explore the references.

APPENDIX—CONSTITUENT MECHANICAL PROPERTIES

As mentioned in the text of this chapter, proper modeling of the individual constituents is critical for approximating the composite behavior. In fact, the particular micromechanics model chosen is, in some instances, of secondary importance. No amount of mathematical complexity in the micromechanics model will improve the solution if the thermoelastic or inelastic material properties of the constituents are inaccurate. Therefore, patiently ensuring the fiber and matrix are individually modeled correctly is time well spent. This appendix presents material properties for the SCS6 fiber and four titanium-based alloy matrix materials. The thermoelastic mechanical properties are presented first followed by the inelastic properties of the various matrix materials. Although there are numerous inelastic material models, only the properties from three unique models are presented. These consist of the classical bilinear elastic-plastic theory with strain hardening [42], the Bodner-Partom unified viscoplastic theory with directional hardening [28–30], and the modified Eisenberg-Yen viscoplastic theory as employed by Bahei-El-Din et al. [12,15].

Appendix A: Thermoelastic Mechanical Properties

Table A1. Thermoelastic properties of
the SCS6 fiber ($v = 0.25$,
$T_{REF} = 23\,°C$) [75].

T (°C)	E (GPa)	α ($\times 10^{-6}\,°C^{-1}$)
21	393	3.564
93	390	3.660
204	386	3.618
316	382	3.638
427	378	3.687
538	374	3.752
649	370	3.826
760	365	3.903
871	361	3.980
1093	354	4.103

Table A2. Thermoelastic properties of TIMETAL®21S ($\nu = 0.34$, $T_{REF} = 23\,°C$) [76].

T (°C)	E (GPa)	$\alpha\ (\times10^{-6}°C^{-1})$
23	112	6.31
260	108	7.26
315	106	7.48
365	104	7.68
415	101	7.88
465	99.1	8.08
482	98.1	8.15
500	97.0	8.22
525	95.5	8.32
550	93.9	8.43
575	92.2	8.53
600	90.4	8.63
650	86.6	8.83
760	77.2	9.27
815	72.0	9.49

Table A3. Thermoelastic properties of Ti-15-3 ($\nu = 0.36$, $T_{REF} = 23\,°C$) [54,77].

T (°C)	E (GPa)	$\alpha\ (\times10^{-6}°C^{-1})$
25	86.3	8.48
315	80.4	9.16
427	77.5	9.40
482	72.2	9.71
538	67.8	9.89
566	64.4	9.98
649	53.0	10.26
900	25.0	10.50

Table A4. Thermoelastic properties of Ti-6-4 ($\nu = 0.3$, $T_{REF} = 900\,°C$) [53].

T (°C)	E (GPa)	$\alpha\ (\times10^{-6}°C^{-1})$
21	113.7	9.44
149	107.5	9.62
315	97.9	9.78
482	81.3	9.83
649	49.6	9.72
900	20.7	9.81

Table A5. Thermoelastic properties of
Ti-24-11 (ν = 0.3, T_{REF} = 1010 °C) [78].

T (°C)	E (GPa)	α (×10^{-6}°C^{-1})
20	84	11.33
168	86	11.63
315	88	11.88
482	68	12.22
649	48	12.73
704	42	12.95
760	36	13.22
885	24	13.93
1010	11	14.97

Appendix B: Inelastic Mechanical Properties of the Matrix Material

Table B1. Classical plastic linear strain
hardening parameters for
TIMETAL®21S [61].

T (°C)	σ_{ys} (MPa)	H' (MPa)
23	1107	459
260	1010	1486
482	810	2000
650	350	0
760	120	0
815	110	0
900	94	0

Table B2. *Bodner-Partom material parameters for TIMETAL®21S (B-P model with directional hardening. Note: Z_0 is the initial value for Z^l) [76].*

T (°C)	n	$Z_0 = Z_2$ (MPa)	Z_3 (MPa)	m_2 (MPa^{-1})	$A_1 = A_2$ (s^{-1})
23	4.8	1550	100	0.35	0
260	3.5	1300	300	0.35	0
315	3.05	1250	390	1.50	0
365	2.65	1205	500	2.55	0.0003
415	2.24	1160	660	3.60	0.0013
465	1.84	1115	960	4.64	0.0050
482	1.7	1100	1100	5.00	0.00764
500	1.5	1089	1300	5.76	0.0116
525	1.28	1074	1670	6.82	0.0203
550	1.1	1059	2100	7.88	0.0342
575	0.97	1045	2600	8.94	0.0559
600	0.82	1030	3700	10.0	0.0887
650	0.74	1000	3800	10.0	0.208
760	0.58	600	4000	15.0	1.01
815	0.55	300	4100	30.0	1.97

$m_1 = 0$ MPa^{-1}	$r_1 = 3$	$r_2 = 3$	$Z_1 = 1600$ MPa	$D_0 = 10{,}000$ s^{-1}

Table B3. *Eisenberg-Yen material parameters for TIMETAL®21S (Bahei-El-Din's modified Eisenberg-Yen equations) [79].*

T (°C)	Y (MPa)	$\mathcal{Y}$ (MPa)	H_0 (MPa)	h (MPa)	m_r	k (MPa^{-p} / s)	p
21	925	1000	72	4024	1.365	2.10×10^{-21}	8.59
150	925	1000	72	4024	1.365	2.10×10^{-21}	8.59
316	645	785	146	5432	1.355	2.92×10^{-16}	6.52
427	630	770	143	5432	1.365	1.66×10^{-14}	5.62
482	505	615	103	3165	1.478	5.38×10^{-18}	6.35
565	190	365	256	6311	1.211	6.07×10^{-18}	5.52
621	85	185	149	2618	1.245	6.05×10^{-13}	3.58
650	53	87	42	1043	1.298	1.76×10^{-13}	3.97
704	24	53	45	829	1.226	2.66×10^{-10}	3.04
760	9	20	15	250	1.268	1.00×10^{-8}	2.40

Italic font denotes equilibrium yield surface constants and script font denotes bounding surface constants.
It is assumed no isotropic hardening occurs, so $q = b_r = 0$.

*Table B4. Classical plastic linear
strain hardening parameters for
Ti-15-3 [54].*

T (°C)	σ_{ys} (MPa)	H' (MPa)
25	763	3.32
482	577	3.67
538	447	2.69
566	287	2.39
650	198	1.12

*Table B5. Bodner-Partom material parameters for Ti-15-3 (B-P model with
directional hardening. Note: Z_0 is the initial value for Z^l) [77].*

T (°C)	n	$Z_0 = Z_2$ (MPa)	Z_3 (MPa)	m_2 (MPa^{-1})	$A_1 = A_2$ (s^{-1})
25	4.5	1200	250	0.005	1×10^{-8}
315	2.9	1070	454	0.04	4.4×10^{-6}
427	2.7	1020	550	0.05	1×10^{-5}
482	1.6	850	1100	5.0	1.0
566	1.05	750	2400	15.0	2.5
650	0.9	650	3000	20.0	3.0

$m_1 = 0$ MPa^{-1}	$r_1 = 3$	$r_2 = 3$	$Z_1 = 1300$ MPa	$D_0 = 10{,}000$ s^{-1}

*Table B6. Eisenberg-Yen material parameters for Ti-15-3
(Bahei-El-Din's modified Eisenberg-Yen equations) [12].*

Material Constant	21°C	482°C	649°C
Y (MPa)	790	45	15.5
H_0 (MPa)	1400	40	50
h (MPa)	21	350	162
$\mathcal{Y}$/(MPa)	915	1100	316
p	3.75	1.85	1.43
k (MPa^{-p} / s)	1.6×10^{-7}	4.2×10^{-7}	3.2×10^{-6}
Q_a MPa	−120	−10	−5
$\mathcal{Q}_a$ MPa	350	100	95
q	800	5.5	2.61
$\mathcal{q}$	800	5.5	2.61
m_r	1.2	1.29	1.35
$\mathcal{m}_r$	1.2	1.29	1.35
c_r (MPa^{-mr+1}/s)	8.0×10^{-5}	5.0×10^{-4}	2.0×10^{-3}
$\mathcal{c}_r$	8.0×10^{-5}	5.0×10^{-4}	2.0×10^{-3}

Italic font denotes equilibrium yield surface constants and script font denotes bounding surface
constants.

Table B7. *Classical plastic linear strain hardening parameters for Ti-6-4 [53].*

T (°C)	σ_{ys} (MPa)	H' (MPa)
21	900	4.6
149	730	4.7
315	517	5.4
482	482	4.8
649	303	1.7
900	35	1.2

Table B8. *Classical plastic linear strain hardening parameters for Ti-24-11 [78].*

T (°C)	σ_{ys} (MPa)	H' (MPa)
20	950	1.0
168	680	2.1
315	410	3.3
482	333	3.7
649	256	4.1
704	214	3.9
760	171	3.7
885	112	3.9
1010	53.8	4.1

REFERENCES

1. Hopkins, D. A., and Chamis, C., "A Unique Set of Micromechanics Equations for High-Temperature Metal Matrix Composites," *Testing Technology of Metal Matrix Composites, ASTM STP 964,* DiGiovanni and Adsit, eds. (1988) 159–176.

2. Dvorak, G. J., and Bahei-El-Din, Y. A., "Plasticity Analysis of Fibrous Composites," *Journal of Applied Mechanics,* **49,** (1982) 327–335.

3. Hecker, S. S., Hamilton, C. H., and Ebert, L. J., "Elastoplastic Analysis of Residual Stresses and Axial Loading in Composite Cylinders," *Journal of Materials, JMLSA,* **5**(4) (1970) 868–900.

4. Aboudi, J., "Micromechanical Analysis of Composites by the Method of Cells," *Applied Mechanics Review,* **42**(7), (1989) 193–221.

5. Bigelow, C. A., Johnson, W. S., and Naik, R. A., "A Comparison of Various Micromechanics Models for Metal Matrix Composites," *Mechanics of Composite Materials and Structures,* ASME, Reddy and Teply, eds. (1989) 21–31.

6. Kroupa, J. L., Neu, R. W., Nicholas, T., Coker, D., Robertson, D. D., and Mall, S., "A Comparison of Analysis Tools for Predicting the Inelastic Cyclic Response of Cross-Ply Titanium Matrix Composites," *Life Predic-*

tion Methodology for Titanium Matrix Composites, ASTM STP 1253, Johnson, Larsen, and Cox, eds. (1996) 297–327.

7. Chamis, C. C., and Hopkins, D. A., "Thermoviscoplastic Nonlinear Constitutive Relationships for Structural Analysis of High-Temperature Metal Matrix Composites," *Testing Technology of Metal Matrix Composites, ASTM STP 964,* Digiovanni and Adsit, eds. (1988) 177–196.

8. Murthy, P. L. N., and Chamis, C. C., "Metal Matrix Composite Analyzer (METCAN): Theoretical Manual," NASA TM 106025, Feb. 1993.

9. Mital, S. K., Murthy, P. L. N., and Chamis, C. C., "Micromechanics for Ceramic Matrix Composites via Fiber Substructuring," *Journal of Composite Materials, 29*(5) (1995) 614–633.

10. Lee, H.-J., Murthy, P. L. N., and Chamis, C. C., "METCAN Updates for High Temperature Composite Behavior: Simulation/Verification," NASA TM 103682.

11. Hopkins, D. A., and Chamis, C., "A Unique Set of Micromechanics Equations for High-Temperature Metal Matrix Composites," NASA-TM 87154 (1985).

12. Bahei-El-Din, Y. A., Shah, R. S., and Dvorak, G. J., "Numerical Analysis of the Rate-Dependent Behavior of High Temperature Fibrous Composites," *Mechanics of Composites at Elevated Temperatures,* AMD Vol-118, Singhal, Jones, Herakovich, and Cruse, eds. (1991) 67–78.

13. Bahei-El-Din, Y. A., and Dvorak, G. J., "Plasticity Analysis of Laminated Composite Plates," *ASME Journal of Applied Mechanics, 49* (1982) 740–746.

14. Bahei-El-Din, Y. A., "Plasticity Analysis of Fibrous Composite Laminates Under Thermomechanical Loads," *Thermal and Mechanical Behavior of Ceramic and Metal Matrix Composites, ASTM STP 1080,* Kennedy, Moeller, and Johnson, eds. (1990) 20–39.

15. Eisenberg, M. A., and Yen, C.-F., "A Theory of Multiaxial Anisotropic Viscoplasticity," *ASME Journal of Applied Mechanics, 48* (1991) 276–284.

16. Pollock, W. D., and Johnson, W. S., "Characterization of Unnotched SCS-6/Ti-15-3 Metal Matrix Composites at 650°C," *Composite Materials: Testing and Design (Tenth Volume), ASTM STP 1120,* Grimes, ed. (1992) 175–191.

17. Pollock, W. D., and Johnson, W. S., "Characterization of Unnotched SCS-6/Ti-15-3 Metal Matrix Composites at 650°C," NASA-TM.

18. Mirdamadi, M., Johnson, W. S., Bahei-El-Din, Y. A., and Castelli, M. G., "Analysis of Thermomechanical Fatigue of Unidirectional Titanium Metal Matrix Composites," *Composite Materials: Fatigue and Fracture, Fourth Volume, ASTM STP 1156,* Stinchcomb and Ashbaugh, eds. (1993) 591–607.

19. Mirdamadi, M., Johnson, W. S., Bahei-El-Din, Y. A., and Castelli, M. G., "Analysis of Thermomechanical Fatigue of Unidirectional Titanium Metal Matrix Composites," NASA-TM 104105 (1991).

20. Johnson, W. S., and Mirdamadi, M., "Modeling and Life Prediction Methodology for Titanium Matrix Composites Subjected to Mission Profiles," NASA TM 109148 (1994).

21. Bakuckas, J. G., Johnson, W. S., and Bigelow, C. A., "Fatigue Damage in Cross-Ply Titanium Metal Matrix Composites Containing Center Holes," *Journal of Engineering Materials and Technology,* **115**(4) (1993) 404–410.

22. Hillberry, B. M., and Johnson, W. S., "Prediction of Matrix Fatigue Crack Initiation in Notched SCS6/Ti-15-3 Metal Matrix Composites," *ASTM Journal of Composites Technology & Research,* **14**(4) (1992) 221–224.

23. Jones, R. M., *Mechanics of Composite Materials,* Hemisphere Publishing Corporation, (1975) 106–107.

24. Coker, D., Ashbaugh, N. E., and Nicholas, T., "Analysis of Thermomechanical Cyclic Behavior of Unidirectional Metal Matrix Composites," *Thermomechanical Fatigue Behavior of Materials, ASTM STP 1186,* Sehitoglu, ed. (1993) 50–69.

25. Coker, D., Ashbaugh, N. E., and Nicholas, T., "Analysis of the Thermomechanical Behavior of [0] and [0/90] SCS6/Timetal 21S Composites," *Thermomechanical Behavior of Advanced Structural Materials,* AD-Vol. 34, Jones, ed. (1993) 1.

26. Neu, R. W., Coker, D., and Nicholas, T., "Cyclic Behavior of Unidirectional and Cross-Ply Titanium Matrix Composites," *International Journal of Plasticity,* **12**(3) (1996) 361–385.

27. Ashbaugh, N. E., "Mechanical Behavior of High Temperature Structural Materials," University of Dayton Research Institute interim report for 5/24/92 to 5/23/93, Materials Directorate, Wright Laboratory, Wright-Patterson AFB, Ohio, WL-TR-94-4059.

28. Bodner, S. R., and Partom, Y., "Constitutive Equations for Elastic Viscoplastic Strain Hardening Materials," *ASME Journal of Applied Mechanics,* **42** (1975) 385–389.

29. Stouffer, D. C., and Bodner, S. R., "A Constitutive Model for the Deformation Induced Anisotropic Plastic Flow of Metals," *International Journal of Engineering Science,* **17** (1979) 757–764.

30. Chan, K. S., and Lindholm, U.S., "Inelastic Deformation Under Nonisothermal Loading," *ASME Journal of Engineering Materials and Technology,* **112** (1990) 15–25.

31. Ashbaugh, N. E., "Mechanical Behavior of High Temperature Structural Materials," University of Dayton Research Institute final report for 5/24/91 to 5/24/94, Materials Directorate, Wright Laboratory, Wright-Patterson AFB, Ohio, WL-TR-95-4003.

32. Paley, M., and Aboudi, J., "The Over-all Instantaneous Properties of Metal-Matrix Composites," *Composites Science and Technology,* **41** (1991) 411–429.

33. Aboudi, J., "A Continuum Theory for Fiber-Reinforced Elastic-Viscoplastic Composites," *International Journal of Engineering Science,* **20**(5) (1982) 605–621.

34. Aboudi, J., "Closed Form Constitutive Equations for Metal Matrix Composites," *International Journal of Engineering Science,* **25**(9) (1987) 1229–1240.

35. Aboudi, J., "Micromechanical Analysis of Composites by the Method of Cells," *ASME Applied Mechanics Review,* **42**(7) (1989) 193–221.

36. Aboudi, J., "Damage in Composites—Modeling of Imperfect Bonding," *Composites Science and Technology,* **28** (1987) 103–128.

37. Aboudi, J., and Cederbaum, G., "Analysis of Viscoelastic Laminated Composite Plates," *Composite Structures,* **12** (1989) 243–256.

38. Arenburg, R. T., and Reddy, J. N., "Analysis of Metal-Matrix Composite Structures—I. Micromechanics Constitutive Theory," *Computers & Structures,* **40**(6) (1991) 1357–1368.

39. Arenburg, R. T., and Reddy, J. N., "Analysis of Metal-Matrix Composite Structures—II. Laminate Analysis," *Computers & Structures,* **40**(6) (1991) 1369–1385.

40. Robertson, D. D., and Mall, S., "Micromechanical Analysis for Thermoviscoplastic Behavior of Unidirectional Fibrous Composites," *Composites Science and Technology,* **50** (1994) 483–496.

41. Aboudi, J., *Mechanics of Composite Materials—A Unified Micromechanical Approach,* Elsevier Science Publishing Company, New York, NY (1991).

42. Skrzypek, J. J., *Plasticity and Creep, Theory, Examples, and Problems,* editor for English version: Hetnarski, R. B., CRC Press, Boca Raton, FL (1993) 93–111.

43. Robertson, D. D., and Mall, S., "Micromechanical Relations for Fiber-Reinforced Composites Using the Free Transverse Shear Approach," *ASTM Journal of Composites Technology & Research,* **15**(3) (1993) 181–192.

44. Paley, M., and Aboudi, J., "Micromechanical Analysis of Composites by the Generalized Cells Model," *Mechanics of Materials,* **14** (1992) 127–139.

45. Aboudi, J., "Constitutive Behavior of Multiphase Metal Matrix Composites with Interfacial Damage by the Generalized Cells Model," *Damage in Composite Materials,* Voyiedjis, ed. Elsevier Science Publishers (1993) 3–22.

46. Arnold, S. A., Wilt, T. E., Saleeb, A. F., and Castelli, M. G., "An Investigation of Macro and Micromechanical Approaches for a Model MMC System," NASA Conf. Pub. 19117, Vol. II, *Proc. of the 6th Annual HITEMP Conf.,* NASA-Lewis Research Ctr, Cleveland, (1995) 52-1–52-12.

47. Robertson, D. D., and Mall, S., "Micromechanics-Based Inelastic Analysis of Fiber Reinforced Titanium Metal Matrix Composite Laminates," *Key Engineering Materials, Vols 104–107,* Newaz, Neber-Aeschbacher, and Wohlbier, eds., Trans Tech Publications, Switzerland, (1995) 799–808.

48. Robertson, D. D., and Mall, S., "A Nonlinear Micromechanics Based Analysis of Metal Matrix Composite Laminates," *Composites Science and Technology,* **52**(3) (1994) 319–331.

49. Aboudi, J., Hidde, J. S., and Herakovich, C. T., "Thermo-Mechanical Response Predictions for Metal Matrix Composite Laminates," *Mechanics of Composites at Elevated and Cryogenic Temperatures,* ASME, AMD-Vol. 118 (1991) 1–8.

50. Aboudi, J., Mirzadeh, F., Herakovich, C. T., "Response of Metal Matrix Laminates with Temperature Dependent Properties," *ASTM Journal of Composites Technology & Research,* **16**(1) (1994) 68–76.

51. Johnson, W. S., Lubowinski, S. J., and Highsmith, A. L., "Mechanical Characterization of Unnotched SCS6/Ti-15-3 Metal Matrix Composites at Room Temperature," *Thermal and Mechanical Behavior of Metal Matrix and Ceramic Matrix Composites,* Kennedy, Moeller, and Johnson, eds, ASTM STP 1080 (1990) 193–218.

52. Jones, J. P., and Whittier, J. S., "Waves at a Flexibly Bonded Interface," *ASME Journal of Applied Mechanics,* (1967) 905–909.

53. Nimmer, R. P., Bankert, R. J., Russel, E. S., Smith, G. A., and Wright, P. K., "Micromechanical Modeling of Fiber/Matrix Interface Effects in Transversely Loaded SiC/Ti-6-4 Metal Matrix Composites," *ASTM Journal of Composites Technology & Research, 13*(1) (1991) 3–13.

54. Santhosh, U., Ahmad, J., and Nagar, A., "Non-Linear Micromechanics Analysis Prediction of the Behavior of Titanium-Alloy Matrix Composites," *ASME Fracture and Damage,* AD-Vol. 27 (1992) 65–76.

55. Robertson, D. D., and Mall, S., "Fiber-Matrix Interphase Effects Upon Transverse Behavior in Metal Matrix Composites," *ASTM Journal of Composites Technology & Research, 14*(1) (1992) 3–11.

56. Kroupa, J. L., Neu, R. W., and Nicholas, T., "Analysis of a [0/90] Metal Matrix Composite Subjected to Thermomechanical Fatigue," *Fatigue 93,* Engineering Advisory Services Ltd., UK (1993).

57. Sherwood, J. A., and Quimby, H. M., "Micromechanical Modeling of Damage Growth in Titanium Based Metal-Matrix Composites," *Computers & Structures, 56* (1995) 505–514.

58. Achenbach, J. D., and Zhu, H., "Effect of Interfacial Zone on Mechanical Behavior and Failure of Fiber-Reinforced Composites," *Journal of the Mechanics and Physics of Solids, 37*(3) (1989) 381–393.

59. Sanders, B. P., and Mall, S., "Transverse Fatigue Response of a Metal Matrix Composite under Strain Controlled Mode at Elevated Temperature. Part II. Analysis," *ASTM Journal of Composites Technology & Research, 18*(1) (1996) 22–29.

60. Lerch, B. A., Melis, M. E., and Tong, M., "Experimental and Analytical Analysis of Stress-Strain Behavior in a [90/0]$_{2S}$ SiC/Ti-15-3 Laminate," NASA TM 104470 (1991).

61. Kroupa, J. L., and Ashbaugh, N. E., "Stress-Free Edge Effects on the Transverse Response of a Unidirectional Metal Matrix Composite," *Composites Engineering, 5*(6) (1995) 569–582.

62. Gayda, J., and Gabb, T. P., "Isothermal Fatigue Behavior of a [90]$_8$ SiC/Ti-15-3 Composite at 426°C," *International Journal of Fatigue, 14*(1) (1992) 14–20.

63. Robertson, D. D., and Mall, S., "Micromechanical Analysis of Metal Matrix Composite Laminates with Fiber/Matrix Interfacial Damage," *Composites Engineering, 4*(12) (1994) 1257–1274.

64. Robertson, D. D., and Mall, S., "Analysis of the Thermo-Mechanical Fatigue Response of Metal Matrix Composite Laminates with Interfacial Normal and Shear Failure," *Thermo-Mechanical Fatigue Behavior of Materials: 2nd Volume, ASTM STP 1263,* Verrilli and Castelli, eds. (1995).

65. Needleman, A., "A Continuum Model for Void Nucleation by Inclusion Debonding," *ASME Journal of Applied Mechanics,* **54** (1987) 525–531.

66. Bigelow, C. A., and Johnson, W. S., "Effect of Fiber-Matrix Debonding on Notched Strength of Titanium Metal Matrix Composites," NASA TM 104131 (1991).

67. Lerch, B. A., and Saltsman, J. F., "Tensile Deformation of SiC/Ti-15-3 Laminates," *Composite Materials: Fatigue and Fracture, Fourth Volume, ASTM STP 1156,* Stinchcomb and Ashbaugh, eds. (1993) 161–175.

68. Clyne, T. W., and Watson, M. C., "Interfacial Mechanics in Fibre-Reinforced Metals," *Composites Science and Technology,* **42** (1991) 25–55.

69. Robertson, D. D., and Mall, S., "Incorporating Fiber Damage in a Micromechanical Analysis of Metal Matrix Composite Laminates," *ASTM Journal of Composites Technology & Research,* **18**(4) (1996) 265–273.

70. Sherwood, J. A., and Boyle, M. J., "Investigation of the Thermomechanical Response of a Titanium Aluminide/Silicon Carbide Composite Using a Unified State Variable Model in ADINA," *Computers & Structures,* **40**(2) (1991) 257–269.

71. Kroupa, J. L., and Neu, R. W., "The Nonisothermal Viscoplastic Behavior of a Titanium-Matrix Composite," *Composites Engineering,* **4**(9) (1994) 965–977.

72. Nimmer, R. P., "Fiber-Matrix Interface Effects on the Presence of Thermally Induced Residual Stresses," *ASTM Journal of Composites Technology & Research,* **12**(2) (1990) 65–75.

73. Sun, C. T., Chen, J. L., Sha, G. T., and Koop, W. E., "Mechanical Characterization of SCS6/Ti-6-4 Metal Matrix Composite," *Journal of Composite Materials,* **24** (1990) 1029–1059.

74. Neu, R. W., and Nicholas, T., "Effect of Laminate Orientation on the Thermomechanical Fatigue Behavior of a Titanium Matrix Composite," *ASTM Journal of Composites Technology & Research,* 16(3) (1994) 214.

75. Hillmer, N. J., "Thermal Expansion of High-Modulus Fibers," *International Journal of Thermophysics,* **12** (1991) 741–750.

76. Neu, R. W., "Nonisothermal Material Parameters for the Bodner-Partom Model," *Material Parameter Estimation for Modern Constitutive Equations,* MD-Vol. 43/AMD-Vol. 168, ASME, Bertram, Brown, and Freed, eds. (1993) 211–226.

77. Foringer, M. A., Robertson, D. D., and Mall S., "A Micro-Mechanistic Based Approach to Fatigue Life Modeling of Titanium Matrix Composites," COMPOSITES PART B: ENGINEERING (in press).

78. Revelos, W. C., and Kroupa, J. L., "Stress-Free Edge Influence on Thermal Fatigue Damage in an SCS6/Ti-24Al-11Nb Composite," *Composites Engineering,* **5**(4) (1995) 347–361.

79. Dvorak, G. J., Bahei-El-Din, Y. A., and Nigam, H., "Modeling of Thermomechanical Fatigue in $[0/\pm45/90]_s$ SiC/Timetal 21S Laminates," ASTM Hilton Head Conference (1994).

Index

465